STEAM

ITS GENERATION AND USE

THE BABCOCK & WILCOX COMPANY

NEW YORK

Thirty-sixth Edition
Third Printing

Bartlett Orr Press
New York

THE BABCOCK & WILCOX COMPANY

85 LIBERTY STREET, NEW YORK, U. S. A.

Works

BAYONNE . NEW JERSEY
BARBERTON . . . OHIO

Officers

A. G. PRATT, *President*
E. H. WELLS, *Chairman of the Board*
J. E EUSTIS, *Secretary*
I. HARTER, *Vice-President*
F. L. WARD, *Treasurer*
C. W. MIDDLETON, *Vice-President*

Directors

J. E. EUSTIS
F. J. FAULKS
I. HARTER
C. A. KNIGHT
C. W. MIDDLETON
A. G. PRATT
J. J. RIKER
E. H. WELLS

Branch Offices

ATLANTA	CANDLER BUILDING
BOSTON	80 FEDERAL STREET
CHICAGO	MARQUETTE BUILDING
CINCINNATI	TRACTION BUILDING
CLEVELAND	GUARDIAN BUILDING
DALLAS	2001 MAGNOLIA BUILDING
DENVER	444 SEVENTEENTH STREET
DETROIT	FORD BUILDING
HAVANA, CUBA	CALLE DE AGUIAR 104
HONOLULU	CASTLE & COOKE BUILDING
HOUSTON	1011-1013 ELECTRIC BUILDING
LOS ANGELES	404-406 CENTRAL BUILDING
NEW ORLEANS	344 CAMP STREET
PHILADELPHIA	PACKARD BUILDING
PHOENIX	HEARD BUILDING
PITTSBURGH	FARMERS DEPOSIT BANK BUILDING
PORTLAND, OREGON	805 FAILING BUILDING
SALT LAKE CITY	405-406 KEARNS BUILDING
SAN FRANCISCO	SHELDON BUILDING
SAN JUAN, PORTO RICO	ROYAL BANK BUILDING
SEATTLE	L. C SMITH BUILDING

Export Department, New York: Alberto de Verastegui, Director

TELEGRAPH ADDRESS: FOR NEW YORK, "*GLOVEBOXES*"
FOR HAVANA, "*BABCOCK*"

Works of The Babcock & Wilcox Company at Bayonne, New Jersey

Works of The Babcock & Wilcox Company at Barberton, Ohio

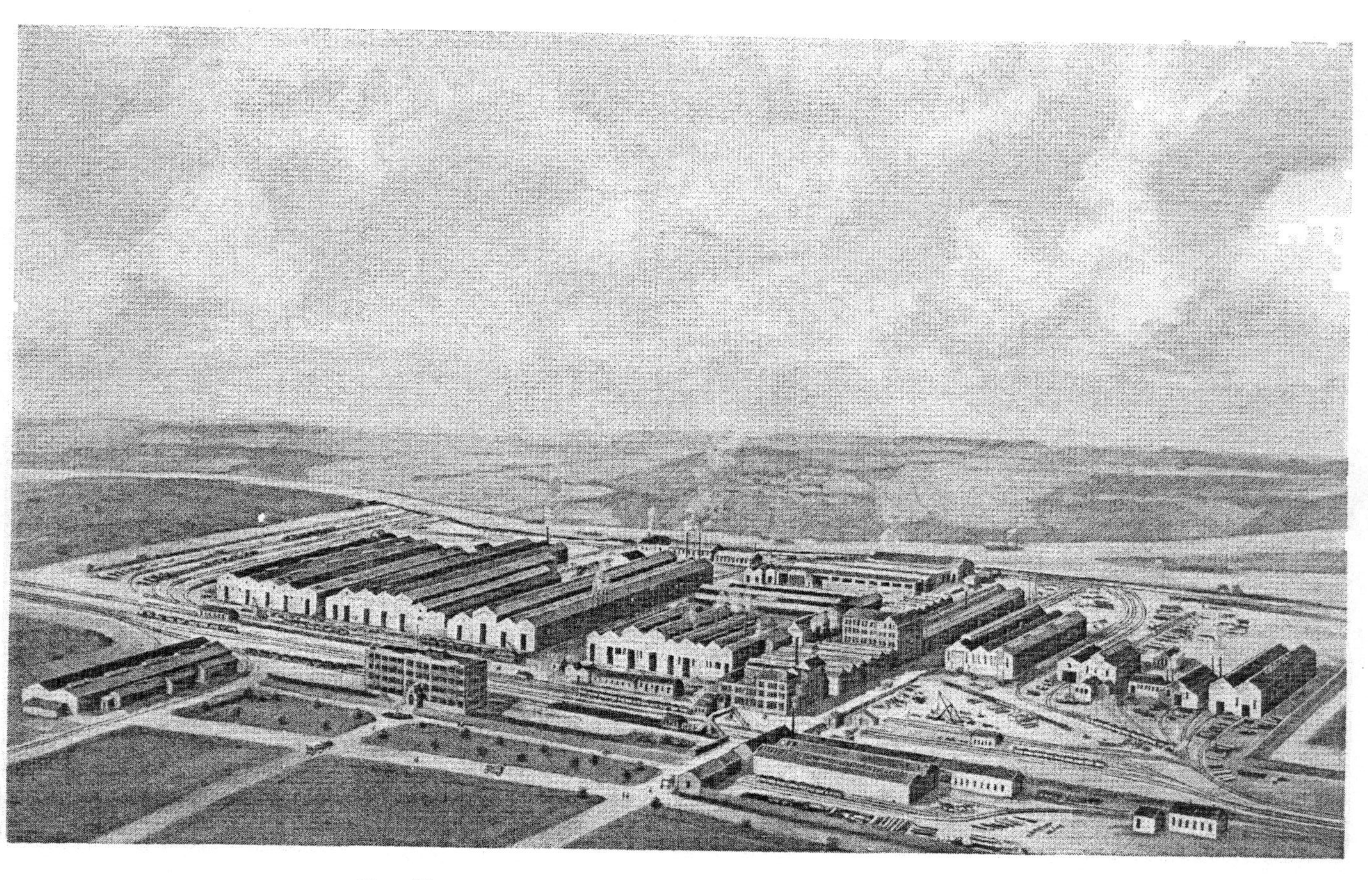

Main Works of Babcock & Wilcox, Limited, at Renfrew, Scotland

BABCOCK & WILCOX, LIMITED

BABCOCK HOUSE, FARRINGDON STREET, LONDON, E. C. 4

MAIN WORKS: RENFREW, SCOTLAND

BRANCH WORKS:

DUMBARTON (SCOTLAND); OLDBURY AND LINCOLN (ENGLAND)
AUSTRALIA AND JAPAN

Directors

SIR JOHN DEWRANCE, *Chairman*
SIR COLES CHILD, BART.
ARTHUR T. SIMPSON
H. W. KOLLE
E. H. WELLS
WALTER COLLS, *Secretary*
CHARLES A. KNIGHT

Branch Offices in Great Britain

BIRMINGHAM
CARDIFF
GLASGOW
LIVERPOOL
MANCHESTER
MIDDLESBOROUGH
NEWCASTLE
SHEFFIELD

Branch Offices Elsewhere

BOMBAY
BRISBANE
BRUSSELS
CALCUTTA
JOHANNESBURG
LIMA
MELBOURNE
MEXICO
SHANGHAI
RIO DE JANEIRO
SYDNEY
TOKYO
WARSAW

District Office

BELFAST: BEDFORD BUILDINGS, BEDFORD STREET

Representatives and Licensees

ADELAIDE
ATHENS
AUCKLAND
BANGKOK
BUENOS AYRES
CAIRO
CAPE TOWN
CHRISTCHURCH
COLOMBO
COPENHAGEN
DUNEDIN
GOTHENBURG
HELSINGFORS
HENGELO
OSLO
PERTH
RANGOON
VALPARAISO
WELLINGTON

TELEGRAPHIC ADDRESS FOR ALL OFFICES EXCEPT LIVERPOOL, BOMBAY AND CALCUTTA: *"BABCOCK"*

FOR LIVERPOOL, BOMBAY AND CALCUTTA: *"BOILER"*

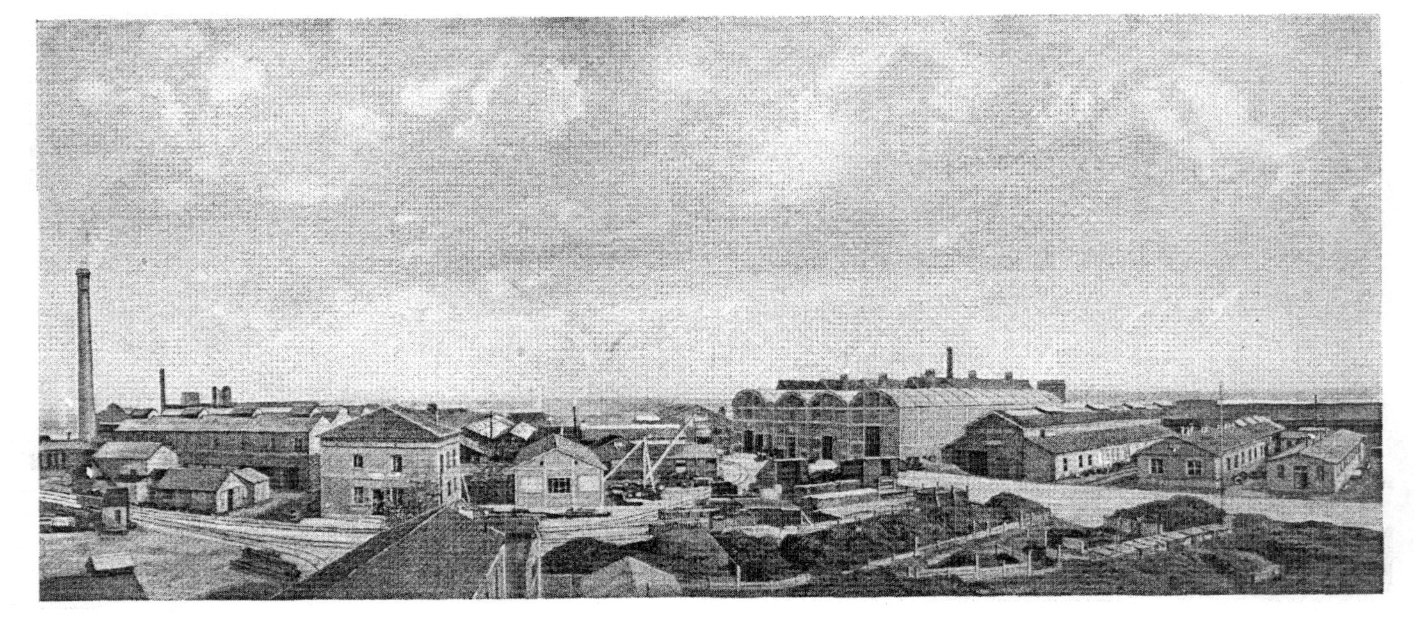

Works of the French Babcock & Wilcox Company at La Courneuve, Seine

SOCIÉTÉ FRANCAISE DES CONSTRUCTIONS BABCOCK ET WILCOX

48 Rue La Boetie, Paris

Directors

Edmund Dupuis, Etienne Besson, Irenee Chavanne, Charles A. Knight, Charles Roszak

Branch Offices

Bordeaux, Lille, Lyon, Marseille, Montpellier, Nancy, Reims, Rouen

Representative for Switzerland: Spoerri & Company, Zurich

Telegraphic Address, "BABCOCK-PARIS"

Works of the Spanish Babcock & Wilcox Company at Galindo, Bilbao

SOCIEDAD ESPAÑOLA DE CONSTRUCCIONES BABCOCK & WILCOX

1 Calle de Ercilla Bilbao, Spain

Directors

Marquis de Triano President, Marquis de Mac-Mahon. Vice-President, José Luis de Oriol, Gabriel Maria de Ibarra Cesar de la Mora: Count de los Gaitanes: Marquis de Casa Palacio, Juan L. Prado, Enrique Ocharan, Ernesto Ugalde: Fidel Alonso Allende; Tomás Urquijo, Count de Zubiria, Marquis de Arriluce de Ibarra; Charles A. Knight: Federico Echevarría, Francisco G. Cowlrick, Counsel

Branch Offices

Barcelona. Madrid Seville and Lisbon

Telegraphic Address "BABCOCK"

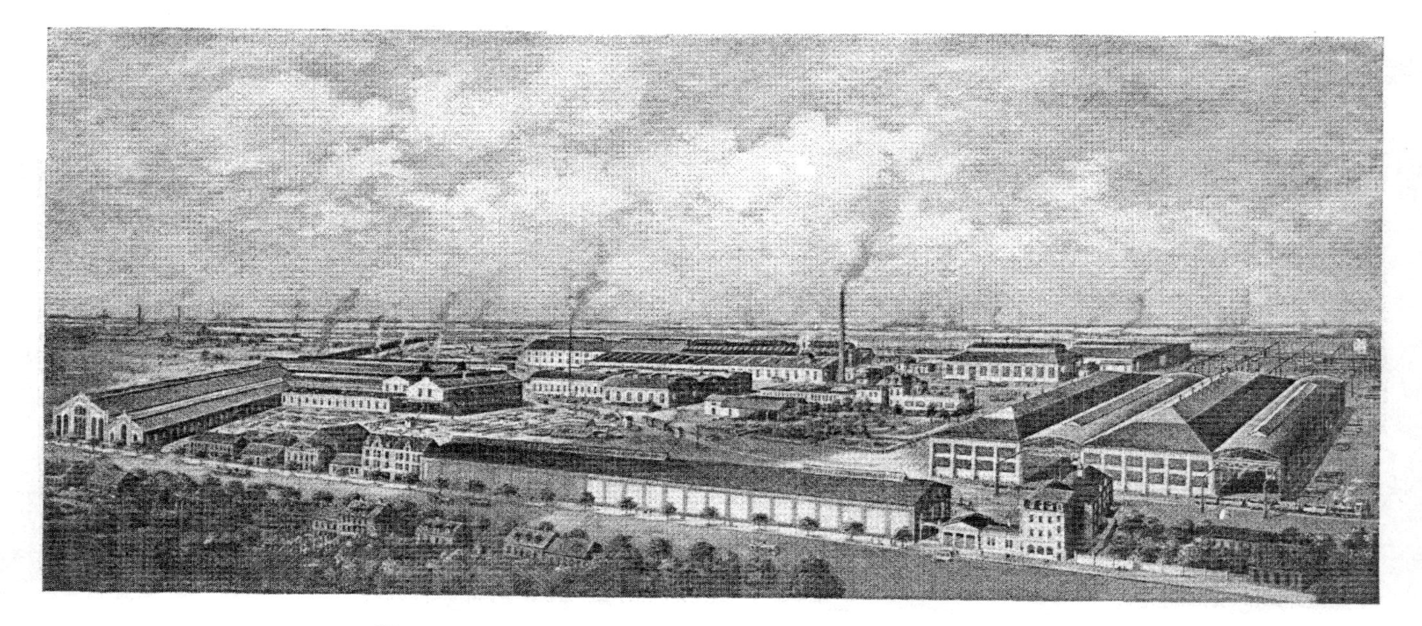

Works of the German Babcock & Wilcox Company at Oberhausen

DEUTSCHE BABCOCK & WILCOX DAMPFKESSELWERKE ACTIEN-GESELLSCHAFT

Oberhausen Rheinland Germany

Branch Works Gleiwitz, Prussian Silesia

Directors

Friedrich Kirchhoff, Eugen Landau, Adelbert Sudhaus, Dr Georg Hahn, Dr Robert Jurenka, Manager

Branch Offices

Berlin, Breslau, Cottbus, Danzig, Dusseldorf, Essen

Frankfort-on-Main, Halle-on Saale, Hamburg, Leipzig, Munich,

Saarbrucken, Stettin, Stuttgart

Telegraphic Address "Babcockwerke"

WORKS AT BEAVER FALLS PENNSYLVANIA

THE BABCOCK & WILCOX TUBE COMPANY

(Pittsburgh Seamless Tube Company)

General Offices and Works: Beaver Falls Pennsylvania

Directors

A. G Pratt President, I Harter, Vice-President C W Middleton, F L Ward E H Wells

Branch Offices

New York, 85 Liberty Street
Pittsburgh, Farmers Deposit Bank Building
Atlanta Candler Building
Cincinnati, Traction Building
Denver, 444 Seventeenth Street

Chicago, Marquette Building
Philadelphia Packard Building
Boston, 8[illegible] Federal Street
Cleveland, Guardian Building
New Orleans, 344 Camp Street

San Francisco, Sheldon Building

The works of this company were established in 1901 In addition to making all the tubes used by The Babcock & Wilcox Company in boilers, superheaters economizers and air heaters, the mills supply seamless steel tubes and pipe up to 5.5 inches outside diameter for general uses.

THE EARLY HISTORY OF THE GENERATION AND USE OF STEAM

WHILE the time of man's first knowledge and use of the expansive force of the vapor of water is unknown, records show that such knowledge existed earlier than 150 B. C. In a treatise of about that time entitled "Pneumatica," Hero, of Alexandria, described not only existing devices of his predecessors and contemporaries, but also an invention of his own which utilized the expansive force of steam for raising water above its natural level. He clearly describes three methods in which steam might be used directly as a source of power: raising water by its elasticity, elevating a weight by its expansive power, and producing a rotary motion by its reaction on the atmosphere. The third method, which is known as "Hero's engine," is described as a hollow sphere supported over a caldron or boiler by two trunnions, one of which was hollow and connected the interior of the sphere with the steam space of the caldron. Two pipes, open at the ends and bent at right angles, were inserted at opposite poles of the sphere, forming a connection between the caldron and the atmosphere. Heat being applied to the caldron, the steam generated passed through the hollow trunnion to the sphere and thence into the atmosphere through the two pipes. By the reaction incidental to its escape through these pipes, the sphere was caused to rotate and here is the primitive steam reaction turbine.

Hero makes no suggestions as to application of any of the devices he describes to a useful purpose. From the time of Hero until the late Sixteenth and early Seventeenth Centuries, there is no record of progress, though evidence is found that such devices as were described by Hero were sometimes used for trivial purposes, the blowing of an organ or the turning of a spit.

Matthesius, a German author, in 1571; Besson, a philosopher and mathematician at Orleans; Ramelli, in 1588; Battista Della Porta, a Neapolitan mathematician and philosopher, in 1601; De Caus, a French engineer and architect, in 1615; and Branca, an Italian architect, in 1629, all published treatises bearing on the subject of the generation of steam.

To the next contributor, Edward Somerset, second Marquis of Worcester, is apparently due the credit of proposing, if not of making, the first useful steam engine. In the "Century of the Names and Scantlings of Inventions by Me Already Practised," published in London in 1663, he describes devices showing that he had in mind the raising of water not only by forcing it from two receivers by direct steam pressure, but also for some sort of reciprocating piston actuating one end of a lever, the other operating a pump. His descriptions are rather obscure and no drawings are extant, so that it is difficult to say whether there were any distinctly novel features to his devices aside from the double action. While there is no direct authentic record that any of the devices he described were actually constructed, it is claimed by many that he really built and operated a steam engine containing pistons.

In 1675, Sir Samuel Morland was decorated by King Charles II for a demonstration of "a certain powerful machine to raise water." Though there appears to be no record of the design of this machine, a mathematical dictionary, published in 1822, credits Morland with the first account of a steam engine, on which subject he wrote a treatise that is still preserved in the British Museum.

A Battery of Two 400 Horse-power Babcock & Wilcox Boilers and Babcock & Wilcox Superheaters in Course of Erection at the Danvers Bleachery of the Naumkeag Steam Cotton Mills, Peabody, Mass.

Dr. Denis Papin, an ingenious Frenchman, invented in 1680 "a steam digester for extracting marrowy, nourishing juices from bones by enclosing them in a boiler under heavy pressure," and finding danger from explosion, added a contrivance which is the first safety valve on record.

The steam engine first became commercially successful with Thomas Savery. In 1699, Savery exhibited before the Royal Society of England (Sir Isaac Newton was President at the time) a model engine which consisted of two copper receivers alternately connected by a three-way hand-operated valve, with a boiler and a source of water supply. When the water in one receiver had been driven out by the steam, cold water was poured over its outside surface, creating a vacuum through condensation and causing it to fill again while the water in the other reservoir was being forced out. A number of machines were built on this principle and placed in actual use as mine pumps.

The serious difficulty encountered in the use of Savery's engine was the fact that the height to which it could lift water was limited by the pressure the boiler and vessels could bear. Before Savery's engine was entirely displaced by its successor, Newcomen's, it was considerably improved by Desaguliers, who applied the Papin safety valve to the boiler and substituted condensation by a jet within the vessel for Savery's surface condensation.

In 1690, Papin suggested that the condensation of steam should be employed to make a vacuum beneath a piston which had previously been raised by the expansion of steam. This was the earliest cylinder and piston steam engine, and his plan took practical shape in Newcomen's atmospheric engine. Papin's first engine was unworkable owing to the fact that he used the same vessel for both boiler and cylinder. A small quantity of water was placed in the bottom of the vessel and heat was applied. When steam formed and raised the piston, the heat was withdrawn and the piston did work on its down stroke under pressure of the atmosphere. After hearing of Savery's engine, Papin developed an improved form. Papin's engine of 1705 consisted of a displacement chamber in which a floating diaphragm or piston on top of the water kept the steam and water from direct contact. The water, delivered to a closed tank by the downward movement of the piston under pressure, flowed in a continuous stream against the vanes of a water wheel. When the steam in the displacement chamber had expanded, it was exhausted to the atmosphere through a valve instead of being condensed. The engine was, in fact, a non-condensing, single action steam pump with the steam and pump cylinders in one. A curious feature of this engine was a heater placed in the diaphragm. This was a mass of heated metal for the purpose of keeping the steam dry or preventing condensation during expansion. This device might be called the first superheater.

Among the various inventions attributed to Papin was a boiler with an internal fire box, the earliest record of such construction.

While Papin had neglected his earlier suggestion of a steam and piston engine to work on Savery's ideas, Thomas Newcomen, with his assistant, John Cawley, put into practical form Papin's suggestion of 1690. Steam admitted from the boiler to a cylinder raised a piston by its expansion, assisted by a counter-weight on the other end of a beam actuated by the piston. The steam valve was then shut and the steam condensed by a jet of cold water. The piston was then forced downward by atmospheric pressure and did work on a pump. The condensed water in the cylinder was

THE CAPITOL POWER PLANT, WASHINGTON, D. C., EQUIPPED WITH SIXTEEN [illegible] HORSE POWER BABCOCK & WILCOX CROSS DRUM BOILERS

expelled through an escapement valve by the next entry of steam. This engine used steam having a pressure but little, if any, above that of the atmosphere.

In 1711, this engine was introduced into mines for pumping purposes. Whether its action was originally automatic or whether dependent upon the hand operation of the valves is unknown. The story commonly believed is that a boy, Humphrey Potter, in 1713, whose duty it was to open and shut such valves of an engine he attended, by suitable cords and catches attached to the beam, caused the engine to manipulate these valves automatically. This device was simplified in 1718 by Henry Beighton, who suspended from the beam a rod called the plug-tree, which actuated the valves by tappets. By 1725, this engine was in common use in collieries and was changed but little for a matter of sixty or seventy years. Compared with Savery's engine, from the aspect of a pumping engine, Newcomen's was a distinct advance, in that the pressure in the pumps was in no manner dependent upon the steam pressure. In common with Savery's engine, the losses from the alternate heating and cooling of the steam cylinder were enormous. Though obviously this engine might have been modified to serve many purposes, its use seems to have been limited almost entirely to the pumping of water.

The rivalry between Savery and Papin appears to have stimulated attention to the question of fuel saving. Dr. John Allen, in 1730, called attention to the fact that owing to the short length of time of the contact between the gases and the heating surfaces of the boiler, nearly half of the heat of the fire was lost. With a view to overcoming this loss, at least partially, he used an internal furnace with a smoke flue winding through the water in the form of a worm in a still. In order that the length of passage of the gases might not act as a damper on the fire, Dr. Allen recommended the use of a pair of bellows for forcing the sluggish vapor through the flue. This is probably the first suggested use of forced draft. In forming an estimate of the quantity of fuel lost up the stack, Dr. Allen probably made the first boiler test.

Toward the end of the period of use of Newcomen's atmospheric engine, John Smeaton, who, about 1770, built and installed a number of large engines of this type, greatly improved the design in its mechanical details.

A patent taken out by William Blakey in 1766, covering an improvement in Savery's engine, included a novel form of steam generator. This boiler, described in the following chapter, was probably the first attempt toward the development of a water-tube boiler.

The improvements in boiler and engine design of Smeaton, Newcomen and their contemporaries, were followed by those of the great engineer, James Watt, an instrument maker of Glasgow. In 1763, while repairing a model of Newcomen's engine, he was impressed by the great waste of steam to which the alternating cooling and heating of the engine gave rise. His remedy was the maintaining of the cylinder as hot as the entering steam, and, with this in view, he added a vessel separate from the cylinder, into which the steam should pass from the cylinder and be there condensed, either by the application of cold water outside or by a jet from within. To preserve a vacuum in his condenser, he added an air pump which should serve to remove the water of condensation and air brought in with the injection water or due to leakage. As the cylinder no longer acted as a condenser, he could maintain it at a high temperature by covering it with non-conducting material and, in particular, by the use of a steam jacket. Further and with the same object in view, he covered the top of

Woolworth Building, New York City, Operating [illegible] Horse Power of Babcock & Wilcox Boilers

the cylinder and introduced steam above the piston to do the work previously accomplished by atmospheric pressure. After several trials with an experimental apparatus based on these ideas, Watt patented his improvements in 1769. Aside from their historical importance, Watt's improvements, as described in his specification, are to this day a statement of the principles which guide the scientific development of the steam engine. His words are:

"My method of lessening the consumption of steam, and consequently fuel, in fire engines, consists of the following principles:

"First: That vessel in which the powers of steam are to be employed to work the engine, which is called the cylinder in common fire engines, and which I call the steam vessel, must, during the whole time the engine is at work, be kept as hot as the steam that enters it; first, by enclosing it in a case of wood, or any other materials that transmit heat slowly; secondly, by surrounding it with steam or other heated bodies; and, thirdly, by suffering neither water nor any other substance colder than the steam to enter or touch it during that time.

"Secondly: In engines that are to be worked wholly or partially by condensation of steam, the steam is to be condensed in vessels distinct from the steam vessels or cylinders, although occasionally communicating with them; these vessels I call condensers; and, whilst the engines are working, these condensers ought at least to be kept as cold as the air in the neighborhood of the engines, by application of water or other cold bodies.

"Thirdly: Whatever air or other elastic vapor is not condensed by the cold of the condenser, and may impede the working of the engine, is to be drawn out of the steam vessels or condensers by means of pumps, wrought by the engines themselves, or otherwise.

"Fourthly: I intend in many cases to employ the expansive force of steam to press on the pistons, or whatever may be used instead of them, in the same manner in which the pressure of the atmosphere is now employed in common fire engines. In cases where cold water cannot be had in plenty, the engines may be wrought by this force of steam only, by discharging the steam into the air after it has done its office. . . . .

"Sixthly: I intend in some cases to apply a degree of cold not capable of reducing the steam to water, but of contracting it considerably, so that the engines shall be worked by the alternate expansion and contraction of the steam.

"Lastly: Instead of using water to render the pistons and other parts of the engine air and steam tight, I employ oils, wax, resinous bodies, fat of animals, quicksilver and other metals in their fluid state."

The fifth claim was for a rotary engine, and need not be quoted here.

The early efforts of Watt are typical of those of the poor inventor struggling with insufficient resources to gain recognition, and it was not until he became associated with the wealthy manufacturer, Matthew Boulton of Birmingham, that he met with the success upon which his present fame is based. In partnership with Boulton, the business of the manufacture and the sale of his engines was highly successful in spite of vigorous attacks on the validity of his patents.

Though the fourth claim of Watt's patent describes a non-condensing engine which would require high pressures, his aversion to such practice was strong. Notwithstanding his entire knowledge of the advantages through added expansion

THE CALUMET STATION OF THE COMMONWEALTH EDISON COMPANY, CHICAGO, OPERATING 28,[illegible]6 HORSE POWER OF
BABCOCK & WILCOX BOILERS AND SUPERHEATERS

under high pressure, he continued to use pressures not more than 7 pounds per square inch above the atmosphere. To overcome such pressures, his boilers were fed through a stand-pipe of sufficient height to have the column of water offset the pressure within the boiler. Watt's attitude toward high pressure made his influence felt long after his patents had expired.

In 1782, Watt patented two other features which he had invented as early as 1769. These were the double-acting engine, that is, the use of steam on both sides of the piston, and the use of steam expansively, that is, the shutting off of steam from the cylinder when the piston had made but a portion of its stroke, the power for the completion of the stroke being supplied by the expansive force of the steam already admitted.

He further added a throttle valve for the regulation of steam admission, invented the automatic governor and the steam indicator, a mercury steam gauge and a glass water column.

It has been the object of this brief history of the early developments in the use of steam to cover such developments only through the time of James Watt. The progress of the steam engine from this time through the stages of higher pressures, combining of cylinders, the application of steam engines to vehicles and steamboats, the adding of third and fourth cylinders, to the invention of the turbine with its development and the accompanying development of the reciprocating engine to hold its place, is one long attestation to the inventive genius of man.

While little is said in the biographies of Watt as to the improvement of steam boilers, all the evidence indicates that Boulton and Watt introduced the first "wagon boiler," so called because of its shape. In 1785, Watt took out a number of patents for variations in furnace construction, many of which contain the basic principles of some of the modern smoke-preventing furnaces. Until the early part of the Nineteenth Century, the low steam pressures used caused but little attention to be given to the form of the boiler operated in connection with the engines above described. About 1800, Richard Trevithick, in England, and Oliver Evans, in America, introduced non-condensing, and for that time, high-pressure steam engines. To the initiative of Evans may be attributed the general use of high-pressure steam in the United States, a feature which for many years distinguished American from European practice. The demand for light weight and economy of space following the beginning of steam navigation and the invention of the locomotive, required boilers designed and constructed to withstand heavier pressures and forced the adoption of the cylindrical form of boiler. There are in use today many examples of every step in the development of steam boilers from the first plain cylindrical boiler to the most modern type of multi-tubular locomotive boiler, which stands as the highest type of fire-tube boiler construction.

The early attempts to utilize water-tube boilers were few. A brief history of the development of the boilers in which this principle was employed is given in the following chapter. From this history it will be clearly indicated that the first commercially successful utilization of water tubes in a steam generator is properly attributed to George H. Babcock and Stephen Wilcox.

WORKS OF THE PENNSYLVANIA SALT MANUFACTURING COMPANY AT WYANDOTTE, MICH., FOR WHICH 7738 HORSE-POWER OF BABCOCK & WILCOX BOILERS HAS BEEN SUPPLIED

BRIEF HISTORY OF WATER-TUBE BOILERS*

As stated in the previous chapter, the first water-tube boiler was built by William Blakey and was patented by him in 1766. Several tubes alternately inclined at opposite angles were arranged in the furnace. the adjacent tube ends being connected by small pipes. The first successful user of water-tube boilers, however, was James Rumsey, an American inventor, celebrated for his early experiments in steam navigation, and it is he who may be truly classed as the originator of the water-tube boiler. In 1788. he patented. in England. several forms of boilers. some of which were of the water-tube type. One had a fire box with flat top and sides, with horizontal tubes across the fire box connecting the water spaces. Another had a cylindrical fire box surrounded by an annular water space and a coiled tube was placed within the box connecting at its two ends with the water space. This was the first of the "coil boilers." Another form in the same patent was the vertical tubular boiler. practically as made at the present time.

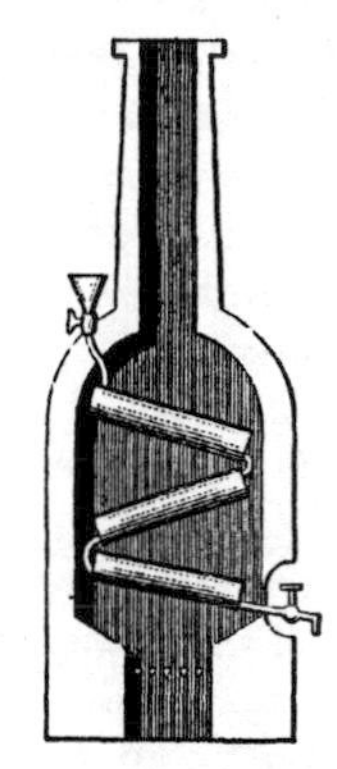

BLAKEY, 1766

The first boiler made of a combination of small tubes, connected at one end to a reservoir, was the invention of another American, John Stevens, in 1804. This boiler was actually employed to generate steam for running a steamboat on the Hudson River, but like all the "porcupine" boilers. of which type it was the first, it did not have the elements of a continued success. Another form of water tube was patented in 1805 by John Cox Stevens, a son of John Stevens. This boiler consisted of twenty vertical tubes, 1¼ inches internal diameter and 40½ inches long, arranged in a circle, the outside diameter of which was approximately 12 inches, connecting a water chamber at the bottom with a steam chamber at the top. The steam and water chambers were annular spaces of small cross section and contained approximately 33 cubic inches. The illustration shows the cap of the steam chamber secured by bolts. The steam outlet pipe "A" is a pipe of one inch diameter. the water entering through a similar aperture at the bottom. One of these boilers was for a long time at the Stevens Institute of Technology at Hoboken, and is now in the Smithsonian Institution at Washington.

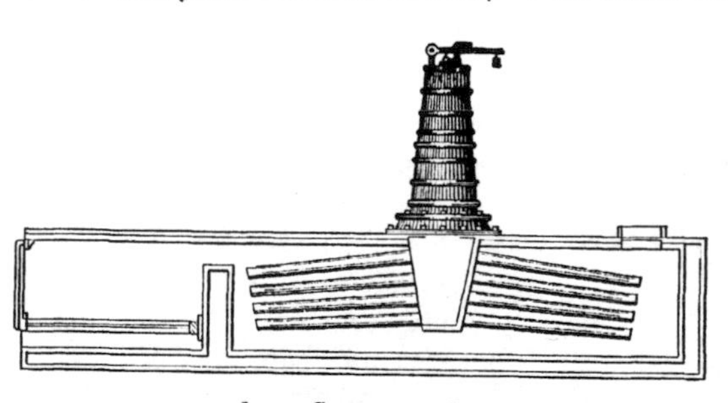

JOHN STEVENS, 1804

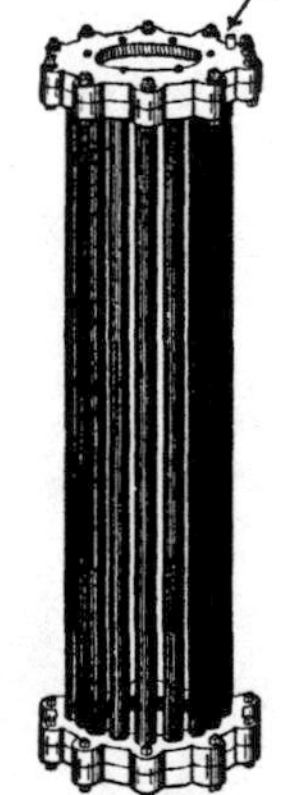

JOHN COX STEVENS 1805

About the same time, Arthur Woolf built a boiler of large horizontal tubes, extending across the furnace and connected at the ends to a longitudinal drum above. The first purely sectional

*See discussion by George H Babcock, of Stirling's paper on "Water-tube and Shell Boilers," in Transactions American Society of Mechanical Engineers, Volume VI., Page 601.

water-tube boiler was built by Julius Griffith, in 1821. In this boiler, a number of horizontal water tubes were connected to vertical side pipes, the side pipes were connected to horizontal gathering pipes, and these latter in turn to a steam drum.

In 1822, Jacob Perkins constructed a flash boiler for carrying what was then considered a high pressure. A number of cast-iron bars having 1½-inch annular holes through them and connected at their outer ends by a series of bent pipes, outside of the furnace walls, were arranged in three tiers over the fire. The water was fed slowly to the upper tier by a force pump, and steam in the superheated state was discharged from the lower tiers into a chamber from which it was taken to the engine.

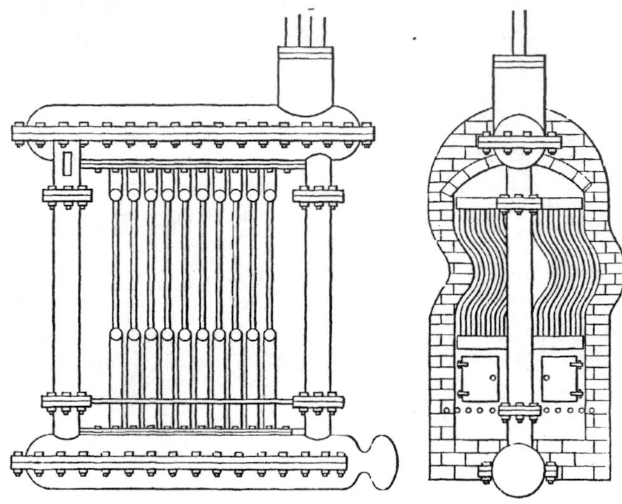
JOSEPH EVE, 1825

The first sectional water-tube boiler, with a well-defined circulation, was built by Joseph Eve, in 1825. The sections were composed of small tubes with a slight double curve, but being practically vertical, fixed in horizontal headers, which headers were in turn connected to a steam space above and a water space below formed of larger pipes. The steam and water spaces were connected by outside pipes to secure a circulation of the water up through the sections and down through the external pipes. In the same year, John M'Curdy of New York, built a "Duplex Steam Generator" of "tubes of wrought or cast iron or other material" arranged in several horizontal rows, connected together alternately at the front and rear by return bends. In the tubes below the water line were placed interior circular vessels closed at the ends in order to expose a thin sheet of water to the action of the fire.

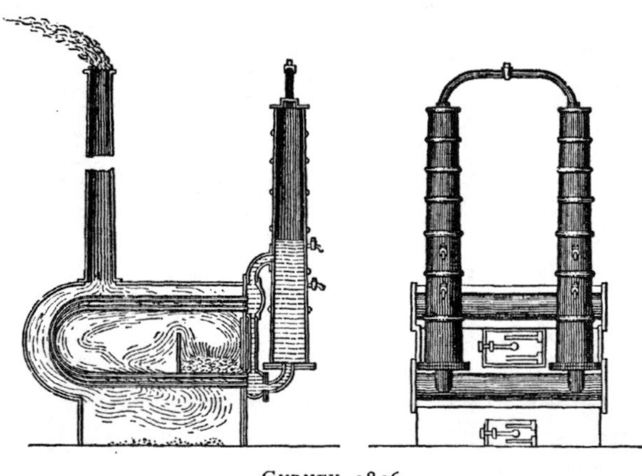
GURNEY, 1826

In 1826, Goldsworthy Gurney built a number of boilers, which he used on his steam carriages. A number of small tubes were bent into the shape of a "U" laid sidewise and the ends were connected with larger horizontal pipes. These were connected by vertical pipes to permit of circulation and also to a vertical cylinder which served as a steam and water reservoir. In 1828, Paul Steenstrup made the first shell boiler with vertical water tubes in the large flues, similar to the boiler known as the "Martin," and suggesting the "Galloway."

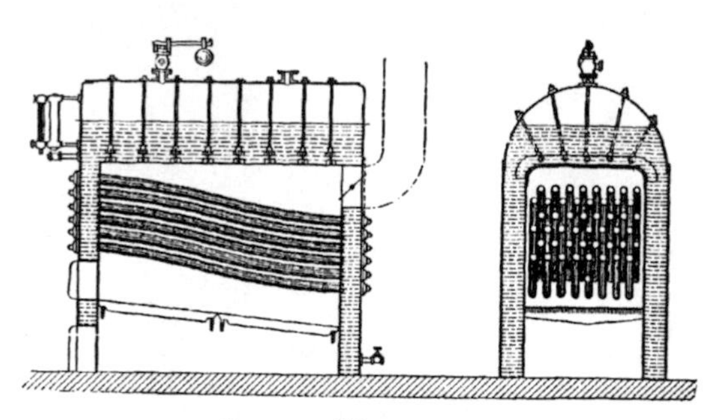

STEPHEN WILCOX, 1856

The first water-tube boiler having fire tubes within water tubes was built in 1830 by Summers & Ogle. Horizontal connections at the top and bottom were connected by a series of vertical water tubes, through which were fire tubes extending through the horizontal connections, the fire tubes being held in place by nuts, which also served to make the joint.

Stephen Wilcox, in 1856, was the first to use inclined water tubes connecting water spaces at the front and rear with a steam space above. The first to make such inclined tubes into a sectional form was Twibill, in 1865. He used wrought-iron tubes connected at the front and rear with standpipes through intermediate connections. These standpipes carried the steam to a horizontal cross drum at the top, the entrained water being carried to the rear.

Clarke, Moore, McDowell, Alban and others worked on the problem of constructing water-tube boilers, but because of difficulties of construction involved, met with no practical success.

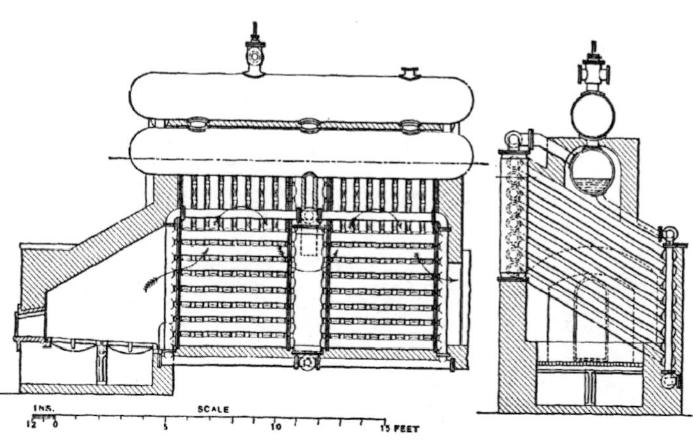

TWIBILL, 1865

It may be asked why water-tube boilers did not come into more general use at an early date, that is, why the number of water-tube boilers built was so small in comparison to the number of shell boilers. The reason for this is found in the difficulties involved in the design and construction of water-tube boilers, which design and construction required a high class of engineering and workmanship, while the plain cylindrical boiler is comparatively easy to build. The greater skill required to make a successful water-tube boiler is readily shown by the great number of failures in the attempts to make them.

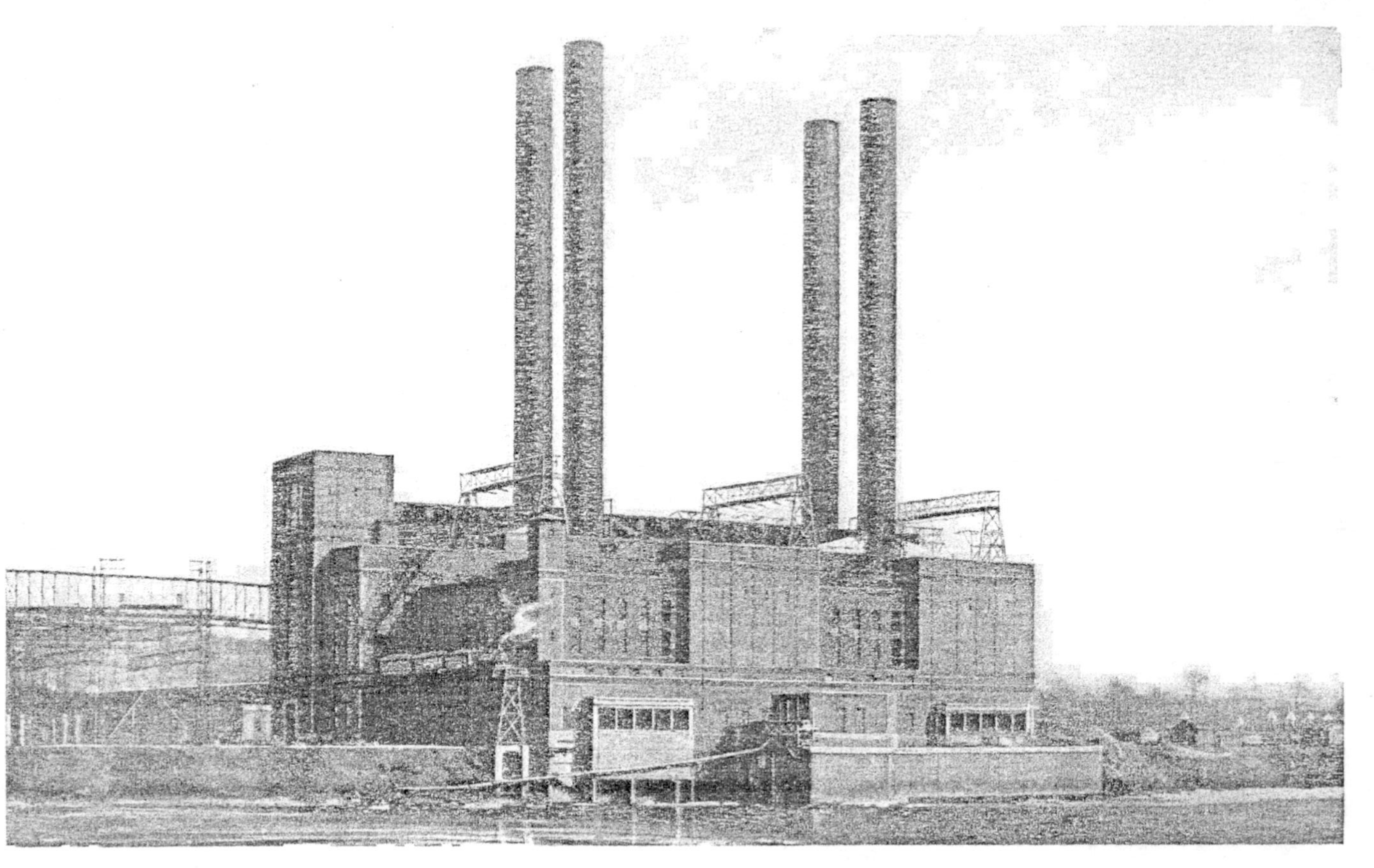

The Colfax Station of the Duquesne Light Company of Pittsburgh in which 30,650 Horse Power of Babcock & Wilcox Boilers Have Been Installed

REQUIREMENTS OF STEAM BOILERS

SINCE the first appearance in " Steam " of the following " Requirements of a Perfect Steam Boiler," the list has been copied many times, either word for word or clothed in different language, and applied to some specific type of boiler design or construction. In most cases, although full compliance with one or more of the requirements was structurally impossible, the reader was left to infer that the boiler under consideration possessed all the desirable features. It is noteworthy that this list of requirements, as prepared by George H. Babcock and Stephen Wilcox, in 1875, represents the best practice of today. Moreover, coupled with their boiler itself, which is used in the largest and most important steam generating plants throughout the world, the list forms a fitting monument to the foresight and genius of the inventors.

REQUIREMENTS OF A PERFECT STEAM BOILER

1st. Proper workmanship and simple construction, using materials which experience has shown to be the best, thus avoiding the necessity of early repairs.

2nd. A mud drum to receive all impurities deposited from the water, and so placed as to be removed from the action of the fire.

3rd. A steam and water capacity sufficient to prevent any fluctuation in steam pressure or water level.

4th. A water surface for the disengagement of the steam from the water, of sufficient extent to prevent foaming.

5th. A constant and thorough circulation of water throughout the boiler, so as to maintain all parts at the same temperature.

6th. The water space divided into sections so arranged that, should any section fail, no general explosion can occur and the destructive effects will be confined to the escape of the contents. Large and free passages between the different sections to equalize the water line and pressure in all.

7th. A great excess of strength over any legitimate strain, the boiler being so constructed as to be free from strains due to unequal expansion, and, if possible, to avoid joints exposed to the direct action of the fire.

8th. A combustion chamber so arranged that the combustion of the gases started in the furnace may be completed before the gases escape to the chimney.

9th. The heating surface as nearly as possible at right angles to the currents of heated gases, so as to break up the currents and extract the entire available heat from the gases.

10th. All parts readily accessible for cleaning and repairs. This is a point of the greatest importance as regards safety and economy.

11th. Proportioned for the work to be done, and capable of working to its full rated capacity with the highest economy.

12th. Equipped with the very best gauges, safety valves and other fixtures.

THE SPRINGDALE POWER STATION OF THE WEST PENN POWER COMPANY, PITTSBURGH, PA., EQUIPPED WITH [illegible] 28 HORSE-POWER OF BABCOCK & WILCOX BOILERS AND SUPERHEATERS

In the light of the performance of the Babcock & Wilcox boiler under the exacting conditions of present-day power-plant practice, the foregoing list of requirements reveals the insight of the inventors of the boiler into the fundamental principles of steam generator design and construction.

Since the Babcock & Wilcox boiler became thoroughly established as a durable and efficient steam generator, many types of water-tube boilers have appeared on the market. Most of them, failing to meet enough of the requirements of a perfect boiler, have fallen by the wayside, while a few, failing to meet all of the requirements, have only a limited field of usefulness. None has been superior, and in the most cases the most ardent admirers of other boilers have been satisfied in looking up to the Babcock & Wilcox boiler as a standard and in claiming that the newer boilers were "just as good."

Records of recent performances, under the most severe conditions of service on land and sea, show that the Babcock & Wilcox boiler can be run continuously and regularly at higher overloads, with higher efficiency, and lower upkeep cost, than any other boiler on the market. It is especially adapted for power-plant work where it is necessary to use a boiler in which steam can be raised quickly and the boiler placed on the line, either from a cold state or from a banked fire, in the shortest possible time, and with which the capacity, with clean feed water, will be largely limited by the quantity of coal that can be burned in the furnace.

The distribution of the circulation through the separate headers and sections and the action of the headers in forcing a maximum and continuous circulation in the lower tubes, permit the operation of the Babcock & Wilcox boiler without objectionable priming, with a higher degree of concentration of salts in the water than is possible in any other type of boiler.

In the development of electrical power stations it becomes more and more apparent that it is economical to run a boiler at high ratings during the times of peak loads, as by so doing the lay-over losses are diminished and the economy of the plant as a whole is increased.

The number and importance of the large electric lighting and power stations constructed during the last ten years that are equipped with Babcock & Wilcox boilers, is a most gratifying demonstration of the merit of the apparatus, especially in view of their satisfactory operation under conditions which are perhaps more exacting than those of any other service.

Time, the test of all, results with boilers as with other things, in the survival of the fittest. When judged on this basis, the Babcock & Wilcox boiler stands preeminent in its ability to cover the whole field of steam generation with the highest commercial efficiency obtainable. Year after year, the Babcock & Wilcox boiler has become more firmly established as the standard of excellence in the boiler-making art.

1456 Horse power Installation of Babcock & Wilcox Boilers at the Raritan Woolen Mills, Raritan, N. J. The first of these Boilers Was Installed in 1878 and Is Still Operated at 80 Pounds Pressure

EVOLUTION OF THE BABCOCK & WILCOX WATER-TUBE BOILER

QUITE as much may be learned from the records of failure as from those of success. Where a device has been once fairly tried and found to be imperfect or impracticable, the knowledge of that trial is of advantage in further investigation. Regardless of the lesson taught by failure, however, it is an almost every-day occurrence that some device or construction which has been tried and found wanting, if not worthless, is again introduced as a great improvement upon a device which has been shown by its survival to be the fittest.

The success of the Babcock & Wilcox boiler is due to many years of constant adherence to one line of research, in which an endeavor has been made to introduce improvements with the view to producing a boiler which would most effectively meet the demands of the times. During the period that this boiler has been built, other companies have placed on the market more than thirty water-tube or sectional water-tube boilers, most of which, though they may have attained some distinction and sale, have now entirely disappeared. The following incomplete list will serve to recall the names of some of the boilers that have had a vogue at various times, but which are now practically unknown: Dimpfel, Howard, Griffith & Wundram, Dinsmore, Miller Fire Box, Miller American, Miller Internal Tube, Miller Inclined Tube, Phleger, Weigant, the Lady Verner, the Allen, the Kelly, the Anderson, the Rogers & Black, the Eclipse or Kilgore, the Moore, the Baker & Smith, the Renshaw, the Shackleton, the Duplex, the Pond & Bradford, the Whittingham, the Bee, the Hazelton or Common Sense, the Reynolds, the Suplee or Luder, the Babbit, the Reed, the Smith, the Standard, the Atlas, etc.

It is with the object of protecting our customers and friends from loss through purchasing discarded ideas that there is given on the following pages a brief history of the development of the Babcock & Wilcox boiler as it is built today. The illustrations and brief descriptions indicate clearly the various designs and constructions that have been used and that have been replaced, as experience has shown in what way improvement might be made. They serve as a history of the experimental steps in the development of the present Babcock & Wilcox boiler, the value and success of which, as a steam generator, is evidenced by the fact that the largest and most discriminating users continue to purchase them after years of experience in their operation.

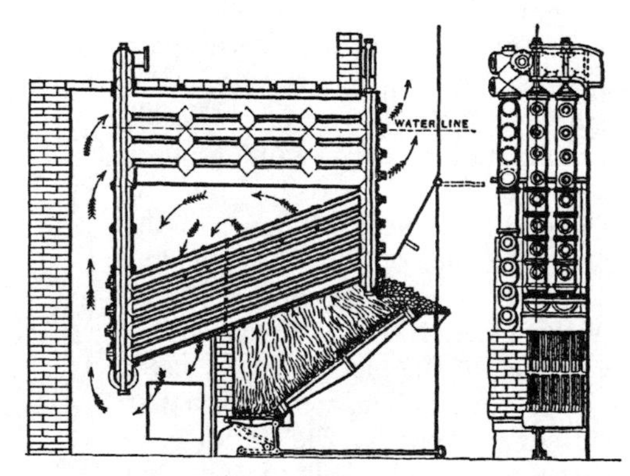

No. 1

No. 1. The original Babcock & Wilcox boiler was patented in 1867. The main idea in its design was safety, to which all other features were sacrificed wherever they conflicted. The boiler consisted of a nest of horizontal tubes, serving as a steam and water reservoir, placed above and connected at each end by bolted joints to a second nest of inclined heating tubes filled with water. The tubes were placed one above the other in vertical rows, each row and its connecting end forming a single casting. Handholes were placed at each end for cleaning. Internal tubes were placed within the inclined tubes with a view to aiding circulation.

No. 2. This boiler was the same as No. 1, except that the internal circulating tubes were omitted, as they were found to hinder rather than help the circulation.

Nos. 1 and 2 were found to be faulty in both material and design, cast metal proving unfit for heating surfaces placed directly over the fire, as it cracked as soon as any scale formed.

No. 3. Wrought-iron tubes were substituted for the cast-iron heating tubes, the ends being brightened, laid in moulds, and the headers cast on.

The steam and water capacities in this design were insufficient to secure regularity of action, there being no reserve upon which to draw during firing or when the water was fed intermittently. The attempt to dry the steam by superheating it in the nest of tubes forming the steam space was found to be impracticable. The steam delivered was either wet, dry or superheated, according to the rate at which it was being drawn from the boiler. Sediment was found to lodge in the lowermost point of the boiler at the rear end and the exposed portions cracked off at this point when subjected to the furnace heat.

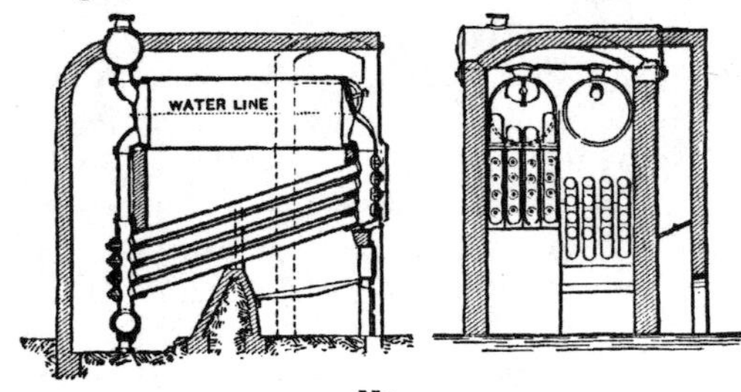

No. 4

No. 4. A plain cylinder, carrying the water line at its center and leaving the upper half for steam space, was substituted for the nest of tubes forming the steam and water space in Nos. 1, 2 and 3. The sections were made as in No. 3 and a mud drum added to the rear end of the sections at the point that was lowest and farthest removed from the fire. The gases were made to pass off at one side and did not come into contact with the mud drum. Dry steam was obtained through the increase of separating surface and steam space, and the added water capacity furnished a storage for heat to tide over irregularities of firing and feeding. By the addition of the drum, the boiler became a serviceable and practical design, retaining all of the features of safety. As the drum was removed from the direct action of the fire, it was not subjected to excessive strain due to unequal expansion, and its diameter, if large in comparison with that of the tubes formerly used, was small when compared with that of cylindrical boilers. Difficulties were encountered in this boiler in securing reliable joints between the wrought-iron tubes and the cast-iron headers.

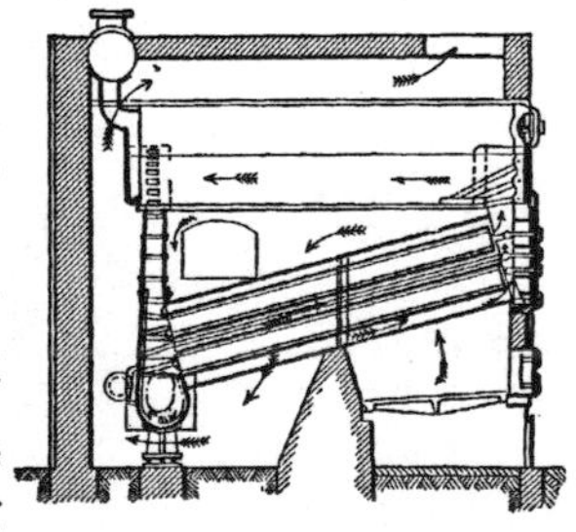
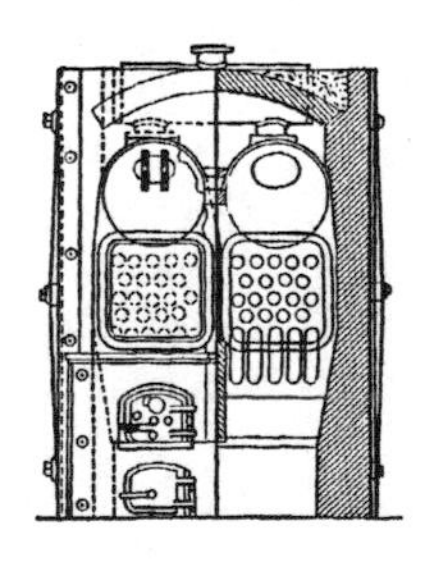
No. 5

No. 5. In this design, wrought-iron water legs were substituted for the cast-iron headers, the tubes being expanded into the inside sheets and a large cover placed opposite the front end of the tubes for cleaning. The tubes were staggered one above the other, an arrangement found to be more efficient in the absorption of heat than where they were placed in vertical rows. In other respects, the boiler was similar to No. 4, except that it had lost the important element of safety through the introduction of the very objectionable feature of flat-stayed surfaces. The large doors for access to the tubes were also a cause of weakness.

An installation of these boilers was made at the plant of the Calvert Sugar Refinery in Baltimore, and while they were satisfactory in their operation, were never duplicated.

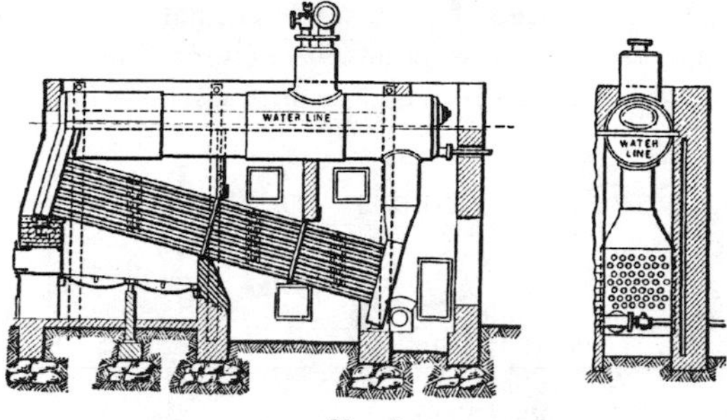

No. 6

No. 6. This was a modification of No. 5, in which longer tubes were used and over which the gases were caused to make three passes with a view of better economy. In addition, some of the stayed surfaces were omitted and handholes substituted for the large access doors. A number of boilers of this design were built, but their excessive first cost, the lack of adjustability of the structure under varying temperatures, and the inconvenience of transportation, led to No. 7.

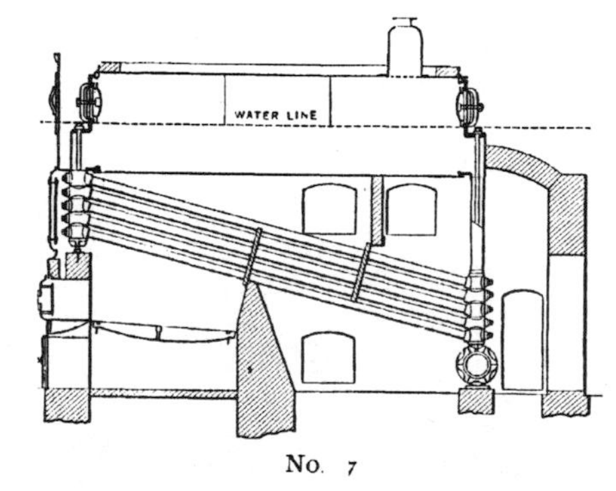

No. 7

No. 7. In this boiler, the headers and water legs were replaced by T-heads screwed to the ends of the inclined tubes. The faces of these Ts were milled and the tubes placed one above the other with the milled faces metal to metal. Long bolts passed through each vertical section of the T-heads and through connecting boxes on the heads of the drums holding the whole together. A large number of boilers of this design were built and many were in successful operation for over twenty years. In most instances, however, they were altered to later types.

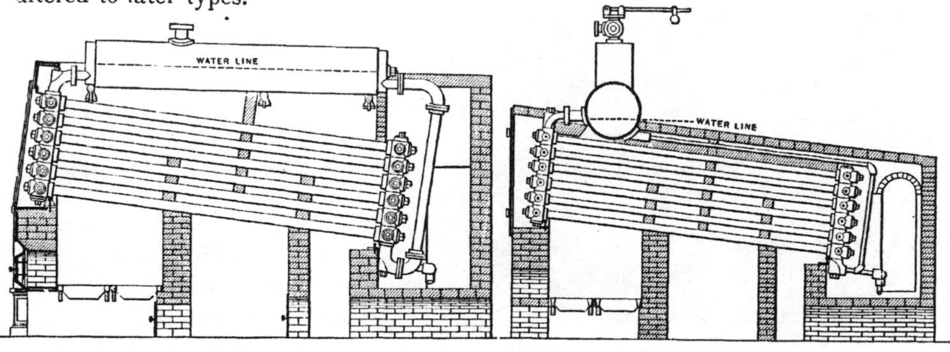

No. 8 No. 9

Nos. 8 and 9. These boilers were known as the Griffith & Wundram type, the concern which built them being later merged in The Babcock & Wilcox Company. Experiments were made with this design with four passages of the gases across the tubes and the downward circulation of the water at the rear of the boiler was carried to the bottom row of tubes. In No. 9, an attempt was made to increase the safety and reduce the cost by reducing the amount of steam and water capacity. A drum at right angles to the line of tubes was used, but as there was no provision made to secure dry steam, the results were not satisfactory. The next move in the direction of safety was the employment of several drums of small diameter instead of a single drum.

This is shown in No. 10. A nest of small horizontal drums, 15 inches in diameter, was used in place of the single drum of larger diameter. A set of circulation tubes was placed at an intermediate angle between the main bank of heating tubes and the horizontal drums forming the steam reservoir. These circulators were to return to the rear end of the circulating tubes the water carried up by the circulation, and in this way were to allow only steam to be delivered to the small drums above. There was no improvement in the action of this boiler over that of No. 9.

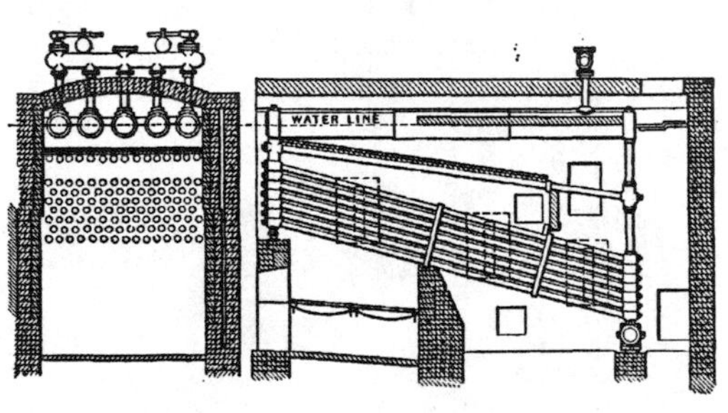

No. 10

The four passages of the gas over the tubes tried in Nos. 8, 9 and 10 were not found to add to the economy of the boiler.

No. 11. A trial was next made of a box coil system, in which the water was made to traverse the furnace several times before being delivered to the drum above. The tendency here, as in all similar boilers, was to form steam in the middle of the coil and blow the water from each end, leaving the tubes practically dry until the steam found an outlet and the water returned. This boiler had, in addition to a defective circulation, a decidedly geyser-like action and produced wet steam.

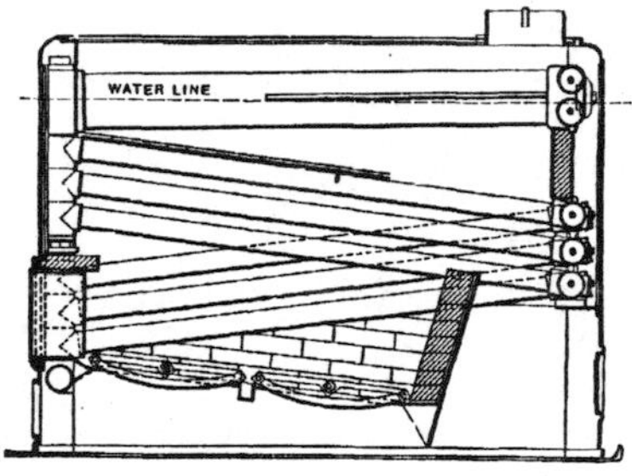

No. 11

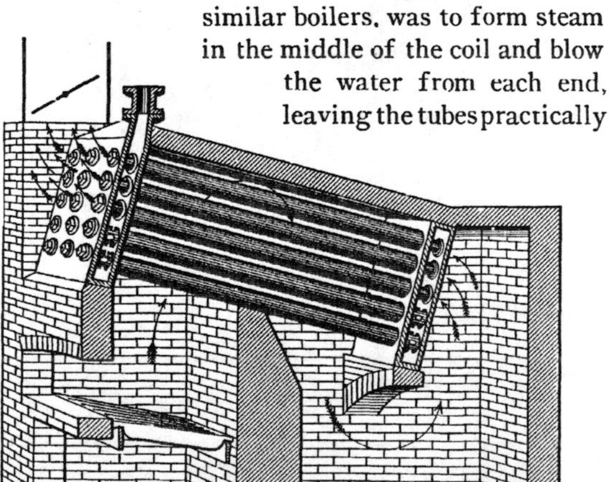
No. 12

All of the types mentioned, with the exception of Nos. 5 and 6, had between their several parts a large number of bolted joints which were subjected to the action of the fire. When these boilers were placed in operation, it was demonstrated that as soon as any scale formed on the heating surfaces, leaks were caused, due to unequal expansion.

No. 12. With this boiler, an attempt was made to remove the joints from the fire and to increase the heating surface in a given space. Water tubes were expanded into both sides of wrought-iron boxes, openings being made for the admission of water and the exit of steam. Fire tubes were placed inside the water tubes to increase the heating surface. This design was abandoned because of the rapid stopping up of the tubes by scale and the impossibility of cleaning them.

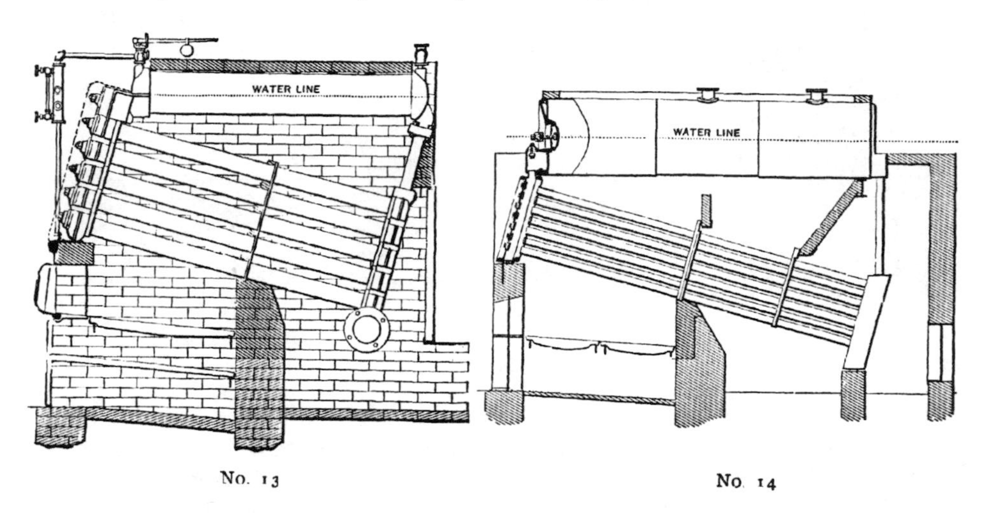

No. 13 No. 14

No. 13. Vertical straight line headers of cast iron, each containing two rows of tubes, were bolted to a connection leading to the steam and water drum above.

No. 14. A wrought-iron box was substituted for the double cast-iron headers. In this design, stays were necessary and were found, as always, to be an element to be avoided wherever possible. The boiler was an improvement on No. 6, however. A slanting bridge wall was introduced underneath the drum to throw a larger portion of its heating surface into the combustion chamber under the bank of tubes. This bridge wall was found to be difficult to keep in repair and was of no particular benefit.

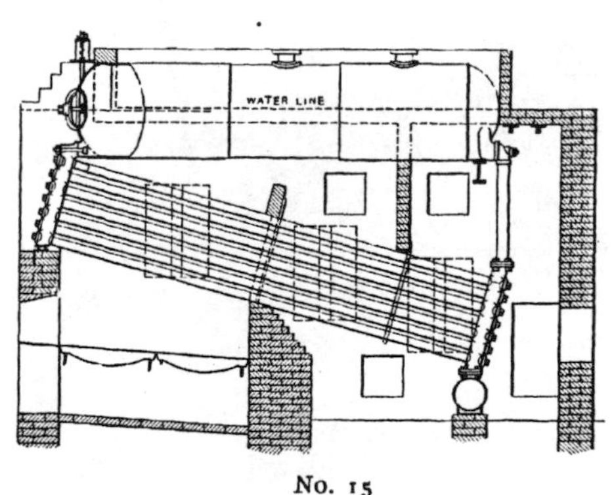

No. 15

No. 15. Each row of tubes was expanded at each end into a continuous header, cast of car wheel metal. The headers had a sinuous form so that they would lie close together and admit of a staggered position of the tubes when assembled. While other designs of header form were tried later, experience with Nos. 14 and 15 showed that the style here adopted was the best for all purposes and it has not been changed materially since. The drum in this design was supported by girders resting on the brickwork. Bolted joints were discarded, with the exception of those connecting the headers to the front and rear ends of the drums and the bottom of the rear headers to the mud drum. Even such joints, however, were found objectionable and were superseded in subsequent construction by short lengths of tubes expanded into bored holes.

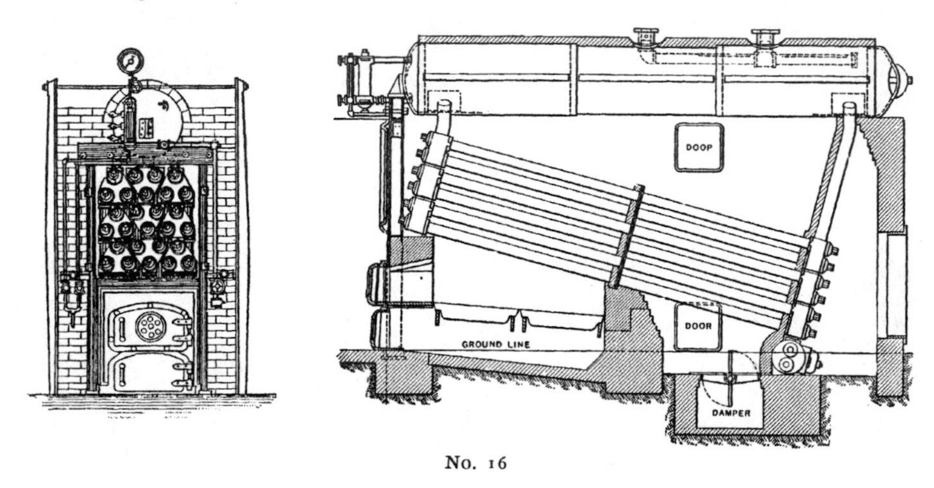

No. 16

No. 16. In this design, headers were tried which were made in the form of triangular boxes, in each of which there were three tubes expanded. These boxes were alternately reversed and connected by short lengths of expanded tubes, being connected to the drum by tubes bent in a manner to allow them to enter the shell normally. The joints between headers introduced an element of weakness and the connections to the drum were insufficient to give adequate circulation.

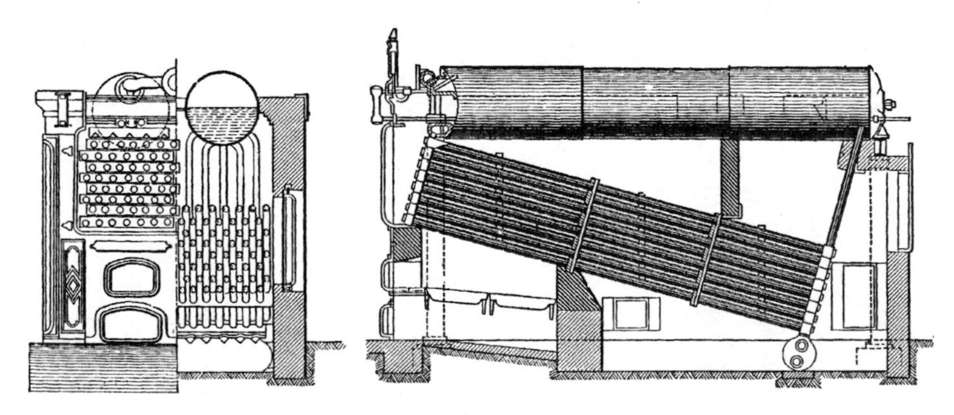

No. 17

No. 17. Straight horizontal headers were next tried, alternately shifted right and left to allow a staggering of tubes. These headers were connected to each other and to the drums by expanded nipples. The objections to this boiler were almost the same as those to No. 16.

Nos. 18 and 19. These boilers were designed primarily for fire-protection purposes, the requirements demanding a small, compact boiler with ability to raise steam

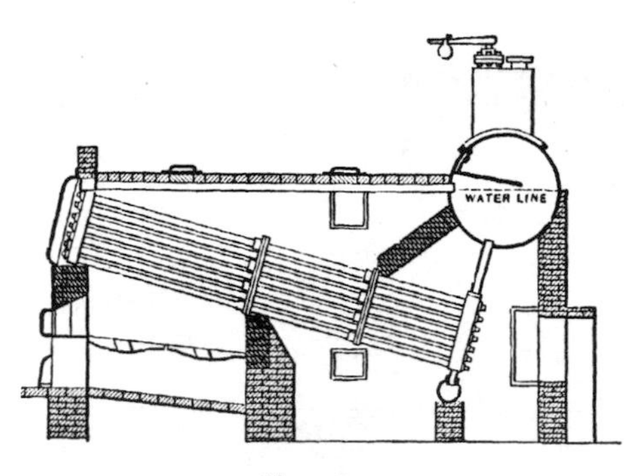

No. 18

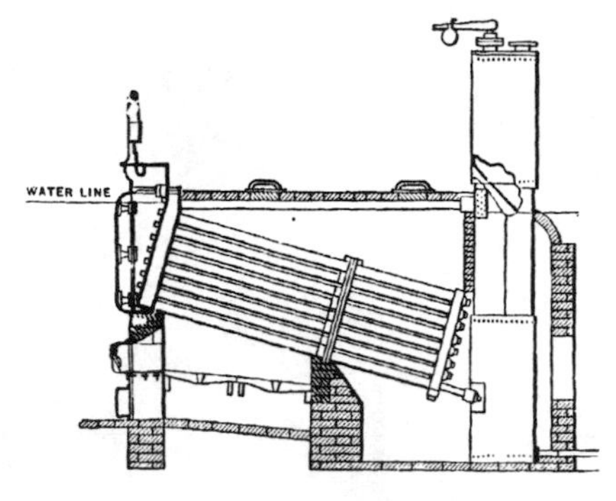

No. 19

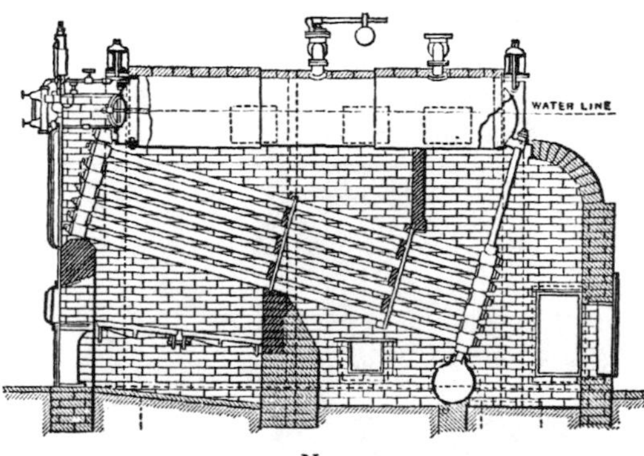

No. 20A

quickly. These both served the purpose admirably, but, as in No. 9, the only provision made for the securing of dry steam was the use of the steam dome, shown in the illustration. This dome was found inadequate and has since been abandoned in nearly all forms of boiler construction. No other remedy being suggested at the time, these boilers were not considered as desirable for general use as Nos. 21 and 22. In Europe, however, where small size units were more in demand, No. 18 was modified somewhat and used largely with excellent results. These experiments, as they may now be called, although many boilers of some of the designs were built, clearly demonstrated that the best construction and efficiency required adherence to the following elements of design:

1st. Sinuous headers for each vertical row of tubes.

2nd. A separate and independent connection with the drum, both front and rear, for each vertical row of tubes.

No. 20B

Printed from Original Engraving in "Steam," Second Edition, 1879

THE 150 HORSE-POWER BOILER EXHIBITED BY BABCOCK & WILCOX AT THE CENTENNIAL EXHIBITION, PHILADELPHIA, IN 1876

The consistently high economy under test, good mechanical design and exceptional safety of this boiler were studied by most of the leading mechanical engineers of that period and the rapid adoption of the Babcock & Wilcox boiler followed. This boiler was used from 1877 to 1911 at the Brooklyn Sugar Refinery of Decastro & Donner, one of the companies consolidated into The American Sugar Refining Company. It was bought by The Babcock & Wilcox Company in 1911 and formed part of the Company's exhibit at the Panama-Pacific Exposition at San Francisco in 1915. It is now preserved at the Bayonne Works of the Company as one of the most important landmarks in the history of steam generation.

3rd. All joints between parts of the boiler proper to be made without bolts or screw plates.

4th. No surfaces to be used which necessitate the use of stays.

5th. The boiler supported independently of the brickwork so as to allow freedom for expansion and contraction as it is heated or cooled.

6th. Ample diameter of steam and water drums, these not to be less than 30 inches except for small size units.

7th. Every part accessible for cleaning and repairs.

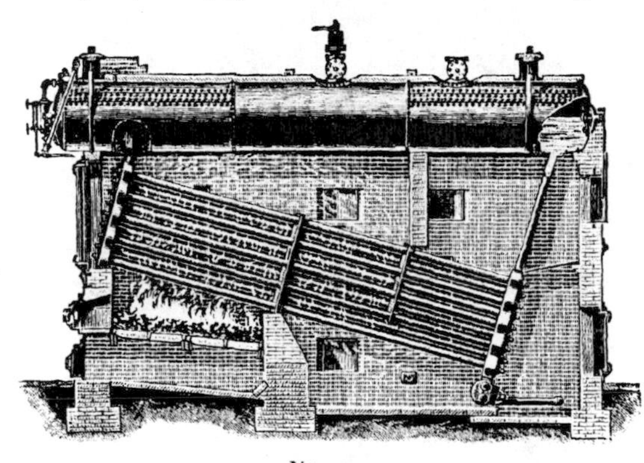

No 21

These desirable features having been determined, No. 20 was designed. This boiler had all the features just enumerated, together with a number of improvements in details of construction. The general form of No. 15 was adhered to, but the bolted connections between the sections and the drum and between the sections and the mud drum were discarded in favor of connections made by short lengths of boiler tube expanded into the adjacent parts. This boiler was suspended from girders, like No. 15, but these in turn were carried on vertical supports, leaving the pressure parts entirely free from the brickwork, the mutually deteriorating strains present where one was supported by the other being in this way overcome. Hundreds of thousands of horse-power of this design were built, giving great satisfaction. The boiler was known as the "C. I. F." (cast-iron front) style, an ornamental cast-iron front having been usually furnished.

The next step, and the one which connects the boilers as described above to the boiler as it is built today, was design No. 21. These boilers were known as the "W. I. F." style, the fronts furnished as part of the equipment being constructed largely of wrought iron. The cast-iron drum heads used in No. 20 were replaced by wrought-steel flanged and "bumped" heads. The drums were made longer and the sections connected to wrought-steel cross boxes riveted to the bottom of the drums. The boilers were supported by girders and columns as in No. 20.

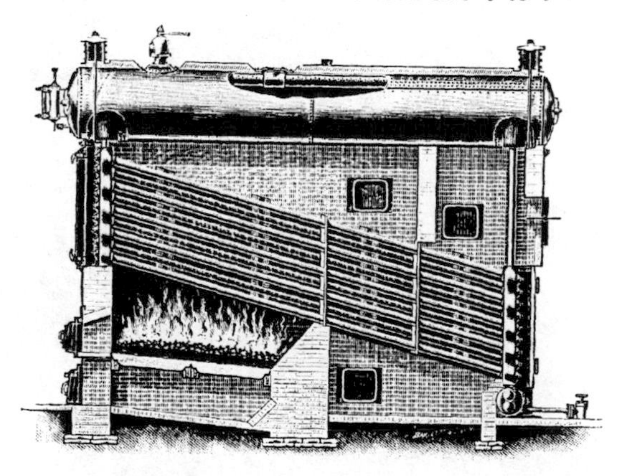

No. 22

U. S. S. "Tennessee," with 13,925 Horse-power of Babcock & Wilcox Marine Boilers. The Cross-drum Type of Water-tube Boiler Was Widely Adopted for Marine Service Before It Had Any Considerable Use on Land. Up to the Close of 1922 Over 8,000,000 Horse-power of Babcock & Wilcox Marine Boilers Had Been Built

No. 22. This boiler, which is designated as the "Vertical Header" type, has the same general features of construction as No. 21, except that the tube sheet side of the headers is "stepped" to allow the headers to be placed vertically and at right angles to the drum and still maintain the tubes at the angle used in Nos. 20 and 21.

No. 23, or the cross-drum design of boiler, is a development of the Babcock & Wilcox marine boiler, in which the cross-drum is used exclusively. The experience of the Glasgow Works of Babcock & Wilcox, Ltd., with No. 18 proved that proper attention to details of construction would make it a most desirable form of boiler where headroom was limited. A large number of this design have been successfully installed and are giving satisfactory results under widely varying conditions. The cross-drum boiler is also built in a vertical header design.

No. 23

Boilers Nos. 21, 22 and 23, with some modification, mostly in construction details, are representative of the standard forms as now offered. These designs as now constructed are illustrated on pages 30, 34, 36 and 42.

As originally constructed, the boilers represented by Nos. 21, 22 and 23 were made with cast-iron headers. With the increased boiler pressures demanded, and the demands of certain laws originally promulgated in Europe and now coming into general use in this country, the last step in the development of the water-tube boiler was taken. This consists in the use, for all pressure parts of the boiler, including headers, cross boxes, nozzles, etc., of wrought steel when the pressure exceeds 160 pounds. The Babcock & Wilcox Company has at the present time plants producing steel forgings that have been pronounced by *The Engineer*, of London, to be "a perfect triumph of the forger's art."

The various designs of these all-wrought-steel boilers are fully illustrated in the following pages.

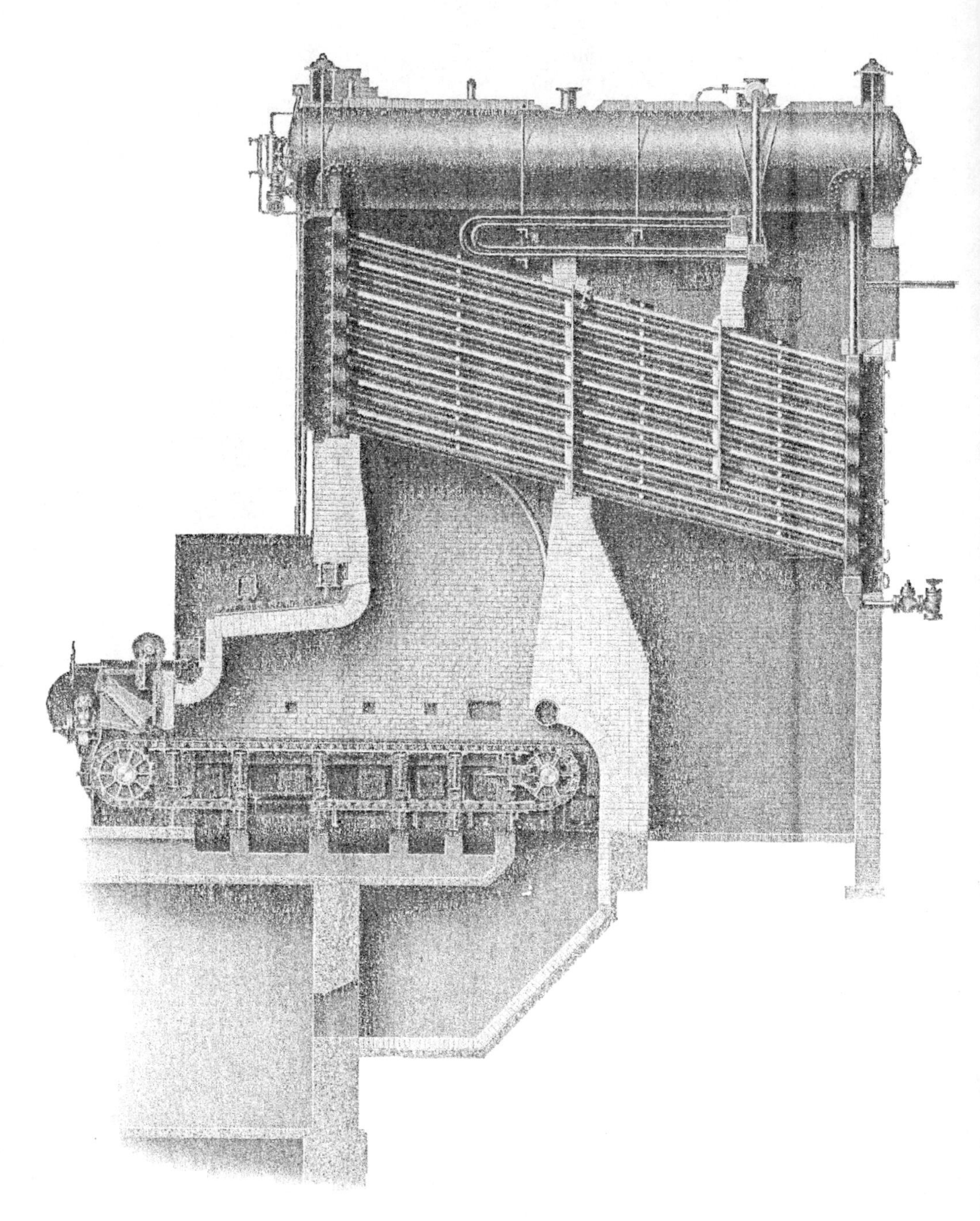

Babcock & Wilcox Longitudinal-drum Boiler with Babcock & Wilcox Superheater and Babcock & Wilcox Blast Chain Grate Stoker

THE BABCOCK & WILCOX BOILER

THE following description of the Babcock & Wilcox boiler will clearly indicate the manner in which it fulfills the requirements of the perfect steam boiler as previously given.

The Babcock & Wilcox boiler is built in two general classes, the longitudinal-drum type and the cross-drum type. While the latter was originally designed to meet certain conditions of headroom, it has become popular in numerous classes of work where headroom is not a factor, and is particularly adaptable for large units. Either the longitudinal or the cross-drum design may be constructed with vertical headers or with inclined headers perpendicular to the tubes, and the headers in turn may be of wrought steel or of cast iron, depending upon the pressure for which the boiler is constructed.

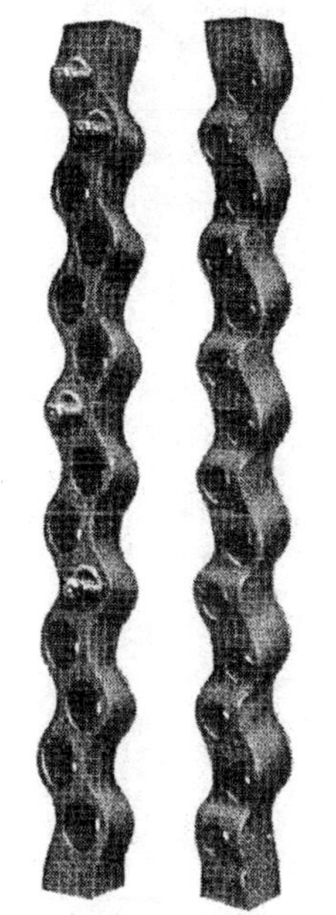

Wrought-steel Vertical Header with Inside Fittings

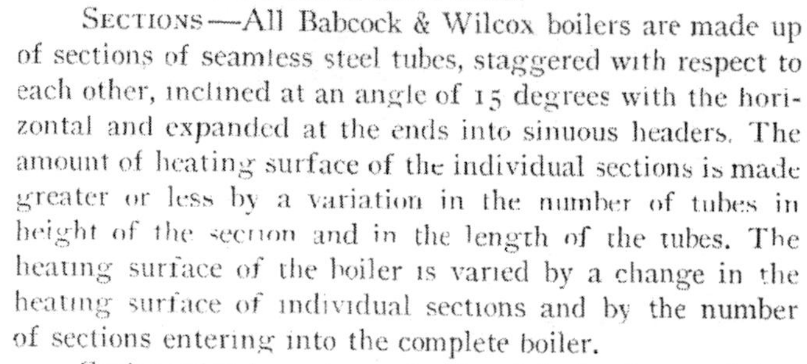

Sections—All Babcock & Wilcox boilers are made up of sections of seamless steel tubes, staggered with respect to each other, inclined at an angle of 15 degrees with the horizontal and expanded at the ends into sinuous headers. The amount of heating surface of the individual sections is made greater or less by a variation in the number of tubes in height of the section and in the length of the tubes. The heating surface of the boiler is varied by a change in the heating surface of individual sections and by the number of sections entering into the complete boiler.

Each section is made up of a downtake header supplying water to the tubes, and an uptake header discharging water and steam from the tubes.

Sections may be single deck, formed with a single downtake and a single uptake header, or double or triple deck, using more than one downtake and more than one uptake header connected by expanded nipples.

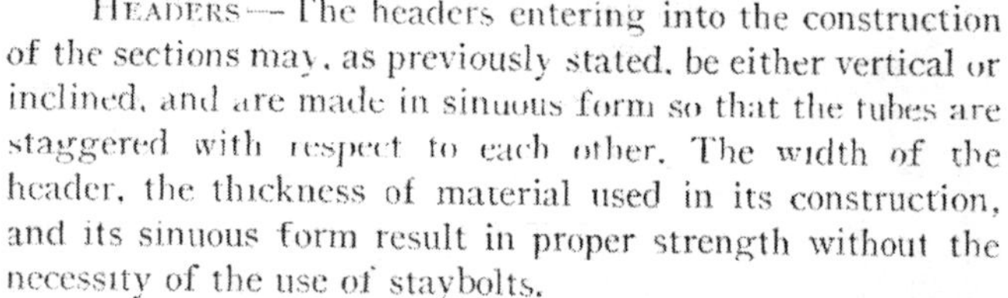

Headers—The headers entering into the construction of the sections may, as previously stated, be either vertical or inclined, and are made in sinuous form so that the tubes are staggered with respect to each other. The width of the header, the thickness of material used in its construction, and its sinuous form result in proper strength without the necessity of the use of staybolts.

Headers are either of cast iron or of wrought steel, depending upon the pressure for which the boiler is constructed and the service in which it is to be used. For all pressures in excess of 160 pounds, wrought-steel headers are used

Handholes and Fittings—Opposite each tube end in the headers there is placed a handhole of sufficient size to permit the cleaning, removal, and renewal of the tube.

These handhole openings in the wrought-steel vertical header are elliptical in shape, machine-faced, and milled to a true plane back from the edge a sufficient distance to form a gasket seat. The openings are closed by inside-fitting, forged-steel plates, shouldered to center in the opening, their flanged seats being milled to a true plane and the fittings being milled to fit the opening. These plates are held in position by studs and forged-steel binders and nuts, as well as by the pressure within the boiler. The joints between plates and header seats are made with thin gaskets

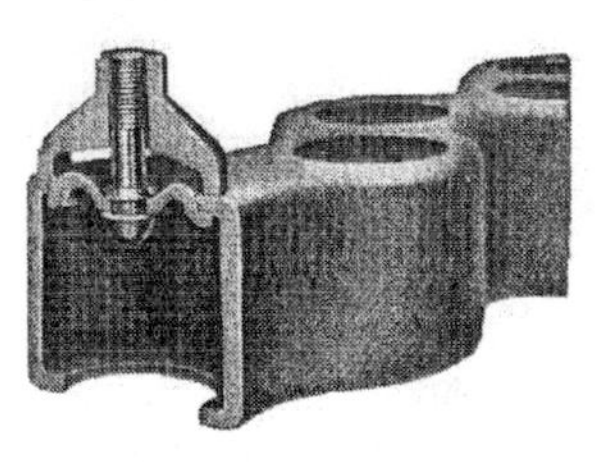

Inside Handhole Fitting Wrought-steel Inclined and Vertical Header

In wrought-steel inclined headers the handholes may be closed with inside fittings similar to those described for wrought-steel vertical headers or by forged-steel. circular, outside fittings, which engage a raised milled seat, machined to a true plane on the header, the fittings being held in place with a forged-steel safety clamp with a central bolt and an outside cast-iron cap nut The surfaces between the handhole fitting and header and between the fitting and cap nut, are lapped to be tight without gaskets

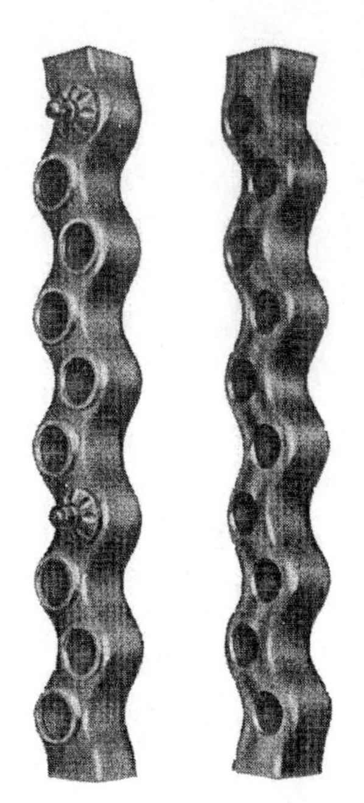

Wrought-steel Inclined Header with Outside Fittings

Cast-iron vertical headers have outside oval fittings of cast iron; otherwise they are similar to the outside wrought-steel circular fittings described for inclined wrought-steel headers.

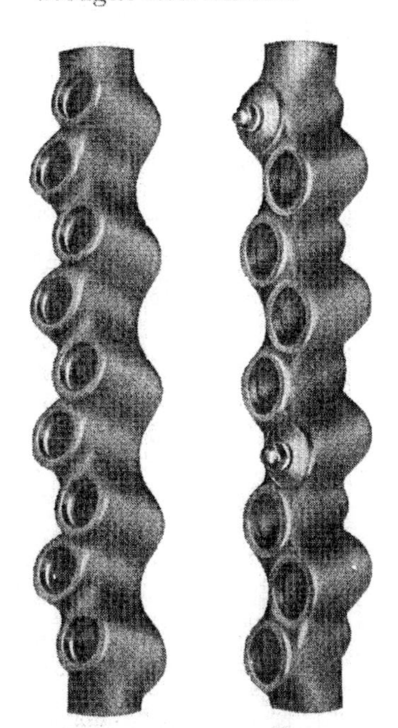

Cast-iron Vertical Header

Cast-iron inclined headers have similar outside handhole fittings, which are circular.

All outside handhole fittings are held in position by forged-steel safety clamps which close the opening from the inside and are held in position by ball-headed bolts to assure proper alignment.

Mud Drums—The lower ends of all downtake headers are connected by nipples to a transverse mud drum. which is a forged-steel box, 7¼ inches square. This mud drum is machined for receiving blow-off connections, and has handholes for cleaning and fittings that are the same as those described for wrought-steel vertical headers

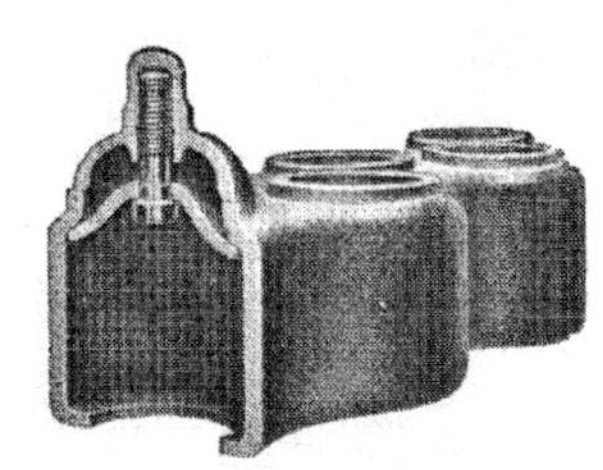

Outside Handhole Fitting Wrought-steel Inclined Header

Tubes and Connections—The tubes entering into the construction of the sections are, as stated, of seamless steel, No. 10 B. W G., for 4-inch tubes up to and including 160 pounds pressure; No. 9 gauge up to and including 240 pounds pressure; and for higher pressures in accordance with the "Report of the Boiler Code Committee of the American Society of Mechanical Engineers."

All connections between sections and drums, and between sections and mud drum, are made of seamless-steel nipples expanded into the parts to be connected. These nipples are of No. 9 B. W G. for pressures of 240 pounds or less and above this pressure of the same gauge as the boiler tubes

DRUM CONNECTIONS—The construction of the sections as above described is the same for longitudinal and cross-drum Babcock & Wilcox boilers.

In longitudinal-drum boilers the connections between the sections and the drum or drums are made by nipples from the sections to wrought-steel cross boxes near the forward and rear ends of the drum, over each group ten sections or less wide. Due to the inclination of the tubes the downtake header connections are relatively long, and, as distinguished from short connecting nipples, are known as vertical circulating tubes.

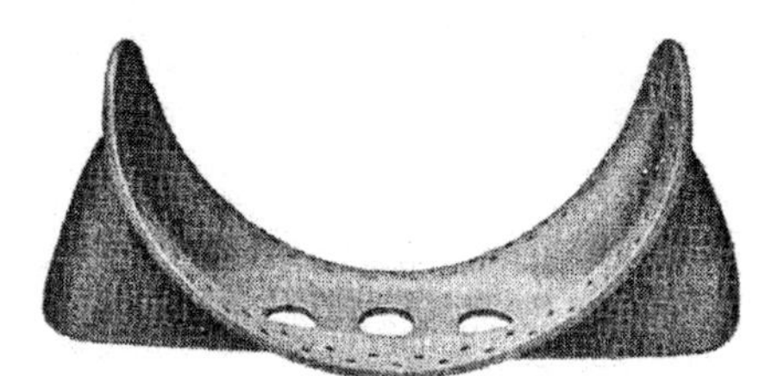

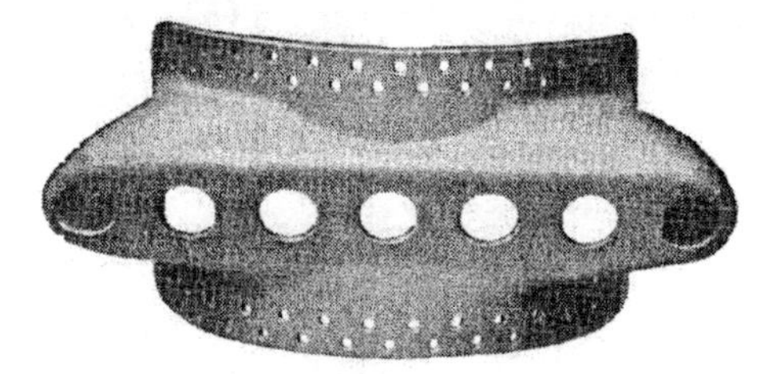

FORGED-STEEL CROSS BOX

In cross-drum boilers the upper ends of the downtake headers are connected below the water line to the lower side of a transverse steam-and-water drum located above the downtake headers, and the upper ends of the uptake headers are connected by one or more rows of tubes, called horizontal circulating tubes, to the forward face of the cross drum, at or above the water level.

DRUM CONSTRUCTION—All drums, both longitudinal and cross drums, have shell plates made with longitudinal seams formed of inner and outer butt straps. Longitudinal drums of 36 and 42 inches diameter are constructed of a single plate with but the one longitudinal butt-strap joint, while drums from 48 to 60 inches in diameter are constructed with a single longitudinal course, consisting of two plates, the butt-strap joints of which are above the water level in the drums. The use of such drum construction eliminates all circumferential seams except those of the head seams In longitudinal-drum boilers this construction results in the absence of drum courses with a corresponding increase in difficulty of manufacture, due to the fitting of such courses to each other Cross-drum Babcock & Wilcox boilers have never been made with circumferential seams other than head seams, but the advantages resulting from the absence of such seams in manufacture would be the same. The primary advantage in the use of a single longitudinal sheet or sheets in drum construction

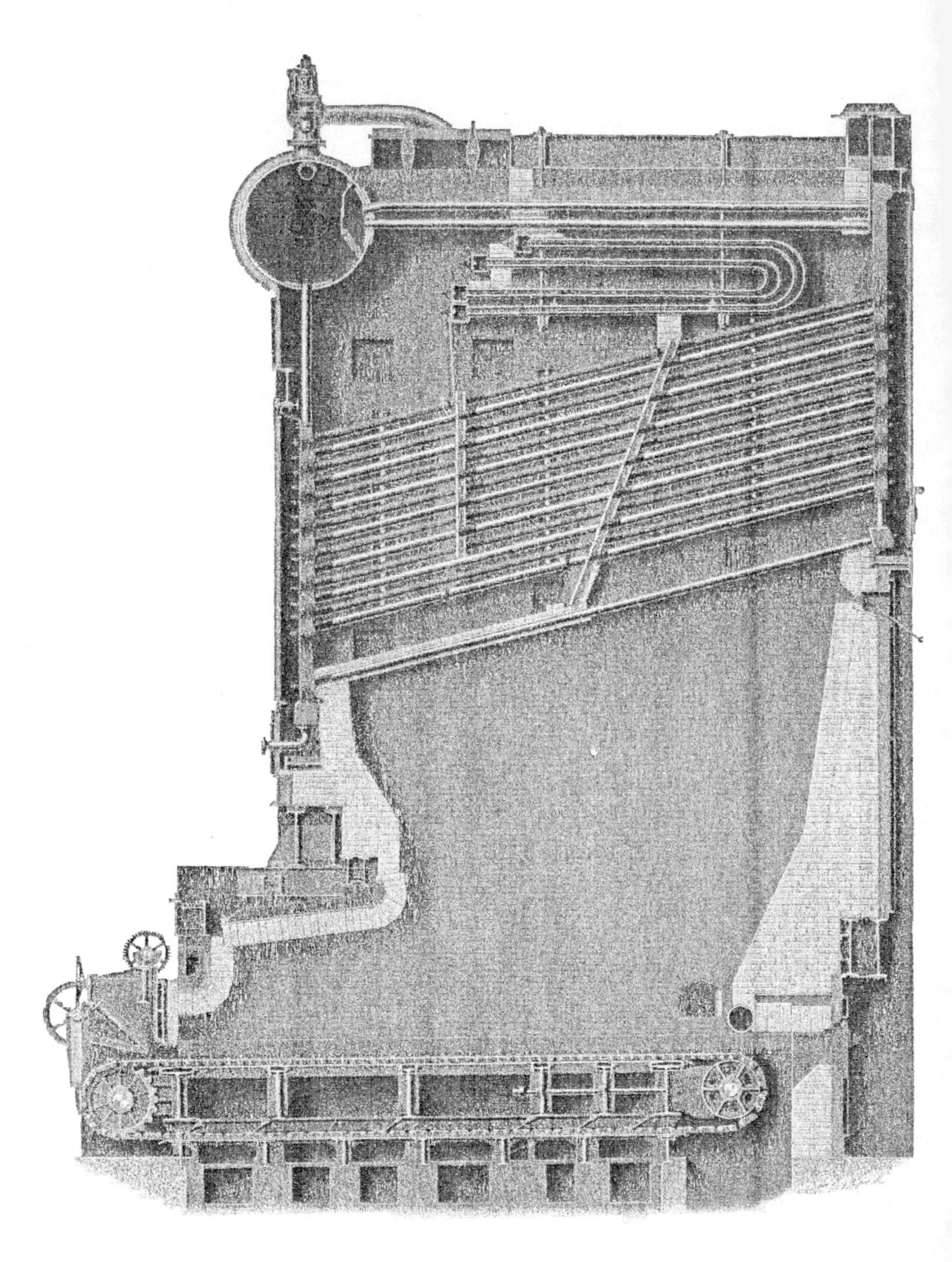

A Babcock & Wilcox Cross-drum Boiler with Babcock & Wilcox Blast Chain Grate Stoker and a Babcock & Wilcox Interposed Superheater

Forged-steel Drum Nozzle

to the boiler user is, however, not dependent upon manufacturing factors, but upon the increased strength of the drum or drums considered as beams.

All shell plates and inner and outer butt straps of all drums are bent to a true circle to the extreme edge of the plates. All edges of plates are planed The edges of the shell plates at the ends of all drums are electrically welded for a distance from the edge of the shells under the outer butt strap. Such construction results in tight drums with a minimum of calking.

All rivet holes are drilled from the solid plate; those holes used for tack bolts are drilled in the flat plate, while all other rivet holes are drilled through the various pieces entering into the joint after the drums are assembled. After the rivet holes

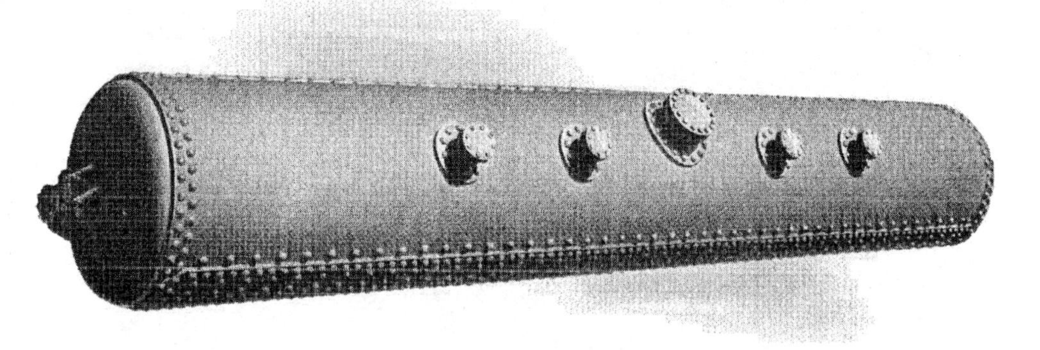

Drum of a Large Cross-drum Boiler

are drilled the drum is taken apart, the burrs are removed from the holes, and the pieces then reassembled and the rivets driven.

All rivets are driven under hydraulic pressure and held until black.

Drum heads are of wrought steel. forged at a single heat. and during the forging process an elliptical manhole opening is cut in the shell, with the edges of the opening flanged inward to give the necessary strength to the opening, and affording a proper seat for the gasket between the drum head and manhole fitting.

The flange of the drum head is given a cylindrical form of the exact size to fit within the drum shell into which it is forced under a heavy hydraulic pressure. The head is then riveted to the drum shell with one or more rows of rivets, depending upon the pressure for which the drum is built, the method of riveting being as described above.

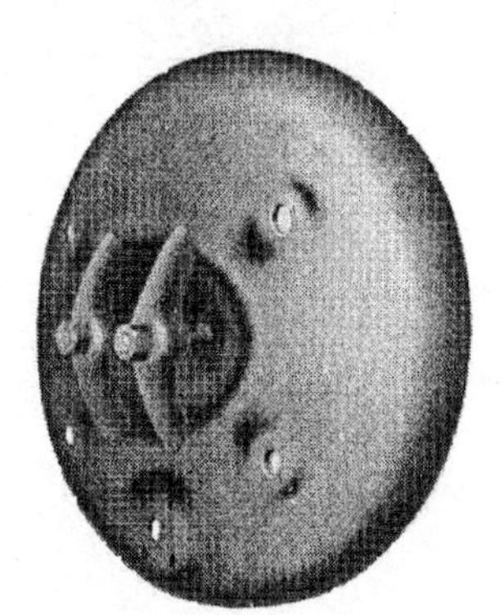

Forged-steel Drum Head with Manhole Plate in Position

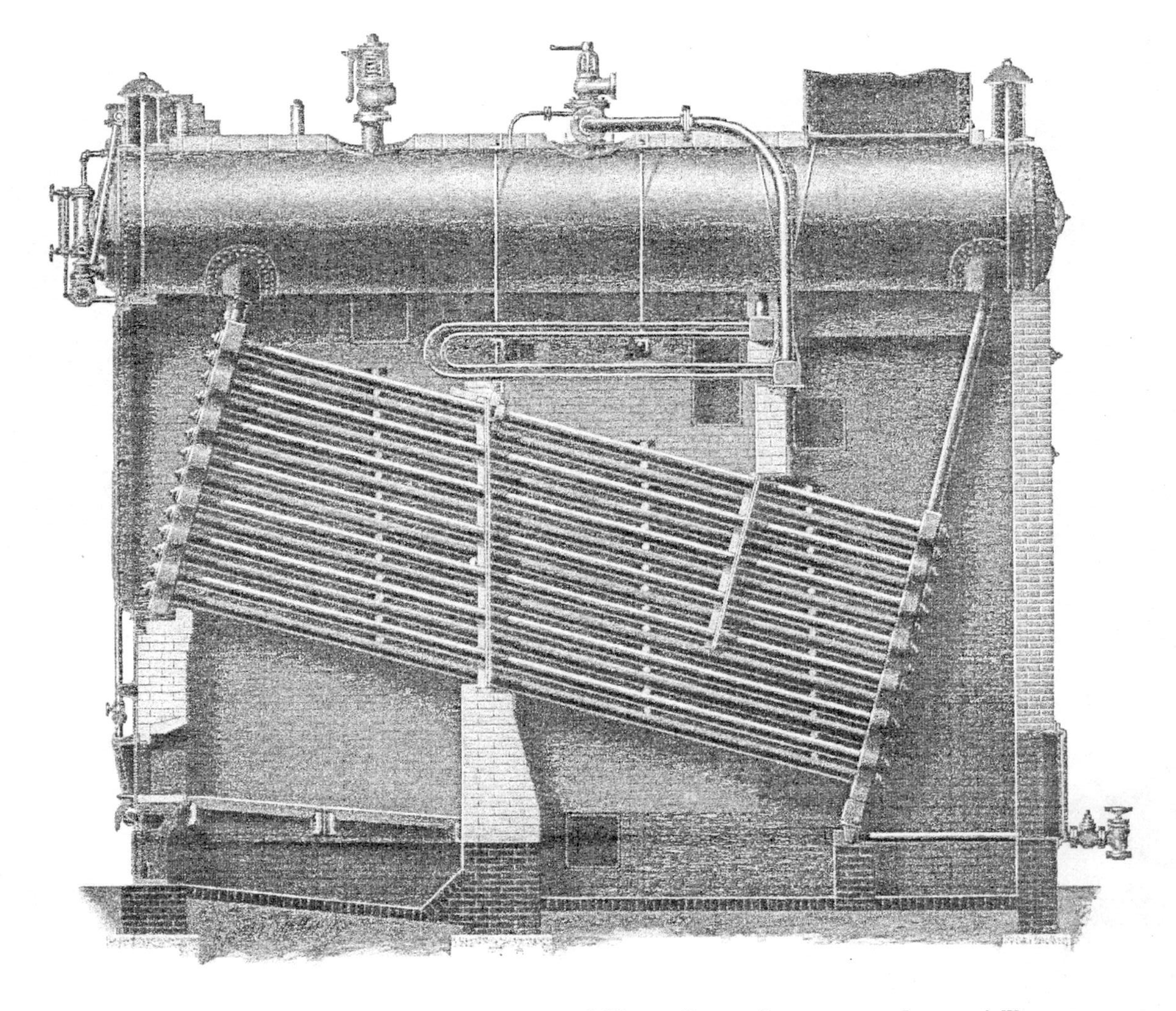

Wrought-steel Inclined Header Longitudinal-drum Babcock & Wilcox Boiler, Equipped with Babcock & Wilcox Superheater

The inner edges of the flanges of the drum head are machined for calking, and the inner end of the flanges at the manhole opening is machined to form a gasket seat.

The manhole fitting is a pressed-steel forging, milled to fit the manhole opening, and its flange is milled to form a seat on the manhole gasket.

In longitudinal-drum Babcock & Wilcox boilers, the connections between sections and drums are made, as stated, through cross boxes. These are forged from a single sheet.

Saturated steam outlets and safety valve connections for all drums are made with wrought-steel high nozzles adapted to receive through bolts, and for high-pressure, cross-drum, power-house boilers all connections to the boilers are made in this way.

All steam drums are fitted with internal feed pipes, baffle plates, and standard dry pipes. These dry pipes, as explained elsewhere, are not dry pipes in the sense of restricting the steam flow but, in reality, collecting pipes.

In longitudinal-drum boilers having more than a single drum, the saturated steam outlets from individual drums are connected by cross pipes. Such cross pipes, for pressures above 160 pounds, are of steel, flanged, while below this pressure flanged cast-iron cross pipes are used.

SUPPORTS—The drums of longitudinal-drum boilers are hung by suspension straps front and rear. The sections are supported by the connections to the drum. The suspension straps ordinarily hang from horizontal members of the boiler suspension frames, though where conditions require it, horizontal supporting members may be supported, or suspension straps may be hung, from the customer's building structure.

In cross-drum boilers the individual sections are hung at the uptake end from individual suspension straps that engage horizontal members which may be boiler supports or a part of the building structure. The downtake end of cross-drum boilers may be supported in one of three ways: by cast-iron chairs from the foundation below the mud drum; by columns extending from below each end of the steam drum to the foundation; or by suspension straps from the customer's overhead building structure.

Any of the methods of suspension described leave the boiler pressure parts independent of the brickwork and perfectly free to expand and contract with temperature changes without stressing the parts.

TUBE DOORS—Front and rear metal tube doors are provided, which, when opened, give full access to headers and handhole fittings.

With small units, these doors are hung on hinges with vertical pins and are held in place with latches.

With large, wide units, removable tube-door panels are hung by landing hooks from upstanding flanges of the tube-door lintels. These panels are held in place by forged-steel clamps, studs and nuts. Such doors are handled with standard chain hoists and I-beam trolleys from transverse I-beams above the tube doors, such beams being ordinarily supplied as a part of the boiler equipment, but may be embodied in the building steelwork. The use of these landing hooks gives a safety factor, in that the tube-door panels cannot fall when the clamps are loosened, and the landing hooks facilitate placing the panels in position.

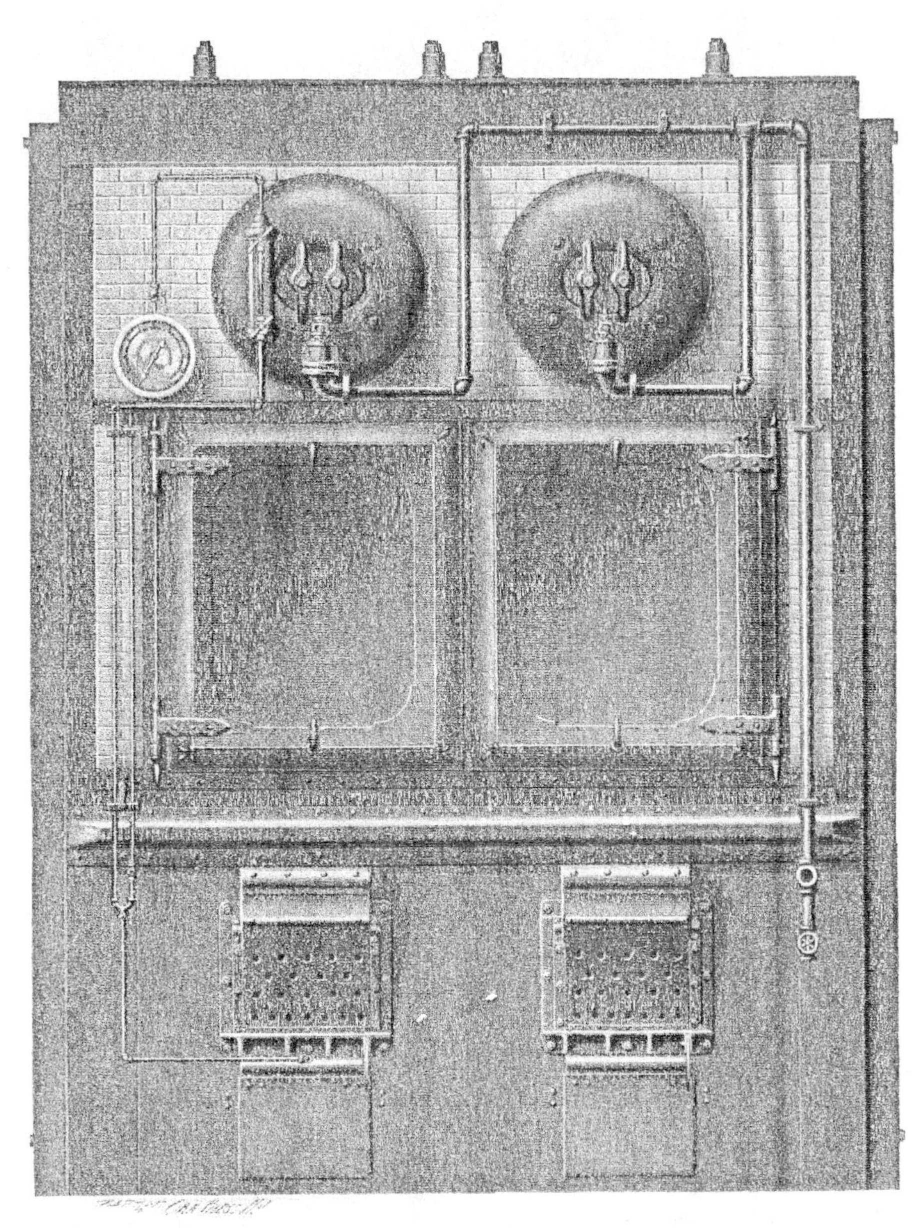

Longitudinal-drum Boiler — Front View

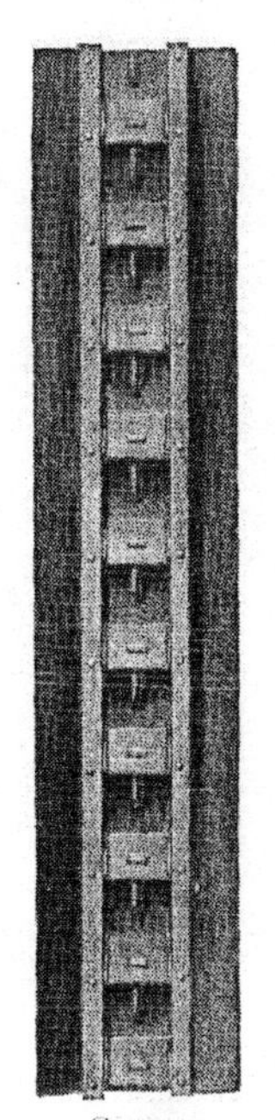

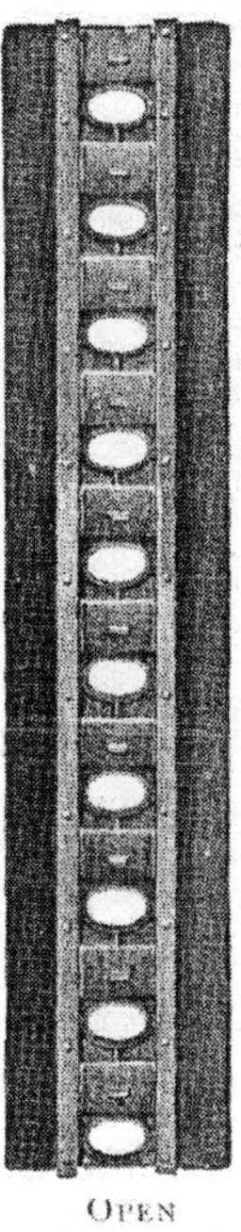

Closed Open

Patented Side Dusting Doors

The removable tube-door panels have an asbestos packing strip between them and the tube-door frame. Tube doors may be lined with insulating material, or not, as conditions warrant

Fronts—Construction below the tube-door sills at the firing end of the boiler is varied to suit the form of the furnace.

Where hand-fired grates are used for coal, a wrought-steel fire front is used, having in-swinging fire and ash-pit doors, which automatically close in the event of the rupture of a tube within the setting, thus affording protection to the firemen. Above this front a steel cross panel makes the connection between the fire front and the tube-door sill. A heavy pressed steel horizontal buckstay forms a joint between the cross panel and fire front, and adds stiffness to the front.

Access and Cleaning Doors—A full set of access and cleaning doors is provided, making all portions of the setting readily accessible for cleaning and inspection. Small dusting doors equipped with self-closing shutters are supplied for the side walls, and through them all portions of the heating surface may be cleaned with a steam lance. These doors are covered by patent.

In batteries of wide boilers, additional cleaning doors are supplied at the top, at the center of the setting, to insure ease in reaching all portions of the setting.

Fixtures—For hand-fired boilers, cast-iron dead plates and supports are supplied, adapted to receive a fire-brick lining. A full set of grate bars and grate bar bearers is furnished, the latter being fitted with expansion sockets for side walls.

Flame baffle plates, with necessary fastenings, and special fire-brick for protecting the plates, are supplied. The first flame baffle is made of a double row of diagonal overlapping tube brick, held in position by horizontal cast-iron overlapping flame plates ground to fit the tubes, and with diagonal cast-iron

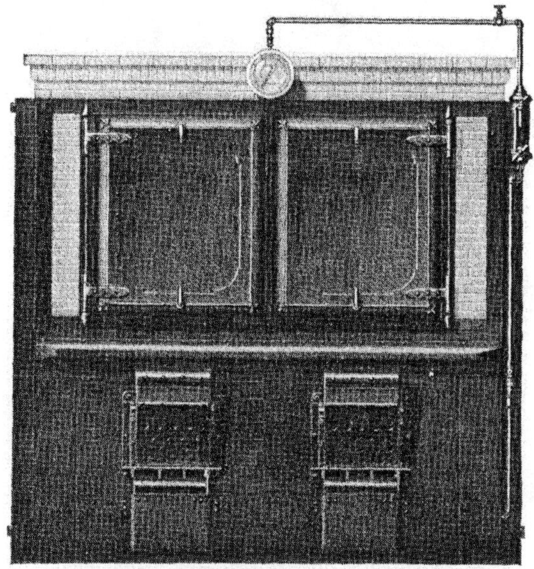

Cross-drum Boiler Front

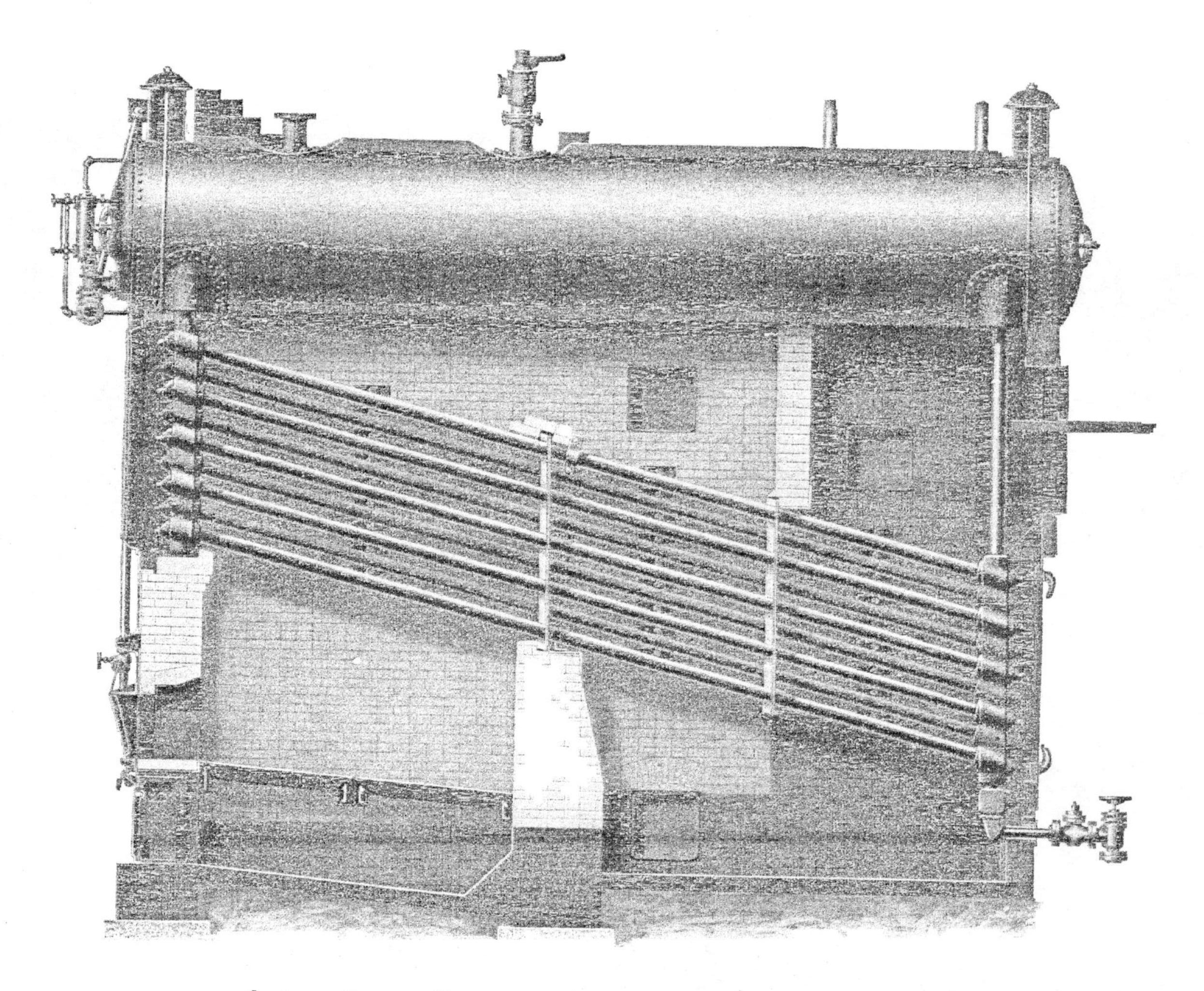

Cast-iron Vertical Header Longitudinal-drum Babcock & Wilcox Boiler

STOP AND CHECK VALVE

backing bars and tube clamps on the upper and lower tubes. This first flame bridge may be vertical or at an angle of 35 degrees with the normal to the tubes. The second flame bridge is like that just described, but without the tube brick.

Where no individual economizers are installed, damper frames at the rear or top of the setting are fitted with a damper with damper-operating rig arranged for convenient operation from the boiler-room floor

Pressed-steel vertical buckstays with tie rods are supplied for each side wall.

FITTINGS—Each boiler is equipped with the following fittings as a part of the standard equipment:

Blow-off connections and valves attached to the mud drum, terminating with companion flanges for attaching the customer's blow-off main.

Safety valves placed on nozzles on the steam drum.

Water column connected to nozzles on the drum.

Steam gauge attached to the front of the boiler.

Feed-water connections and valves, including a flanged stop and check valve of extra heavy pattern, attached to a flange on the drum head, arranged to close automatically in case of a rupture in the feed line

All valves and fittings are substantially built and are designs which, by their successful service for many years, have become standard with The Babcock & Wilcox Company.

There are also supplied with the boilers:

A wrench for handhole nuts.

A wrench for manhole nuts.

A water-driven turbine tube cleaner.

A set of fire tools.

Where soot blowers are not used, a metal steam hose and cleaning pipe equipped with special nozzle for blowing soot and dust from the tubes.

With the methods of suspension described the furnace may be made of any height, while the depth of the furnace may be made as great as the depth of the boiler, and by the use of an extension furnace, greater. Further, the boiler may be fired under the uptake end, or by the use of a horizontal baffle on one of the lower rows of tubes in connection with the vertical or inclined baffles, under the downtake end, or it may be fired under both ends. These factors make possible any desired form of furnace and any desired furnace volume, and it is obvious that by taking advantage of the possible flexibility of furnace form, a proper furnace can be designed to give the best results for any fuel that can be burned.

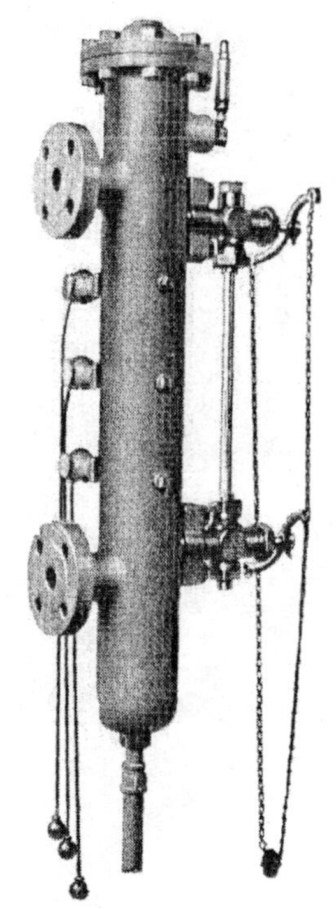

BABCOCK & WILCOX
WATER COLUMN

Babcock & Wilcox Wrought Steel Vertical Header Cross-drum Boiler

The gases of combustion leaving the furnace are led over the heating surface by the two baffles described, giving the gases three passes. These gases are discharged either through the rear circulating tubes or at the top of the third pass, between the drums in longitudinal-drum boilers, or between the horizontal circulators in the case of cross-drum boilers.

The gas passage areas are ample and are so proportioned as to give a maximum heat absorption from the gases without an undue frictional resistance and corresponding high draft loss.

The method of introduction of the feed water to the drums of longitudinal-drum boilers is clearly indicated by the illustration, the feed connections being made to the front drum heads, and the water on entering the drums passing to the rear of the drums. In cross-drum boilers the feed is also through the head, and in the case of wide units a feed connection is supplied on both drum heads. The water passes downward through the rear circulating tubes to the rear headers, upward through the tubes of the sections to the front headers and front circulating tubes to the drum, or, in cross-drum boilers, to the horizontal circulating tubes which deliver it to the drum. The steam formed in the passage through the tubes is liberated as it reaches the drum in longitudinal-drum boilers, and, to an extent, in the horizontal circulating tubes in cross-drum units. The steam so formed is stored in the steam space above the water line in the drums, from which it is drawn through a so-called "dry pipe." The dry pipe as ordinarily supplied with Babcock & Wilcox boilers is misnamed in that it fulfills none of the functions ordinarily attributed to such a device, particularly that of restricting the flow of steam from a boiler, with a view to avoid priming. In the Babcock & Wilcox boiler, since the total area of the holes in the dry pipe is ordinarily greater than that of the steam outlet, there is no restriction, and in reality the pipe is a collecting pipe, taking steam evenly from a great proportion of the length of the steam space of the drum.

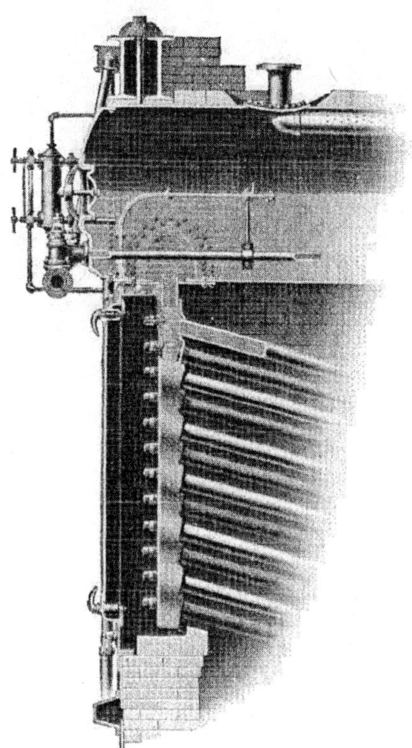

Partial Vertical Section Showing Method of Introducing Feed Water

All boilers manufactured by The Babcock & Wilcox Company are built in accordance with the Report of the Boiler Code Committee of the American Society of Mechanical Engineers. The materials entering into the construction of the Babcock & Wilcox boiler are the best obtainable for the special purpose for which they are used and are subjected to rigid inspection and tests. The boilers are manufactured by means of the most modern shop equipment and appliances in the hands of an old and well-tried organization of skilled mechanics under the supervision of experienced engineers.

Longitudinal-Drum Vertical Header Boiler, Showing Access Doors to Rear Headers. Rear View

ADVANTAGES OF THE BABCOCK & WILCOX BOILER

THE advantages of the Babcock & Wilcox boiler may perhaps be most clearly set forth by a consideration, first, of water-tube boilers as a class as compared with shell and fire-tube boilers; and, secondly, of the Babcock & Wilcox boiler specifically as compared with the other designs of water-tube boilers.

WATER-TUBE *VERSUS* FIRE-TUBE BOILERS

SAFETY—The most important requirement of a steam boiler is that it shall be safe in so far as danger from explosion is concerned. If the energy in a large shell boiler under pressure is considered, the thought of the destruction possible in the case of an explosion is appalling. The late Dr. Robert H. Thurston, Director of Sibley College, Cornell University, and past president of the American Society of Mechanical Engineers, estimated that there is sufficient energy stored in a plain cylinder boiler under 100 pounds steam pressure to project it in case of an explosion to a height of over 3½ miles; a locomotive boiler at 125 pounds pressure, from one-half to one-third of a mile; and a 60 horse-power return tubular boiler under 75 pounds pressure, somewhat over a mile. To quote: "A cubic foot of heated water under a pressure of from 60 to 70 pounds per square inch has about the same energy as one pound of gunpowder." From such a consideration, it may be readily appreciated how the advent of high pressure steam was one of the strongest factors in forcing the adoption of water-tube boilers. A consideration of the thickness of material necessary for cylinders of various diameters under a steam pressure of 200 pounds, and assuming an allowable stress of 12,000 pounds per square inch, will perhaps best illustrate this point. Table 1 gives such thicknesses for various diameters of cylinders, not taking into consideration the weakening effect of any joints which may be necessary. The rapidity with which the plate thickness increases with the diameter is apparent, and in practice, due to the fact that riveted joints must be used, the thicknesses as given in the table, with the exception of the first, must be increased from 30 to 40 per cent.

TABLE 1

PLATE THICKNESS REQUIRED FOR VARIOUS CYLINDER DIAMETERS

ALLOWABLE STRESS, 12,000 POUNDS PER SQUARE INCH, 200 POUNDS GAUGE PRESSURE, NO JOINTS

Diameter Inches	Thickness Inches	Diameter Inches	Thickness Inches
4	0.033	72	0.600
36	0.300	108	0.900
48	0.400	120	1.000
60	0.500	144	1.200

In a water-tube boiler, the drums seldom exceed 60 inches in diameter, and the thickness of plate required, therefore, is never excessive. The thinner metal can be rolled to a more uniform quality, the seams admit of better proportioning. and the joints can be more easily and perfectly fitted than is the case where thicker plates are necessary. All of these joints contribute toward making the drums of water-tube boilers better able to withstand the stress which they will be called upon to endure.

The essential constructive difference between water-tube and fire-tube boilers lies in the fact that the former is composed of parts of relatively small diameter as against the large diameters necessary in the latter.

The factor of safety of boiler parts which come in contact with the most intense heat in water-tube boilers can be made much higher than would be practicable in a shell boiler. Under the assumptions considered in connection with the thickness of plates required, a No. 9 gauge tube (0.148 inch), which is standard in Babcock & Wilcox boilers for pressures up to 240 pounds under the same allowable stress as was used in computing Table 1, the safe working pressure for the tubes is 888 pounds per square inch, indicating the very large margin of safety of such tubes as compared with that possible with the shell of a boiler.

A further advantage in the water-tube boiler as a class is the elimination of all compressive stresses. Cylinders subjected to external pressures, such as fire tubes or the internally fired furnaces of certain types of boilers, will collapse under a pressure much lower than that which they could withstand if it were applied internally. This is due to the fact that if there exists any initial distortion from its true shape, the external pressure will tend to increase such distortion and collapse the cylinder, while an internal pressure tends to restore the cylinder to its original shape.

Stresses due to unequal expansion have been a fruitful source of trouble in fire-tube boilers.

In boilers of the shell type, the riveted joints of the shell, with their consequent double thickness of metal exposed to the fire, give rise to serious difficulties. Upon these points are concentrated all strains of unequal expansion, giving rise to frequent leaks and oftentimes to actual ruptures. Moreover, in the case of such rupture, the whole body of contained water is liberated instantaneously and a disastrous and usually fatal explosion results.

Further, unequal strains result in shell or fire-tube boilers due to the difference in temperature of the various parts. This difference in temperature results from the lack of positive well-defined circulation. While such a circulation does not necessarily accompany all water-tube designs, in general, the circulation in water-tube boilers is much more defined than in fire-tube or shell boilers.

A positive and efficient circulation assures that all portions of the pressure parts will be at approximately the same temperature, and in this way strains resulting from unequal temperatures are obviated.

If a shell or fire-tube boiler explodes, the apparatus as a whole is destroyed. In the case of water-tube boilers, the drums are ordinarily so located that they are protected from intense heat and any rupture is usually in the case of a tube. Tube failures, resulting from blisters or burning, are not serious in their nature. Where a tube ruptures because of a flaw in the metal, the result may be more severe, but there cannot be the disastrous explosion such as would occur in the case of the explosion of a shell boiler.

To quote Dr. Thurston, relative to the greater safety of the water-tube boiler: "The stored available energy is usually less than that of any of the other stationary boilers and not very far from the amount stored, pound for pound, in the plain tubular boiler. It is evident that their admitted safety from destructive explosion does not come from this relation, however, but from the division of the contents into small portions and especially from those details of construction which make it tolerably certain that any rupture shall be local. A violent explosion can only come from the general disruption of a boiler and the liberation at once of large masses of steam and water."

Economy—The requirement probably next in importance to safety in a steam boiler is economy in the use of fuel. To fulfill such a requirement, the three items of proper grate for the class of fuel to be burned, a combustion chamber permitting complete combustion of gases before their escape to the stack, and the heating surface of such a character and arrangement that the maximum amount of available heat may be extracted, must be co-ordinated.

Fire-tube boilers, from the nature of their design, do not permit the variety of combinations of grate surface, heating surface, and combustion space possible in practically any water-tube boiler.

In securing the best results in fuel economy, the draft area in a boiler is an important consideration. In fire-tube boilers this area is limited to the cross-sectional area of the fire tubes, a condition further aggravated in a horizontal boiler by the tendency of the hot gases to pass through the upper rows of tubes instead of through all of the tubes alike. In water-tube boilers, the draft area is that of the space outside the tubes and is hence much greater than the cross-sectional area of the tubes.

Capacity—Due to the generally more efficient circulation found in water-tube than in fire-tube boilers, rates of evaporation are possible with water-tube boilers that cannot be approached where fire-tube boilers are employed.

Quick Steaming—Another important result of the better circulation ordinarily found in water-tube boilers is their ability to raise steam rapidly in starting and to meet the sudden demands that may be thrown on them.

In a properly designed water-tube boiler steam may be raised from a cold boiler to 200 pounds pressure in less than one-half hour.

For the sake of comparison with the figure above, it may be stated that in the United States Government service the shortest time allowed for getting up steam in Scotch marine boilers is 6 hours, and the time ordinarily allowed is 12 hours. In large double-ended Scotch boilers, such as are generally used in trans-Atlantic service, the fires are usually started 24 hours before the time set for getting under way. This length of time is necessary for such boilers in order to eliminate as far as possible excessive strains resulting from the sudden application of heat to the surfaces.

Accessibility—In the "Requirements of a Perfect Steam Boiler," as stated by Mr. Babcock, he demonstrates the necessity for complete accessibility to all portions of the boiler for cleaning, inspection and repair.

Cleaning—When the great difference is realized in performance, both as to economy and capacity, of a clean boiler and one in which the heating surfaces have been allowed to become fouled, it may be appreciated that the ability to keep heating surfaces clean internally and externally is a factor of the highest importance.

Such results can be accomplished only by the use of a design in boiler construction which gives complete accessibility to all portions. In fire-tube boilers the tubes are frequently nested together with a space between them often less than 1¼ inches and, as a consequence, nearly the entire tube surface is inaccessible. When scale forms upon such tubes it is impossible to remove it completely from the inside of the boiler and if it is removed by a turbine hammer, there is no way of knowing how thorough a job has been done. With the formation of such scale there is danger through overheating, and frequent tube renewals are necessary.

In Scotch marine boilers, even with the engines operating condensing, complete tube renewals at intervals of six or seven years are required, while large replacements

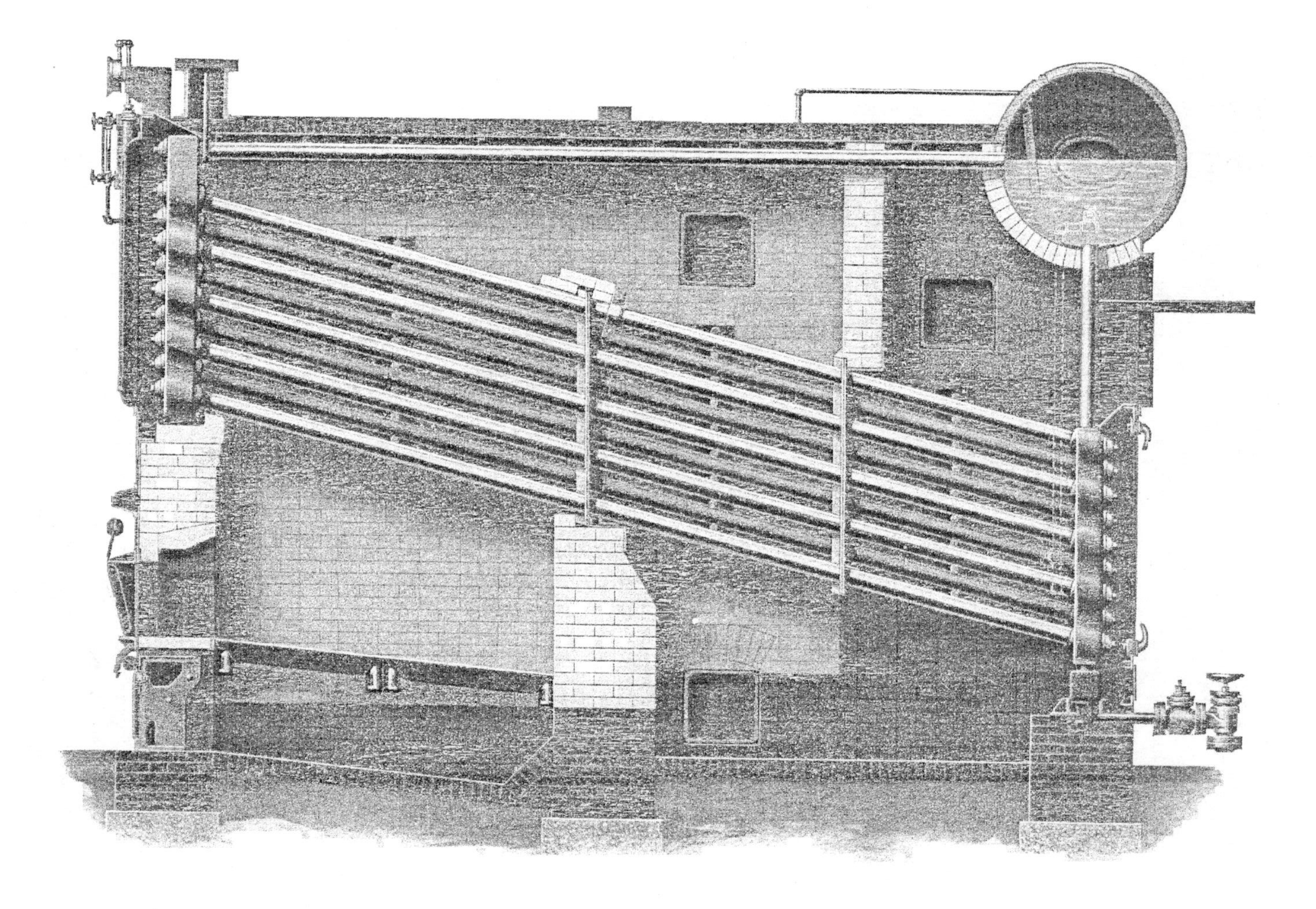

Cast-iron Vertical Header Cross-drum Babcock & Wilcox Boiler

are often necessary in less than one year. In return tubular boilers operated with bad feed water, complete tube renewals annually are not uncommon. In this type of boiler much sediment falls on the bottom sheets where the intense heat to which they are subjected bakes it to such an excessive hardness that the only method of removing it is to chisel it out. This can be done only by omitting tubes enough to leave a space into which a man can crawl, and the discomforts under which he must work are apparent. Unless such a deposit is removed, a burned and buckled plate will invariably result, and if neglected too long an explosion will follow.

In vertical fire-tube boilers using a water-leg construction, a deposit of mud in such legs is an active agent in causing corrosion, and the difficulty of removing such a deposit through handholes is well known. A complete removal is practically impossible, and as a last resort to obviate corrosion in certain designs, the bottoms of the water legs in some cases have been made of copper. A thick layer of mud and scale is also liable to accumulate on the crown sheet of such boilers and may cause the sheet to crack and lead to an explosion.

The soot and fine coal swept along with the gases by the draft will settle in fire tubes and unless removed promptly, must be cut out with a special form of scraper. It is not unusual where soft soal is used to find tubes half filled with soot, which renders useless a large portion of the heating surface and so restricts the draft as to make it difficult to burn sufficient coal to develop the required power from such heating surface as is not covered by soot.

Water-tube boilers in general are, from the nature of their design, more readily accessible for cleaning than are fire-tube boilers.

INSPECTION—The objections given above, in the consideration of the inability to clean fire-tube boilers properly, hold as well for the inspection of such boilers.

REPAIRS—The lack of accessibility in fire-tube boilers further leads to difficulties where repairs are required.

In fire-tube boilers, tube renewals are a serious undertaking. The accumulation of hard deposit on the exterior of the surfaces so enlarges the tubes that it is oftentimes difficult, if not impossible, to draw them through the tube sheets, and it is usually necessary to cut out such tubes as will allow access to the one which has failed and remove them through the manhole.

When a tube sheet blisters, the defective part must be cut out by hand-tapped holes drilled by ratchets, and as it is frequently impossible to get space in which to drive rivets, a " soft patch " is necessary. This is but a makeshift at best and usually results in either a reduction of the safe working pressure or in the necessity for a new plate. If the latter course is followed, the old plate must be cut out, a new one scribed to place to locate rivet holes and in order to obtain room for driving rivets, the boiler will have to be re-tubed.

The setting must, of course, be at least partially torn out and replaced.

After repairs of such nature in fire-tube boilers, the working pressure of such repaired boilers will frequently be lowered by the insurance companies when the boiler is again placed in service.

In the case of a rupture in a water-tube boiler, the loss will ordinarily be limited to one or two tubes which can readily be replaced. The fire-tube boiler will be so completely demolished that the question of repairs will be shifted from the boiler to the surrounding property, the damage to which will usually exceed many times the cost of a boiler of a type which would have eliminated the possibility of a disastrous explosion. In considering the proper repair cost of the two types of boilers, the fact should not be overlooked that it is poor economy to invest large sums in equipment that, through a possible accident to the boiler, may be wholly destroyed or so damaged that the cost of repairs, together with the loss of time while such repairs are being made, would purchase boilers of absolute safety and leave a large margin besides. The possibility of loss of human life should also be considered, though this may seem a far cry from the question of repair costs.

SPACE OCCUPIED—The space required for the boilers in a plant often exceeds the requirements for the remainder of the plant equipment. Any saving of space in a boiler room will be a large factor in reducing the cost of real estate and of the building. Even when the boiler plant is comparatively small, the saving in space frequently will amount to a considerable percentage of the cost of the boilers. Water-tube boilers, comparable in size to the largest fire-tube boilers manufactured, occupy an amount of space appreciably less than that occupied by fire-tube boilers. In large plants, because of the limitations in size to which fire-tube units can be built, and because of the possibility of building water-tube boilers of practically any size, the saving in space occupied is greatly in favor of the latter.

BABCOCK & WILCOX BOILERS AS COMPARED WITH OTHER WATER-TUBE DESIGNS

It must be borne in mind that the simple fact that a boiler is of the water-tube design does not as a necessity indicate that it is a good or safe boiler.

SAFETY—Many of the water-tube boilers on the market are as lacking as are fire-tube boilers in the positive circulation which is so necessary in the requirements of the perfect steam boiler. In boilers using water-leg construction, there is danger of defective circulation, leaks are common, and unsuspected corrosion may be going on in portions of the boiler that cannot be inspected. Stresses due to unequal expansion of the metal cannot be well avoided, but they may be minimized by maintaining at the same temperature all pressure parts of the boiler. This result is to be secured only by means of a well-defined circulation.

The main feature to which the Babcock & Wilcox boiler owes its safety is the construction made possible by the use of headers, by which the water in each vertical row of tubes is separated from that in the adjacent rows. This construction results in the very efficient circulation produced through the breaking up of the steam and water in the front headers. The use of a number of sections, thus composed of headers and tubes, has a distinct advantage over the use of a common chamber at the outlet ends of the tubes. In the former case, the circulation of water on one vertical row of tubes cannot interfere with that in the other rows, while in the latter construction there will be downward as well as upward currents and such downward currents tend to neutralize any good effect there might be through the diminution of the density of the water column by the steam.

Further, the circulation results directly from the design of the boiler and requires no assistance from "retarders," check valves and the like, within the boiler. All such mechanical devices in the interior of a boiler serve only to complicate the design and should not be used.

This positive and efficient circulation assures that all portions of the pressure parts of the Babcock & Wilcox boiler will be at approximately the same temperature, and in this way strains resulting from unequal temperatures are obviated.

Where the water throughout the boiler is at the temperature of the steam contained, a condition to be secured only by proper circulation, danger from internal pitting is minimized, or at least limited only to effects of the water fed the boiler. Where the water in any portion of the boiler is lower than the temperature of the steam corresponding to the pressure carried, whether the fact that such lower temperatures exist as a result of lack of circulation or because of intentional design, internal pitting or corrosion will almost invariably result.

Dr. Thurston has already been quoted to the effect that the admitted safety of a water-tube boiler is the result of the division of its contents into small portions. In boilers using a water-leg construction, while the danger from explosion will be largely limited to the tubes, there is the danger, however, that such legs may explode due to the deterioration of their stays, and such an explosion might be almost as disastrous as that of a shell boiler. The headers in a Babcock & Wilcox boiler are practically free from any danger of explosion. Were such an explosion to occur, it would still be localized to a much larger extent than in the case of a water-leg boiler, and the header construction thus almost absolutely localizes any danger from such a cause.

Staybolts are admittedly an undesirable element of construction in any boiler. They are wholly objectionable and the only reason for the presence of staybolts in a boiler is to enable a cheaper form of construction to be used than if they were eliminated.

In boilers utilizing in their design flat-stayed surfaces, or staybolt construction under pressure, corrosion and wear and tear in service tend to weaken some single part subject to continual strain, the result being an increased strain on other parts greatly in excess of that for which an allowance can be made by any reasonable factor of safety. Where the construction is such that the weakening of a single part will produce a marked decrease in the safety and reliability of the whole, it follows of necessity that there will be a corresponding decrease in the working pressure which may be safely carried.

In water-leg boilers, the use of such flat-stayed surfaces under pressure presents difficulties that are practically unsurmountable. Such surfaces exposed to the heat of the fire are subject to unequal expansion, distortion, leakage and corrosion, or in general, to many of the objections that have already been advanced against the fire-tube boilers in the consideration of water-tube boilers as a class in comparison with fire-tube boilers.

Aside from the difficulties that may arise in actual service due to the failure of staybolts, or in general, due to the use of flat-stayed surfaces, constructional features are encountered in the actual manufacture of such boilers that make it difficult if not impossible to produce a first-class mechanical job. It is practically impossible in the building of such a boiler to so design and place the staybolts that all will be under equal strain. Such unequal strains, resulting from constructional difficulties, will be greatly multiplied when such a boiler is placed in service. Much of the riveting in boilers of this design must of necessity be hand work, which is never the equal of machine riveting. The use of water-leg construction ordinarily requires the flanging of large plates, which is difficult, and because of the number of heats necessary and the continual working of the material, may lead to the weakening of such plates.

In vertical or semi-vertical water-tube boilers utilizing flat-stayed surfaces under pressure, these surfaces are ordinarily so located as to offer a convenient lodging place for flue dust, which fuses into a hard mass, is difficult of removal and under which corrosion may be going on with no possibility of detection.

Where stayed surfaces or water legs are features in the design of a water-tube boiler, the factor of safety of such parts must be most carefully considered. In such parts, too, is the determination of the factor most difficult, and because of the "rule-of-thumb" determination frequently necessary, the factor of safety becomes in reality a factor of ignorance. As opposed to such indeterminate factors of safety, in the Babcock & Wilcox boiler, when the factor of safety for the drum or drums has been determined, and such a factor may be determined accurately, the factors for all other portions of the pressure parts are greatly in excess of that of the drum. All Babcock & Wilcox boilers are built with a factor of safety of at least five, and inasmuch as the factor of safety of the tubes and headers is greatly in excess of this figure, it applies specifically to the drum or drums. This factor represents a greater degree of safety than a considerably higher factor applied to a boiler in which the shell or any riveted portion is acted upon directly by the fire, or the same factor applied to a boiler utilizing flat-stayed surface construction, where the accurate determination of the limiting factor of safety is difficult, if not impossible.

That the factor of safety of stayed surfaces is questionable may perhaps be best realized from a consideration of the severe requirements as to such factor called for by the rules and regulations of the Board of Supervising Inspectors, United States Government.

In view of the above, the absence of any stayed surfaces in the Babcock & Wilcox boiler is obviously a distinguishing advantage where safety is a factor. It is of interest to note, in the article on the evolution of the Babcock & Wilcox boiler, that staybolt construction was used in several designs, found unsatisfactory and unsafe, and discarded.

Another feature in the design of the Babcock & Wilcox boiler tending toward added safety is its manner of suspension. This has been indicated in the previous chapter and is of such nature that all of the pressure parts are free to expand or contract under variations of temperature without in any way interfering with any part of the boiler setting. The sectional nature of the boiler allows a flexibility under varying temperature changes that practically obviates internal strain.

In boilers utilizing water-leg construction, on the other hand, the construction is rigid, giving rise to serious internal strains, and the method of support ordinarily made necessary by the boiler design is not only unmechanical but frequently dangerous, due to the fact that proper provision is not made for expansion and contraction under temperature variations.

Boilers utilizing water-leg construction are not ordinarily provided with mud drums. This is a serious defect in that it allows impurities and sediment to collect in a portion of the boiler not easily inspected, and corrosion may result.

Economy—That the water-tube boiler as a class lends itself more readily than does the fire-tube boiler to a variation in the relation of grate surface, heating surface and combustion space has been already pointed out. In economy again, the construction made possible by the use of headers in Babcock & Wilcox boilers appears as a distinct advantage. Because of this construction, there is a flexibility possible, in an unlimited variety of heights and widths, that will satisfactorily meet the special requirements of the fuel to be burned in individual cases.

An extended experience in the design of furnaces best suited for a wide variety of fuels has made The Babcock & Wilcox Company a leader in the field of economy. Furnaces have been built and are in successful operation for burning anthracite and bituminous coals, lignite, crude oil, gas-house tar, wood, sawdust and shavings, bagasse, tan bark, natural gas, blast-furnace gas, by-product coke oven gas and for the utilization of waste heat from commercial processes. The great number of Babcock & Wilcox boilers now in satisfactory operation under such a wide range of fuel conditions constitutes an unimpeachable testimonial to their ability to meet all of the many conditions of service.

The limitations in the draft area of fire-tube boilers as affecting economy have been pointed out. That a greater draft area is possible in water-tube boilers does not of necessity indicate that proper advantage of this fact is taken in all boilers of the water-tube class. In the Babcock & Wilcox boiler, the large draft area taken in connection with the effective baffling allows the gases to be brought into intimate contact with all portions of the heating surfaces and renders such surfaces highly efficient.

In certain designs of water-tube boilers the baffling is such as to render ineffective certain portions of the heating surface, due to the tendency of soot and dirt to collect on or behind baffles, in this way causing the interposition of a layer of non-conducting material between the hot gases and the heating surfaces.

In Babcock & Wilcox boilers the standard baffle arrangement is such as to allow the installation of a superheater without in any way altering the path of the gases from furnace to stack, or requiring a change in the boiler design. In certain water-tube boilers the baffle arrangement is such that if a superheater is to be installed a complete change in the ordinary baffle design is necessary. Frequently to insure sufficiently hot gas striking the heating surfaces, a portion is by-passed directly from the furnace to the superheater chamber without passing over any of the boiler-heating surfaces. Any such arrangement will lead to a decrease in economy and the use of boilers requiring it should be avoided.

CAPACITY—Babcock & Wilcox boilers are run successfully in every-day practice at higher ratings than any other boilers in practical service. The capacities thus obtained are due directly to the efficient circulation already pointed out. Inasmuch as the construction utilizing headers has a direct bearing in producing such circulation, it is also connected with the high capacities obtainable with this apparatus.

Where intelligently handled and kept properly cleaned, Babcock & Wilcox boilers are operated in many plants at from 200 to 225 per cent of their rated evaporative capacity and it is not unusual for them to be operated at 300 per cent of such rated capacity during periods of peak load.

DRY STEAM—In the list of the requirements of the perfect steam boiler, the necessity that dry steam be generated has been pointed out. The Babcock & Wilcox boiler will deliver dry steam under higher capacities and poorer conditions of feed water than any other boiler now manufactured. Certain boilers will, when operated at ordinary ratings, handle poor feed water and deliver steam in which the moisture

A 2[illegible] Horse-power Cross-drum Babcock & Wilcox Boiler Fifty-one Sections Wide

content is not objectionable. When these same boilers are driven at high overloads, there will be a direct tendency to prime and the percentage of moisture in the steam delivered will be high. This tendency is the result of the lack of proper circulation and once more there is seen the advantage of the headers of the Babcock & Wilcox boiler, resulting as it does in securing a positive circulation.

In the design of the Babcock & Wilcox boiler sufficient space is provided between the steam outlet and the disengaging point to insure the steam passing from the boiler in a dry state without entraining or again picking up any particles of water in its passage, even at high rates of evaporation. Ample time is given for a complete separation of steam from the water at the disengaging surface before the steam is carried from the boiler. These two features, which are additional causes for the ability of the Babcock & Wilcox boiler to deliver dry steam, result from the proper proportioning of the steam and water space of the boiler. From the history of the development of the boiler, it is evident that the cubical capacity per horse-power of the steam and water space has been adopted after numerous experiments.

That the "dry pipe" serves in no way the generally understood function of such device has been pointed out. As stated, the function of the "dry pipe" in a Babcock & Wilcox boiler is simply that of a collecting pipe. and this statement holds true regardless of the rate of operation of the boiler.

In certain boilers, "superheating surface" is provided to "dry the steam," or to remove the moisture due to priming or foaming. Such surface is invariably a source of trouble unless the steam is initially dry, and a boiler which will deliver dry steam is obviously to be preferred to one in which surface must be supplied especially for such purpose. Where superheaters are installed with Babcock & Wilcox boilers, they are in every sense of the word superheaters and not driers, the steam being delivered to them in a dry state.

Quick Steaming—The advantages of water-tube boilers as a class over fire-tube boilers in ability to raise steam quickly have been indicated.

Due to the constant and thorough circulation resulting from the sectional nature of the Babcock & Wilcox boiler, steam may be raised more rapidly than in practically any other water-tube design.

In starting up a cold Babcock & Wilcox boiler with either coal or oil fuel, where a proper furnace arrangement is supplied, steam may be raised to a pressure of 200 pounds in less than half an hour. With a Babcock & Wilcox boiler in a test where forced draft was available, steam was raised from an initial temperature of the boiler and its contained water of 72 degrees to a pressure of 200 pounds, in 12½ minutes after lighting the fire. The boiler also responds quickly in starting from banked fires, especially where forced draft is available.

In Babcock & Wilcox boilers, the water is divided into many small streams. which circulate without undue frictional resistance in thin envelopes passing through the hottest part of the furnace, the steam being carried rapidly to the disengaging surface. There is no part of the boiler exposed to the heat of the fire that is not in contact with water internally, and, as a result, there is no danger of overheating on starting up quickly, nor can leaks occur from unequal expansion, such as might be the case where an attempt is made to raise steam rapidly in boilers using water-leg construction.

Bankers Trust Building, New York City Operating 000 Horse-power of Babcock & Wilcox Boilers

Storage Capacity for Steam and Water—Where sufficient steam and water capacity are not provided in a boiler, its action will be irregular, the steam pressure varying over wide limits and the water level being subject to frequent and rapid fluctuation.

Owing to the small relative weight of steam, water capacity is of greater importance in this respect than steam space. With a gauge pressure of 180 pounds per square inch, 8 cubic feet of steam, which is equivalent to one-half cubic foot of water space, are required to supply one boiler horse-power for one minute, and if no heat be supplied to the boiler during such an interval, the pressure will drop to 150 pounds per square inch. The volume of steam space, therefore, may be over-rated, but if this be too small, the steam passing off will carry water with it in the form of spray. Too great a water space results in slow steaming and waste of fuel in starting up; while too much steam space adds to the radiating surface and increases the losses from that cause.

That the steam and water spaces of the Babcock & Wilcox boiler are the result of numerous experiments has previously been pointed out.

Accessibility—Cleaning. That water-tube boilers are more accessible as a class than are fire-tube boilers has been indicated. All water-tube boilers, however, are not equally accessible. In certain designs, due to the arrangement of baffling used, it is practically impossible to remove all deposits of soot and dirt. Frequently, in order to cheapen the product, sufficient cleaning and access doors are not supplied as part of the boiler equipment. The tendency of soot to collect on the crown sheets of certain vertical water-tube boilers has been noted. Such deposits are difficult to remove, and if corrosion goes on beneath such a covering, the sheet may crack and an explosion result.

It is almost impossible to thoroughly clean water legs internally, and in such places also there is a tendency to unsuspected corrosion under deposits that cannot be removed.

In Babcock & Wilcox boilers, every portion of the interior of the heating surfaces can be reached and kept clean, while any soot deposited on the exterior surfaces can be blown off while the boiler is under pressure.

Inspection—The accessibility which makes possible the thorough cleaning of all portions of the Babcock & Wilcox boiler also provides a means for a thorough inspection.

Drums are accessible for internal inspection by the removal of the manhole plates. Front headers may be inspected through large doors furnished for the purpose. Rear headers in the inclined header designs may be inspected from the chamber formed by such headers and the rear wall of the boiler. In the vertical header designs, rear tube doors are furnished, as has been stated. In certain designs of water-tube boilers, in order to assure accessibility for inspection of the rear ends of the tubes, the rear portion of the boiler is exposed to the atmosphere, with resulting excessive radiation losses. In other designs, the means of access to the rear ends of the tubes are of a makeshift and unworkmanlike character.

By the removal of handhole plates, all tubes in a Babcock & Wilcox boiler may be inspected for their full length, either for the presence of scale or for suspected corrosion.

FOUR 208 HORSE-POWER BABCOCK & WILCOX BOILERS EQUIPPED WITH BABCOCK & WILCOX SUPERHEATERS AND STOKERS AT THE GEORGIA INSTITUTE OF TECHNOLOGY, ATLANTA

Repairs—In Babcock & Wilcox boilers, the possession of great strength, the elimination of stresses due to uneven temperatures and of the resulting danger of leaks and corrosion, the protection of the drums from the intense heat of the fire, and the decreased liability of the scale-forming matter to lodge on the hottest tube surfaces, all tend to minimize the necessity for repairs. The tubes of the Babcock & Wilcox boiler are practically the only part which may need renewal and these only at infrequent intervals. When necessary, such renewals may be made cheaply and quickly. A small stock of tubes, 4 inches in diameter, of sufficient length for the boiler used, is all that need be carried to make renewals.

Repairs in water-leg boilers are difficult at best, and frequently unsatisfactory when completed. When staybolt replacements are necessary, in order to get at the inner sheet of the water leg, several tubes must in some cases be cut out. Not infrequently a replacement of an entire water leg is necessary and this is difficult and requires a lengthy shutdown. With the Babcock & Wilcox boiler, on the other hand, even if it is necessary to replace a section, this may be done in a few hours after the boiler is cool.

In the case of certain staybolt failures, the working pressure of a repaired boiler utilizing such construction will frequently be lowered by the insurance companies when the boiler is again placed in service. The sectional nature of the Babcock & Wilcox boiler enables it to maintain its original working pressure over long periods of time, almost regardless of the nature of any repair that may be required.

Durability—Babcock & Wilcox boilers are being operated in every-day service with entirely satisfactory results and under the same steam pressure as that for which they were originally sold, that have been operated over thirty-five years. It is interesting to note in considering the life of a boiler that the length of life of a Babcock & Wilcox boiler must be taken as the criterion of what length of life is possible. This is due to the fact that there are Babcock & Wilcox boilers in operation today that have been in service from a time that antedates by a considerable margin that at which the manufacture of any other water-tube boiler now on the market was started.

Probably the very best evidence of the value of the Babcock & Wilcox boiler as a steam generator and of the reliability of the apparatus, is seen in the sales of the company. Since the company was formed, there have been sold throughout the world over 35,000,000 horse-power.

A feature that cannot be overlooked in the consideration of the advantages of the Babcock & Wilcox boiler is the fact that as a part of the organization back of the boiler there is a body of engineers of recognized ability, ready at all times to assist its customers in every possible way.

POWER STATION OF THE KANSAS CITY TERMINAL COMPANY, OPERATING EIGHT 508 HORSE-POWER BABCOCK & WILCOX BOILERS WITH BABCOCK & WILCOX STOKERS AND SUPERHEATERS

HEAT AND ITS MEASUREMENT

THE usual conception of heat is that it is a form of energy produced by the vibratory motion of the minute particles or molecules of a body. All bodies are assumed to be composed of these molecules, which are held together by mutual cohesion and yet are in a state of continual vibration. The hotter the body or the more heat added to it, the more vigorous is the vibration of the molecules.

The effect of heat on a body may be to change its temperature, its volume, or its state—that is, from solid to liquid or from liquid to gaseous. Where ice is melted to water and evaporated into steam, the various changes are admirably described in the lecture by Mr. Babcock on "The Theory of Steam Making," given in the next chapter.

Maxwell defines the temperature of a body as "its thermal state with reference to its power of communicating heat to other bodies."

From this definition as indicating temperature as purely relative, an arbitrary scale of some sort must be assumed for temperature measurement. The ideal scale is the thermodynamic scale which Kelvin defines as a scale such that "the absolute values of two temperatures are to one another in proportion of the heat taken in to the heat rejected in a reversible thermodynamic engine working with a source and refrigerator at the higher and lower temperatures, respectively." The arbitrary unit degree is defined by giving the temperature interval between the freezing and the boiling points of pure water under standard conditions a value of 100 degrees Centigrade. No method of temperature measurement uses this principle, but scales defined by other means are in agreement with the thermodynamic scale or vary from it by small known amounts.

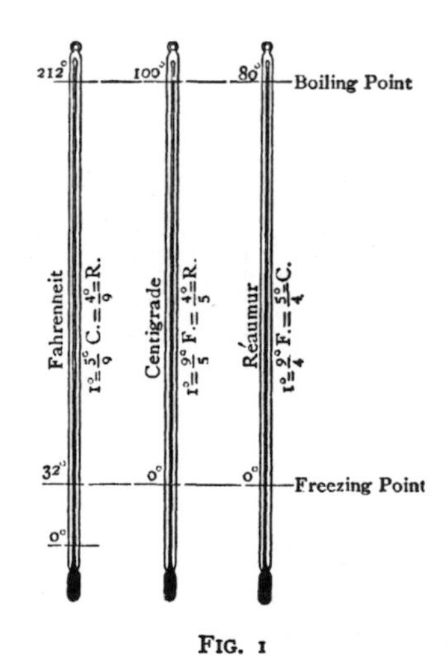

FIG. 1

The high temperature scale is defined by nitrogen expansion in the constant-volume gas thermometer. Such thermometers are not used commercially but serve as fundamental standard instruments. A series of fixed points defined by the boiling or melting points of chemical elements or compounds is determined by the gas thermometer and corrected to agree with the true thermodynamic scale. These points determine the scale as defined by other forms of pyrometers. The standard fixed points used in thermometry are given in Table 2, on page 62.

The temperature interval between the freezing and the boiling points of pure water under standard conditions is the unit of temperature used in all thermometric scales. The division of this interval into smaller units or degrees, however, differs in different scales. There are in use today three thermometric scales, known as the Fahrenheit, the Centigrade or Celsius, and the Réaumur. As shown in Fig. 1, in the Fahrenheit scale the interval between the two fixed points is divided into 180 parts; the boiling point is marked 212 and the freezing point 32, and zero is a temperature which, at the time this thermometer was invented, was incorrectly imagined to be the lowest temperature obtainable. In the Centigrade and Réaumur scales, the fixed interval is divided into 100 and 80 parts or degrees, respectively. In each of these scales, the freezing point is marked zero, and the boiling point in the Centigrade scale 100 and in the Réaumur scale 80.

Table 3 and appended formulæ are useful in converting from one scale to another.

In measuring any temperature it is essential for the measuring instrument to be in such relation to the body whose temperature is desired that the two are in

TABLE 2

STANDARD FIXED POINTS USED IN THERMOMETRY

Substance	Transformation	Temperature	
		Degrees Centigrade	Degrees Fahrenheit
Oxygen	Boiling Point	-182.98	-297.36
Carbon Bisulphide	Melting Point	-112.0	-169.6
Carbon Dioxide	Boiling Point	-78.5	-109.3
Mercury	Melting Point	-38.88	-37.98
Water	Melting Point	0.00	32.00
Water	Boiling Point	100.00	212.00
Naphthalene	Boiling Point	217.95	424.31
Tin	Melting Point	231.9	449.4
Benzophenone	Boiling Point	305.9	582.6
Cadmium	Melting Point	320.9	609.6
Zinc	Melting Point	419.4	786.9
Sulphur	Boiling Point	444.55	832.19
Antimony	Melting Point	630.0	1166.0
Aluminum	Melting Point	658.7	1217.7
Silver	Melting Point	960.2	1760.4
Gold	Melting Point	1062.6	1944.7
Copper	Melting Point	1082.8	1981.0
Lithium Metasilicate	Melting Point	1201.	2293.8
Diopside	Melting Point	1391.5	2536.7
Nickel	Melting Point	1452.6	2646.7
Palladium	Melting Point	1549.5	2821.1
Platinum	Melting Point	1755.	3191.

N. B.—The boiling points are at standard pressure.

complete thermal equilibrium, *i. e.*, that they are at the "same" temperature. Because of variation in temperatures to be measured and of bodies whose temperatures are desired, and because of limitations in the different measuring devices, thermometry is necessarily subdivided into two divisions: The first involves the insertion of the measuring device into the body whose temperature is desired and obtaining results through heat conduction, and the second, the use of a measuring instrument at a distance, involving radiation. Radiation methods are dependent upon the fact that all bodies are radiating energy, that the amount radiated varies with the temperature, and that the energy may be measured.

Low and moderate temperatures are ordinarily measured by thermometers. High temperatures are measured by some form of pyrometer, discussed hereafter under the heading, "High Temperature Measurements." Mercury is the fluid most commonly used in the construction of thermometers. In the United States, the bulbs of high-grade thermometers are usually made of either Jena 59^{III} borosilicate thermometer glass or Jena 16^{III} glass, the stems being made of ordinary glass. The Jena 16^{III} is not suitable for use at temperatures much above 850 degrees Fahrenheit and the harder Jena 59^{III} should be used in thermometers for temperatures higher than this.

TABLE 3

COMPARISON OF THERMOMETER SCALES

	Fahrenheit	Centigrade	Réaumur		Fahrenheit	Centigrade	Réaumur
Absolute Zero .	-459.6	-273.1	-218.5		50	10	8
	0	-17.78	-14.22		75	23.89	19.11
	10	-12.22	-9.78		100	37.78	30.22
	20	-6.67	-5.33		200	93.33	74.67
	30	-1.11	-0.89	Boiling Point . .	212	100	80
Freezing Point .	32	0	0		250	121.11	96.89
Maximum Density					300	148.89	119.11
of Water . . .	39.1	3.94	3.15		350	176.67	141.33

$F = \frac{9}{5} C + 32^\circ = \frac{9}{4} R + 32^\circ$ $\quad$ $C = \frac{5}{9} (F - 32^\circ) = \frac{5}{4} R$ $\quad$ $R = \frac{4}{9} (F - 32^\circ) = \frac{4}{5} C$

Below the boiling point, the hydrogen-gas thermometer is the almost universal standard with which mercurial thermometers may be compared, while above this point the nitrogen-gas thermometer is used. In both of these standards, the change in temperature is measured by the change in pressure of a constant volume of the gas.

In graduating a mercurial thermometer for the Fahrenheit scale, ordinarily a degree is represented as $\frac{1}{180}$ part of the volume of the stem between the readings at the melting point of ice and the boiling point of water. For temperatures above the latter, the scale is extended in degrees of the same volume. For very accurate work, however, the thermometer may be graduated to read true gas-scale temperatures by comparing it with the gas thermometer and marking the temperatures at 25 or 50 degree intervals. Each degree is, then, $\frac{1}{25}$ or $\frac{1}{50}$ of the volume of the stem in each interval.

Every thermometer, especially if intended for use above the boiling point, should be suitably annealed before it is used. If this is not done, the true melting point and also the "fundamental interval," that is, the interval between the melting and the boiling points, may change considerably. After continued use at the higher temperatures, also, the melting point will change, so that the thermometer must be calibrated occasionally to insure accurate readings.

As a general rule, thermometers are graduated to read correctly for total immersion, that is, with bulb and stem of the thermometer at the same temperature, and they should be used in this way when compared with a standard thermometer. If the stem emerges into space either hotter or colder than that in which the bulb is placed, a " stem correction " must be applied to the observed temperature in addition to any correction that may be found in the comparison with the standard. For instance, for a particular thermometer. comparison with the standard with both fully immersed made necessary the following corrections:

Temperature	*Correction*	*Temperature*	*Correction*
40°F.	0.0	300°F.	+2.5
100	0.0	400	—0.5
200	0.0	500	—2.5

When the sign of the correction is positive (+) it must be added to the observed reading, and when the sign is a negative (—) the correction must be subtracted.

The formula for the stem correction is as follows:

$$\text{Stem correction} = 0.000085 \times n\,(T - t) \qquad (1)$$

in which T is the observed temperature, t is the mean temperature of the emergent column. n is the number of degrees of mercury column emergent, and 0.000085 is the difference between the coefficient of expansion of the mercury and that in the glass in the stem.

Suppose the observed temperature is 400 degrees and the thermometer is immersed to the 200-degree mark, so that 200 degrees of the mercury column project into the air. The mean temperature of the emergent column may be found by tying another thermometer on the stem with the bulb at the middle of the emergent mercury column, as in Fig. 2. Suppose this mean temperature is 85 degrees, then

$$\text{Stem correction} = 0.000085 \times 200 \times (400 - 85) = 5.3 \text{ degrees}$$

As the stem is at a lower temperature than the bulb, the thermometer will evidently read too low, so that this correction must be added to the observed reading to find the reading corresponding to total immersion. The corrected reading will therefore be 405.3 degrees. If this thermometer is to be corrected in accordance with the calibrated corrections given above, we note that a further correction of 0.5 must be applied to the observed reading at this temperature, so that the correct temperature is 405.3—0.5=404.8 degrees or 405 degrees.

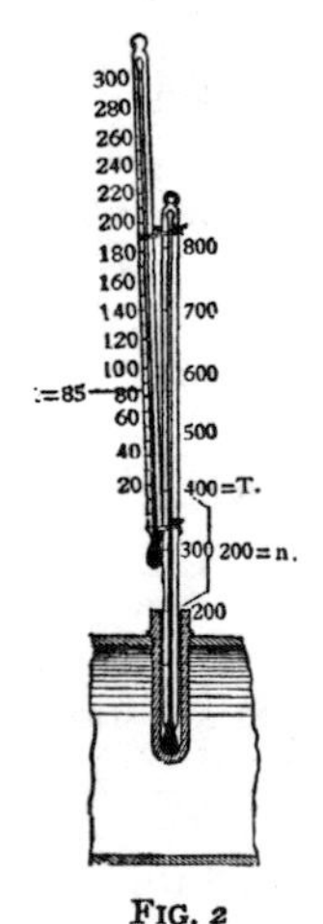

FIG. 2

Fig. 2 shows how a stem correction can be obtained for the case just described.

Fig. 3 affords an opportunity for comparing the scale of a thermometer correct for total immersion with one which will read correctly when submerged to the 300-degree mark, the stem being exposed at a mean temperature of 100 degrees Fahrenheit, a temperature often prevailing when thermometers are used for measuring temperatures in steam mains.

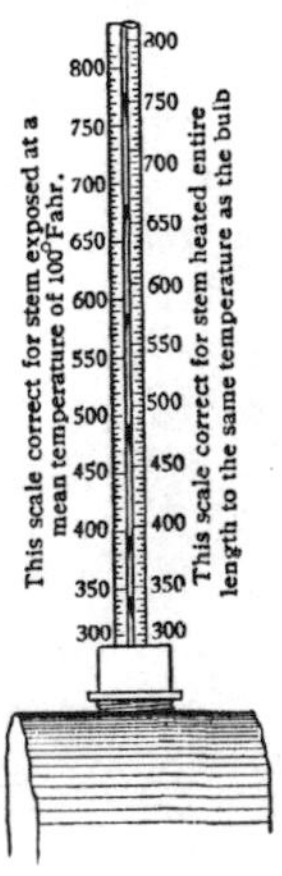

ABSOLUTE ZERO—Experiments show that at 32 degrees Fahrenheit a perfect gas expands $\frac{1}{491.6}$ part of its volume if its pressure remains constant and its temperature is increased one degree. Thus if gas at 32 degrees Fahrenheit occupies 100 cubic feet and its temperature is increased one degree,

FIG. 3

its volume will be increased to $100+\frac{100}{491.6}=100.203$ cubic feet, while for a temperature increase of two degrees the volume will be increased to $100+\frac{100\times2}{491.6}=100.407$ cubic feet. If this rate of expansion per one degree held good at all temperatures, and experiment shows that it does above the freezing point, if the pressure remains constant, the volume would double if raised to a temperature of $32+491.6=523.6$ degrees Fahrenheit, while under a diminution of temperature it would decrease in volume and finally disappear at a temperature of $491.6-32=459.6$ degrees below zero Fahrenheit. While undoubtedly some change in the law would take place before the lower temperature could be reached, there is no reason why the law may not be used over the range of temperature within which it is known to hold good. It is evident from this explanation that under a constant pressure, the volume of a gas will vary directly as the number of degrees between its temperature and —459.6 degrees Fahrenheit. The value, —459.6, is uncertain by approximately two units in the last place, and for all ordinary engineering purposes, such as reductions of gas volumes, the value —460 is more than sufficiently accurate.

To simplify the application of this law, a new thermometric scale is constructed as follows. The point corresponding to —460 degrees Fahrenheit is taken as the zero point on the new scale, and the degrees are identical in magnitude with those of the Fahrenheit scale. Temperatures referred to this scale are called absolute temperatures and the point —460 degrees Fahrenheit is called absolute zero. It is obvious that the relation between the absolute temperature (T°) and the temperature of the Fahrenheit scale (t°) is

$$T^\circ = t^\circ + 460. \qquad (2)$$

Thus if one pound of gas is at 60 degrees Fahrenheit and another pound at 140 degrees Fahrenheit, the pressure at both temperatures being the same, the respective volumes will be in the ratio of 520 to 600.

British Thermal Unit—The quantitative measure of heat is the British thermal unit, ordinarily expressed B.t.u. While this has until recently been ordinarily defined as the amount of heat necessary to raise the temperature of one pound of water at a definite temperature, one degree Fahrenheit, the value now commonly accepted is $\frac{1}{180}$th of the amount of heat required to raise one pound of pure water from the ice point (32 degrees Fahrenheit) to the steam point (212 degrees Fahrenheit), both at standard pressure. In the metric system, the quantitative measure of heat is the calorie, and is $\frac{1}{100}$th of the amount of heat required to raise the temperature of one kilogram of pure water from 0 to 100 degrees Centigrade. The relations of the two units are as follows:

Unit	*Water*	*Temperature Rise*
1 B.t.u.	1 Pound	1 Degree Fahrenheit
1 Calorie	1 Kilogram	1 Degree Centigrade

1 kilogram = 2.2046 pounds; 1 degree Centigrade = $\frac{9}{5}$ degrees Fahrenheit.
Hence 1 calorie = $(2.2046 \times \frac{9}{5}) = 3.968$ B.t.u.

The heat values are ordinarily given in B.t.u. per pound, and in calories per kilogram, in which case the B.t.u. per pound are approximately equivalent to $\frac{9}{5}$ the calories per kilogram.

As stated in the first law of thermodynamics, heat and mechanical energy are mutually convertible. The relation of heat energy to work as determined by Joule's original experiments indicated one B.t.u. as being equivalent to 772 foot-pounds. Joule's later experiments and those of Rowland and others indicated that this value was low. Some of the values more recently determined are: Mollier (1906), 778.23; Goodenough (1914), 777.64; Callendar (1915), 777.8; Marks and Davis (1916), 777.52. The value ordinarily accepted is 778 foot-pounds, the greatest variation from this value in any of those given being insignificant. Inasmuch as this value is uncertain in the last place, the use of figures beyond the decimal point is useless.

One horse-power is equivalent to 550 foot-pounds per second, or 1,980,000 foot-pounds per hour, while one kilowatt-hour is equivalent to 1.3410 horse-power-hours. On this basis, with the mechanical equivalent of one B.t.u. taken as 778 foot-pounds, we have

$$1 \text{ horse-power-hour} = \frac{1{,}980{,}000}{778} = 2545 \text{ B.t.u.}$$

$$1 \text{ kilowatt-hour} = \frac{1{,}980{,}000}{778 \times .7457} = 3413 \text{ B.t.u.}$$

SPECIFIC HEAT—The specific heat of a substance is the quantity of heat expressed in thermal units required to raise unit weight of the substance through one degree of temperature. The specific heat of all substances varies with temperature, and since all substances vary in volume or pressure with temperature changes, it is necessary to distinguish between specific heats at constant volume and at constant pressure, expressed ordinarily as C_v and C_p, respectively.

Because of the low coefficients of expansion of solids and liquids, there is but little difference in the specific heat at constant volume and that at constant pressure. With gases, on the other hand, the difference is appreciable. When heat is added to a gaseous substance its volume may be kept constant, in which case no external work is done, or the gas may be allowed to expand during the addition of heat, the pressure being kept constant. The specific heat at constant volume, therefore, is always less than that at constant pressure by an amount equal to the heat required to do the work of expansion against external pressure.

In both specific heat at constant volume and at constant pressure, it is necessary to distinguish between *instantaneous* and *mean* specific heat. The instantaneous specific heat of a substance is the amount of heat that must be added to unit weight of such substance at some definite temperature to increase its temperature one degree under given conditions of pressure or volume. The mean specific heat of a substance, over a given temperature range, is the value by which such range must be multiplied to determine the quantity of heat required to raise unit weight of the substance through the range under the conditions of pressure or volume which exist.

In practically all boiler and combustion work, mean specific heats are used.

If a pound of pure water is considered a standard, the mean specific heat over a temperature range of 32 to 212 degrees Fahrenheit, from the definition of a B.t.u., is unity. The pound is the unit of weight considered in expressing the specific heat of all substances.

The specific heat of gases is of the greatest importance in combustion work, and is treated more fully in the chapter on "Combustion."

The specific heats of various solids and liquids are given in Table 4.

TABLE 4

SPECIFIC HEAT OF VARIOUS LIQUIDS AND SOLIDS

SOLIDS					
	Temperature* Degrees Fahrenheit	Specific Heat		Temperature* Degrees Fahrenheit	Specific Heat
Copper	59–460	.0951	Glass (normal ther. 16III)	66–212	.1988
Gold	32–212	.0316	Lead	59	0299
Wrought Iron	59–212	.1152	Platinum	68–212	.0319
Cast Iron	68–212	.1189	Silver	32–212	.0559
Steel (soft)	68–208	1175	Tin	−105–64	.0518
Steel (hard)	68–208	.1165	Ice	. . .	.5040
Zinc	32–212	0931	Sulphur (newly fused)	. . .	.2025
Brass (yellow)	32	.0883			
LIQUIDS					
	Temperature* Degrees Fahrenheit	Specific Heat		Temperature* Degrees Fahrenheit	Specific Heat
Water†	32–212	1.0000	Sulphur (melted)	246–297	.2350
Alcohol	32 176	.5475 .7694	Tin (melted)	482	.0578
			Sea Water (sp. gr. 1.0043)	64	.980
Mercury	32	.03346	Sea Water (sp. gr. 1.0463)	64	903
Benzol	50 122	.340 .423	Oil of Turpentine	32	.411
			Petroleum	70–136	.511
Glycerine	59–102	.576	Sulphuric Acid	68–133	.3363
Lead (melted)	500	.0356			

* When one temperature alone is given the "true" specific heat is given; otherwise the value is the "mean" specific heat for the range of temperature given.

† For variation, see Table 11.

Sensible Heat—The heat utilized in raising the temperature of a body is termed "sensible heat." In the case of water, the sensible heat required to raise it from 32 degrees Fahrenheit to the boiling point corresponding to the pressure at which steam is formed is termed the "heat of the liquid."

Latent Heat—Latent heat is the heat which apparently disappears in producing some change in the condition of a body without increasing its temperature. If heat be added to ice at freezing temperature, the ice will melt but its temperature will not be raised. The heat so utilized in changing the condition of the ice is the latent heat, and in this particular case is known as the "latent heat of fusion." If heat be added to water at 212 degrees under atmospheric pressure, the water will not become hotter but will be evaporated into steam, the temperature of which will also be 212 degrees. The heat so utilized is called the "latent heat of evaporation" and is the heat which apparently disappears in causing the substance to pass from a liquid to a gaseous state.

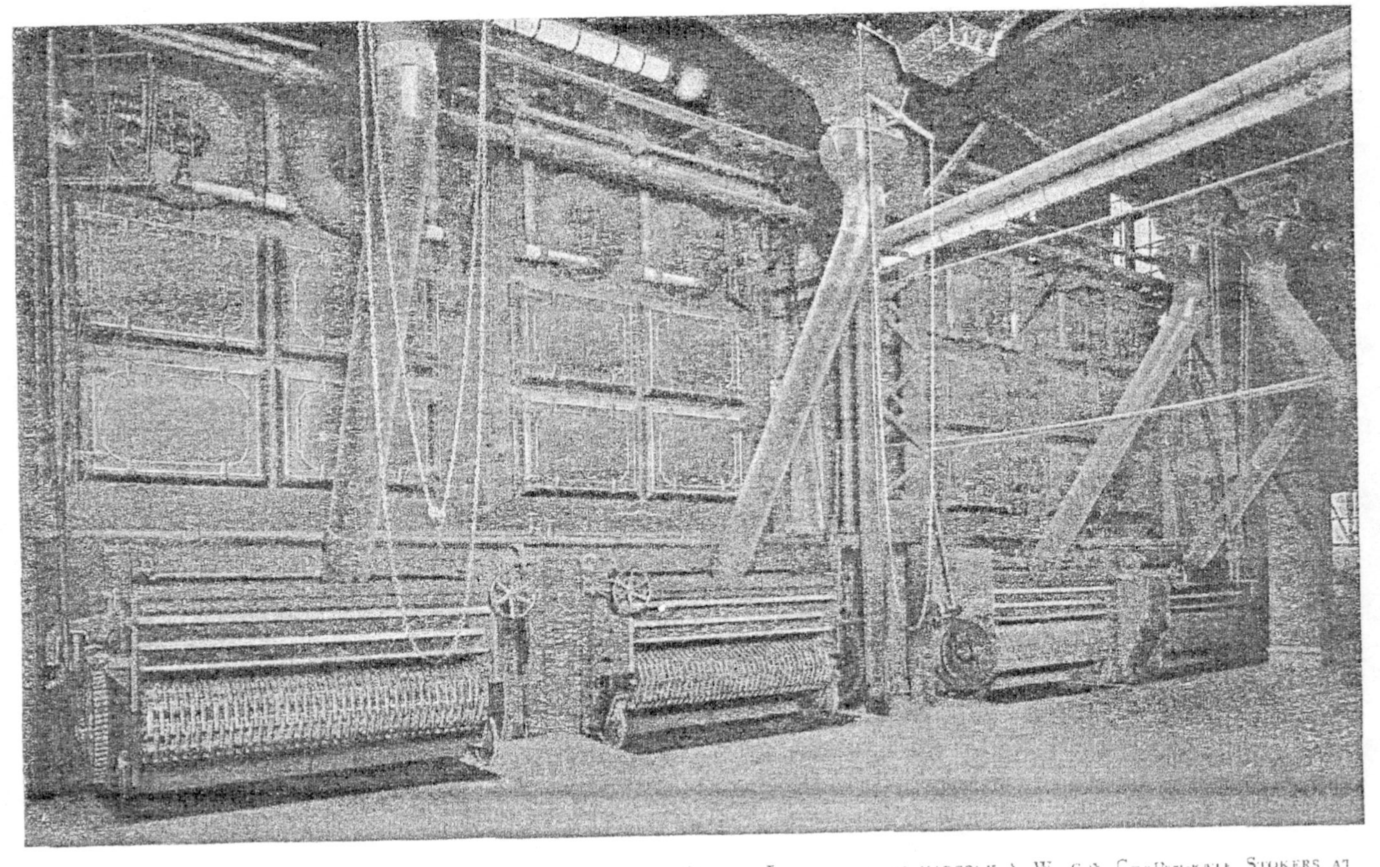

Portion of 3600 Horse-power Installation of Babcock & Wilcox Boilers Equipped with Babcock & Wilcox Chain-grate Stokers at The Loomis Street Plant of the Peoples Gas Light & Coke Co., Chicago, Ill. This Company Has Installed 7700 Horse-power of Babcock & Wilcox Boilers

Latent heat is not lost, but reappears whenever the substances pass through a reverse cycle, from a gaseous to a liquid or from a liquid to a solid state. It may, therefore, be defined, as stated, as the heat which apparently disappears, or is lost to thermometric measurement, when the molecular constitution of a body is being changed. Latent heat is expended in overcoming the molecular cohesion of the particles of the substance and in overcoming the resistance of external pressure to a change of volume of the heated body. Latent heat of evaporation, therefore, may be said to consist of internal and external heat, the former being utilized in overcoming the molecular resistance of the water in changing to steam, while the latter is expended in overcoming any resistance to the increase of its volume during formation. In evaporating a pound of water at 212 degrees to steam at 212 degrees, 897.6 B.t.u. are expended as internal latent heat and 72.8 B.t.u. as external latent heat. For a more detailed description of the changes brought about in water by sensible and latent heat, the reader is again referred to the chapter on "The Theory of Steam Making."

EBULLITION—The temperature of ebullition of any liquid, or its boiling point, may be defined as the temperature which exists where the addition of heat to the liquid no longer increases its temperature, the heat added being absorbed or utilized in converting the liquid into vapor. This temperature is dependent upon the pressure under which the liquid is evaporated, being higher as the pressure is greater.

TOTAL HEAT* OF EVAPORATION—The quantity of heat required to raise a unit of any liquid from the freezing point to any given temperature, and to entirely evaporate it at that temperature, is the total heat of evaporation of the liquid for that temperature. It is the sum of the heat of the liquid and the latent heat of evaporation.

To recapitulate, the heat added to a body is divided as follows:

Total heat = Heat to change the temperature + heat to overcome the molecular cohesion + heat to overcome the external pressure resisting an increase of volume of the body.

Where water is converted into steam, this total heat is divided as follows:

Total heat = Heat to change the temperature of the water + heat to separate the molecules of the water + heat to overcome resistance to increase in volume of the steam,

= Heat of the liquid + internal latent heat + external latent heat,

= Heat of the liquid + total latent heat of steam,

= Total heat of evaporation.

The steam tables given on pages 118 and 119 give the heat of the liquid and the total latent heat through a wide range of temperatures.

GASES—When heat is added to gases there is no internal work done; hence the total heat is that required to change the temperature plus that required to do the external work. If the gas is not allowed to expand but is preserved at constant volume, the entire heat added is that required to change the temperature only.

LINEAR EXPANSION OF SUBSTANCES BY HEAT—To find the increase in the length of a bar of any material due to an increase of temperature, multiply the number of

* The Committee on Definitions and Values of the A. S. M. E. Power Test Codes Committee has recommended that the term "Total Heat" be replaced by the term "Heat Content."

Vesta Coal Co., California, Pa. Operating at This Plant 2160 Horse Power of Babcock & Wilcox Boilers

degrees of increase in temperature by the coefficient of expansion for one degree and by the length of the bar. Where the coefficient of expansion is given for 100 degrees, as in Table 5, the result should be divided by 100. The expansion of metals per one degree rise of temperature increases slightly as high temperatures are reached, but for all practical purposes it may be assumed to be constant for a given metal.

TABLE 5

LINEAL EXPANSION OF SOLIDS AT ORDINARY TEMPERATURES

(Tabular values represent increase per foot per 100 degrees increase in temperature, Fahrenheit or Centigrade)

Substance	Temperature Conditions* Degrees Fahrenheit	Coefficient per 100 Degrees Fahrenheit	Coefficient per 100 Degrees Centigrade
Brass (cast)	32 to 212	.001042	.001875
Brass (wire)	32 to 212	.001072	.001930
Copper	32 to 212	.000926	.001666
Glass (English flint)	32 to 212	.000451	.000812
Glass (French flint)	32 to 212	.000484	.000872
Gold	32 to 212	.000816	.001470
Granite (average)	32 to 212	.000482	.000868
Iron (cast)	104	.000589	.001061
Iron (soft forged)	0 to 212	.000634	.001141
Iron (wire)	32 to 212	.000800	.001440
Lead	32 to 212	.001505	.002709
Mercury	32 to 212	.010141†	.018252†
Platinum	104	.000499	.000899
Limestone	32 to 212	.000139	.000251
Silver	104	.001067	.001921
Steel (Bessemer rolled, hard)	0 to 212	.00056	.00101
Steel (Bessemer rolled, soft)	0 to 212	.00063	.00117
Steel (cast, French)	104	.000734	.001322
Steel (cast annealed, English)	104	.000608	.001095

*Where range of temperature is given, coefficient is mean over range.
†Coefficient of cubical expansion.

HIGH TEMPERATURE MEASUREMENTS*—The temperatures to be dealt with in boiler practice range from those of ordinary air to the temperature of burning fuel. The gases of combustion, originally at the temperature of the furnace, cool as they pass over the boiler-heating surface, to a temperature approximating that due to the pressure within the boiler, resulting in a wide range of temperatures through which definite measurements are frequently required.

Of the different methods devised for ascertaining these temperatures, the most important are as follows:

1st. Mercurial thermometers or pyrometers for temperatures up to 750 or 800 degrees Fahrenheit.

2nd. Expansion thermometers for temperatures up to 1500 degrees Fahrenheit.

3rd. Platinum resistance pyrometers for temperatures up to 1200 or 1400 degrees Fahrenheit.

*For a full discussion of high temperature measurements and the instruments for such measurements, see the Bureau of Standards Technologic Paper No. 170, "Pyrometric Practice."

4th. Base metal thermoelectric pyrometers for temperatures up to 2000 degrees Fahrenheit.

5th. Rare metal thermoelectric pyrometers for temperatures up to 2900 degrees Fahrenheit.

6th. Melting points of metals which flow at various temperatures up to the melting point of platinums, 3159 degrees Fahrenheit.

7th. Radiation pyrometers for temperatures up to 3600 degrees Fahrenheit.

8th. Optical pyrometers capable of measuring temperatures up to 12,600 degrees Fahrenheit.* In ordinary boiler practice their range is from 1000 to 3600 degrees Fahrenheit.

MERCURIAL PYROMETERS—While mercurial pyrometers are ordinarily given a scale up to 1000 degrees Fahrenheit, the accuracy of readings above 800 degrees is questionable.

At atmospheric pressure, mercury boils at 676 degrees Fahrenheit, and even at lower temperatures the mercury in thermometers will be distilled and will collect in the upper part of the stem. Therefore, for temperatures much above 400 degrees Fahrenheit, some inert gas, such as nitrogen or carbon dioxide, must be forced under pressure into the upper part of the thermometer stem. The pressure at 600 degrees Fahrenheit is about 15 pounds, and at 1000 degrees about 300 pounds.

Flue-gas temperatures are nearly always taken with mercurial thermometers, as they are the most accurate and are easy to read and manipulate. Care must be taken that the bulb of the instrument projects into the path of the moving gases, in order that the temperature reading may truly represent the flue-gas temperature. No readings should be considered until the thermometer has been in place long enough to heat it up to the temperature of the gases. Even with mercurial thermometers of known accuracy, the determination of true exit gas temperatures is extremely difficult, and where there is the possibility of stratification of gases, and particularly with large boilers, a number of locations for thermometers should be used to assure that the temperature represents an average.

EXPANSION PYROMETERS—Expansion thermometers may be subdivided into two general classes, bimetallic and pressure thermometers. The former utilizes the principle of the turning moment of a strip formed by brazing together two metals having different coefficients of expansion, or a combination of similar metals with one of them fixed to the other at one end and the different rates of expansion of the two metals indicated by a multiple gear attached to a pointer on a graduated dial.

Pressure thermometers include vapor-pressure, liquid-filled and gas-filled. Liquid and gas-filled pressure thermometer scales are based on approximately linear expansion with temperature. Vapor pressure does not change with temperature in this manner, and the scale of the vapor-pressure thermometer is more open at higher temperatures.

The volatile liquid employed in vapor-pressure thermometers must be stable, readily obtainable, and should not act on the metals with which it is in contact. It should further have a pressure at low temperatures that allows a readable scale. The action of the vapor-pressure thermometer is based on the fact that the pressure is determined solely by the temperature of the free surface of the liquid.

* Le Chatelier's investigation

In liquid and gas-filled expansion thermometers, chemically pure materials are requisite and the highest temperature to be recorded must not exceed the critical temperature of the liquid.

PLATINUM RESISTANCE PYROMETERS—The resistance pyrometer depends in its operation upon the variation of the resistance of an electrical conductor with changes in temperature. The primarily important property of the material to be used in an instrument of this type is constancy of resistance and reproducibility. A simple and convenient relation between temperature and resistance in the material should also exist. For industrial purposes, resistance thermometers are made either of platinum or nickel, the latter being suitable for temperatures up to 575 degrees Fahrenheit. Platinum of the highest degree of purity in resistance pyrometers enables temperatures up to 1900 degrees Fahrenheit to be measured accurately, though this temperature, for precision measurements, requires accurate construction and careful experimental manipulation. The probable limit of temperature in commercial apparatus is approximately 1300 degrees Fahrenheit, and within its limits the platinum resistance pyrometer is the most accurate instrument available.

While the resistance pyrometer is, as stated, remarkably accurate within the limits of its temperature range, because of such limits and because of the precautions necessary in securing accuracy, its importance for industrial, as against laboratory work, is decreasing.

THERMOELECTRIC PYROMETERS—Thermoelectric pyrometry is based upon the principle that if, in a closed circuit of two metals, the two junctions are at different temperatures, an electric current will flow through the circuit. Not only is there an electromotive force between the two metals at the junctions, but in the single leads of homogeneous metal there is, provided the temperature of the two ends is not the same, an electromotive force depending upon the variation in the temperature of the two ends and the metals used in the leads. The total electromotive force, or the sum of the two described, thus depends on the difference in temperature between the two junctions. If the temperature of one junction is fixed, the temperature of the other may be determined by measuring the total electromotive force developed in the circuit, and it is this that is the basic principle of thermoelectric pyrometry.

Since any two dissimilar metals might be employed for thermoelectric couples, it is necessary in the selection of such metals to consider:

1st. The ability to resist corrosion and oxidation.

2nd. The development of a relatively high electromotive force.

3rd. That the temperature-electromotive force relation be such that the latter increases continuously with temperature increase over the range for which the couple is employed.

The metals in general use for thermoelectric couples vary with the temperature range to be covered and with the degree of accuracy desired in temperature measurement. A couple made up of one copper and one constantan wire will give extreme precision of measurement up to 680 degrees Fahrenheit and accuracy within 9 to 18 degrees at 930 degrees. Nichrome-constantan or iron-constantan couples may be used for technical processes up to 1650 degrees Fahrenheit, while for temperatures below 2015 degrees special alloys of chromium and nickel and of aluminum and nickel (chromel-alumel and nichrome-alumel) give satisfactory service even where the use of the thermocouple is continuous. For a temperature range of from 575 to 2725 degrees Fahrenheit, the Le Chatelier thermocouple, composed of platinum and an alloy of 90 per cent platinum and 10 per cent rhodium, should be used. No couples so far developed have been satisfactory at temperatures higher than 2725 degrees Fahrenheit. Several metals and alloys have a melting temperature appreciably above this, but all as yet tried have serious disadvantages.

The decrease in the availability of metals required for high temperature measurements is the cause of the subdivision on page 72 of thermocouples into base and rare metal couples. The principle of the two forms is the same, the difference being in initial cost, durability and accuracy. For accuracy, small wires forming the couple are desirable, and base metal couples do not give satisfactory results as to accuracy where small wires are employed. Consequently some degree of precision is sacrificed in base metal couples to give greater serviceability and lower initial cost.

TABLE 6

APPROXIMATE MELTING POINTS OF METALS

Metal	Temperature Degrees Fahrenheit	Metal	Temperature Degrees Fahrenheit
Wrought Iron	2737	Lead	621
Pig Iron (gray)	2190–2327	Bismuth	520
Cast Iron (white)	2075	Tin	449
Steel	2460–2550	Platinum	3191
Steel (cast)	2500	Gold	1945
Copper	1981	Silver	1760
Zinc	786	Aluminum	1218
Antimony	1166		

Because of greater or less delicacy, it is ordinarily necessary to give thermocouples some form of protection when obtaining temperature readings, and the choice of a proper protection tube is almost as important as that of the selection of material for the couple. Protection tubes should be able to withstand high temperatures and sudden changes in temperature. They should have a low porosity to gases which might attack the couple, a low volatility since the distillation of metal of the tube might affect the calibration of the couple, a sufficient degree of rigidity to withstand any tendency toward plastic deformation, the proper degree of thermal conductivity, which should be high where rapidly changing temperatures are to be measured but ordinarily low to minimize the flow of heat along the tube, and ability to resist corrosion from furnace gases.

The material used in thermocouple protection tubes varies in different industrial processes. The materials in most common use are composed of fused quartz, porcelain, carborundum, nichrome, clay, corundite, alundum and nickel.

Temperature measurements with thermocouples are indicated by readings on a galvanometer or potentiometer, and it is the improved accuracy of this apparatus that has to a large extent caused thermoelectric pyrometry to replace resistance pyrometry in many industrial processes.

Melting Point of Metals—The approximate temperature of a furnace or flue may be determined, if so desired, by introducing certain metals of which the melting points are known. The more common metals form a series in which the respective melting points differ by 100 to 200 degrees Fahrenheit, and by using these in order, the temperature can be fixed between the melting points of some two of them. This method lacks accuracy, but it suffices for determinations where approximate readings are satisfactory.

The approximate melting points of certain metals that may be used for determinations of this nature are given on Table 6.

RADIATION PYROMETERS—This class of pyrometer measures the intensity of all wave lengths emitted by a glowing body, the light and the heat rays combined. Ordinarily this combined radiated energy is focused in some manner upon the hot junction of a small thermocouple. The temperature to which the junction rises is approximately proportional to the rate at which the energy falls upon it, this in turn being proportional to the fourth power of the absolute temperature of the radiation body. In determining temperature as indicated by a radiation pyrometer, the same methods are employed as with thermoelectric pyrometers.

Errors in true temperatures as measured by radiation pyrometers may be as great as 400 degrees Fahrenheit, due to the difference in the power of emission of energy and light of different substances. Since the error is approximately constant for a given substance, however, this fact in connection with heat control in industrial processes is not of particular importance.

OPTICAL PYROMETERS—This class of pyrometer measures the intensity of a narrow spectral band of radiation emitted from a glowing body. Optical pyrometers are of two general classes. In the first, the light emitted from the body whose temperature is to be measured is compared with that from a constant source of light by means of suitable lenses. The intensity of the field from the fixed source of light and that from the light-emitting body are relatively decreased or increased by the manipulation of some portion of the instrument, and when the intensities of the two fields agree, a setting is obtained which is a measure of the temperature. In the second class of optical pyrometer, which is known as the disappearing filament type, the filament of a small electric lamp is placed at the focal point of the objective, and, by ordinary telescopic means, the image of the source viewed is superimposed upon the lamp. In setting the instrument for temperature determination, the current passing through the lamp is adjusted by a rheostat until the filament is indistinguishable from the surrounding field. The current through the lamp is measured on an ammeter and the corresponding temperature obtained from a chart or table.

The radiation and optical pyrometers offer the only pyrometric means of measuring temperatures above 2725 degrees Fahrenheit. While the human element enters into the temperatures as determined

TABLE 7

TEMPERATURE AND APPEARANCE OF FLAME*

Appearance of Flame	Temperature Degrees Fahrenheit
Dark Red	975
Dull Red	1290
Dull Cherry Red	1470
Full Cherry Red	1650
Clear Cherry Red	1830
Deep Orange	2010
White	2370
Bright White	2550
Dazzling White	2730

*C. S. M. Pouillet.

by either type of apparatus, they are rather remarkably accurate. Optical pyrometers, when designed specially for precision are accurate to within 3.6 degrees Fahrenheit, and in commercial forms are accurate to within 10 degrees.

DETERMINATION OF TEMPERATURE FROM CHARACTER OF EMITTED LIGHT — The temperature evolved in combustion may be approximated from the appearance of the fuel mass or the flame. These approximate temperatures are indicated by the figures of Table 7. Such figures are of necessity but the roughest approximations, but, in connection with the flame length, are of some value where apparatus for more accurate determination of the extent and degree of combustion is not available.

THE ESSEX STATION OF THE PUBLIC SERVICE ELECTRIC COMPANY OF NEW JERSEY, EQUIPPED WITH 36,080 HORSE-POWER OF BABCOCK & WILCOX BOILERS WITH BABCOCK & WILCOX SUPERHEATERS

THE THEORY OF STEAM MAKING

[Extracts from a lecture delivered by George H. Babcock, at Cornell University, 1887*]

THE chemical compound known as H_2O exists in three states or conditions—ice, water and steam; the only difference between these states or conditions is in the presence or absence of a quantity of energy exhibited partly in the form of heat and partly in molecular activity, which, for want of a better name, we are accustomed to call "latent heat"; and to transform it from one state to another we have only to supply or extract heat. For instance, if we take a quantity of ice, say one pound, at absolute zero† and supply heat, the first effect is to raise its temperature until it arrives at a point 492 Fahrenheit degrees above the starting point. Here it stops growing warmer, though we keep on adding heat. It, however, changes from ice to water, and when we have added sufficient heat to have made it, had it remained ice, 283 degrees hotter or a temperature of 315 degrees Fahrenheit's thermometer, it has all become water, at the same temperature at which it commenced to change, namely, 492 degrees above absolute zero, or 32 degrees by Fahrenheit's scale. Let us still continue to add heat, and it will now grow warmer again, though at a slower rate—that is, it now takes about double the quantity of heat to raise the pound one degree than it did before—until it reaches a temperature of 212 degrees Fahrenheit, or 672 degrees absolute (assuming that we are at the level of the sea). Here we find another critical point. However much more heat we may apply, the water, as water, at that pressure, cannot be heated any hotter, but changes on the addition of heat to steam; and it is not until we have added heat enough to have raised the temperature of the water 966 degrees, or to 1178 degrees by Fahrenheit's thermometer (presuming for the moment that its specific heat has not changed since it became water), that it has all become steam, which steam, nevertheless, is at the temperature of 212 degrees, at which the water began to change. Thus over four-fifths of the heat which has been added to the water has disappeared, or become insensible in the steam to any of our instruments.

It follows that if we could reduce steam at atmospheric pressure to water, without loss of heat, the heat stored within it would cause the water to be red hot; and if we could further change it to a solid, like ice, without loss of heat, the solid would be white hot, or hotter than melted steel—it being assumed, of course, that the specific heat of the water and ice remain normal, or the same as they respectively are at the freezing point.

After steam has been formed, a further addition of heat increases the temperature again at a much faster ratio to the quantity of heat added, which ratio also varies according as we maintain a constant pressure or a constant volume; and I am not aware that any other critical point exists where this will cease to be the fact until we arrive at that very high temperature, known as the point of dissociation, at which it becomes resolved into its original gases.

The heat which has been absorbed by one pound of water to convert it into a pound of steam at atmospheric pressure is sufficient to have melted 3 pounds of steel or 13 pounds of gold. This has been transformed into something besides heat;

*See Scientific American Supplement, 624, 625, December, 1887. While the values of the physical constants given by Mr. Babcock are not the values accepted today, the theory is not affected.

†460 degrees below the zero of Fahrenheit. This is the nearest approximation in whole degrees to the latest determinations of the absolute zero of temperature.

stored up to reappear as heat when the process is reversed. That condition is what we are pleased to call latent heat, and in it resides mainly the ability of the steam to do work.

The diagram shows graphically the relation of heat to temperature, the horizontal scale being quantity of heat in British thermal units, and the vertical temperature in Fahrenheit degrees, both reckoned from absolute zero and by the usual scale. The dotted lines for ice and water show the temperature which would have been obtained if the conditions had not changed. The lines marked "gold" and "steel" show the relation to heat and temperature and the melting points of these metals. All the inclined lines would be slightly curved if attention had been paid to the changing specific heat, but the curvature would be small. It is worth noting that, with one or two exceptions. the curves of all substances lie between the vertical and that for water. That is to say, that water has a greater capacity for heat than all other substances except two, hydrogen and bromine.

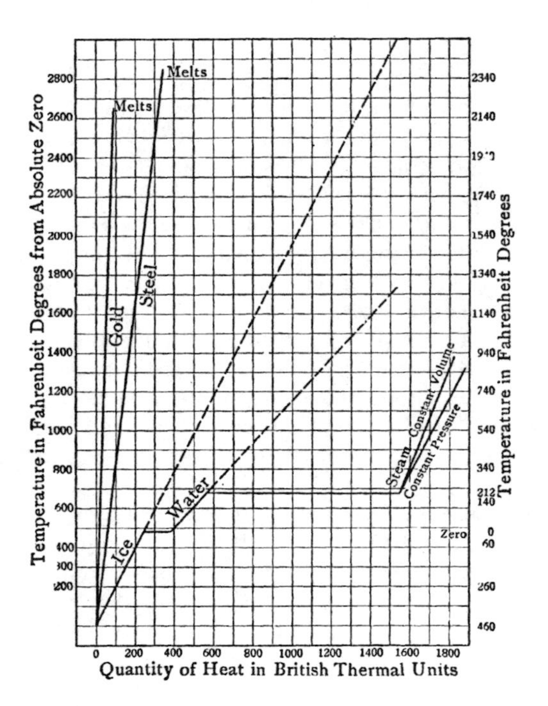

In order to generate steam, then only two steps are required: 1st, procure the heat. and 2nd, transfer it to the water. Now, you have it laid down as an axiom that when a body has been transferred or transformed from one place or state into another, the same work has been done and the same energy expended, whatever may have been the intermediate steps or conditions, or whatever the apparatus. Therefore, when a given quantity of water at a given temperature has been made into steam at a given temperature, a certain definite work has been done, and a certain amount of energy expended. from whatever the heat may have been obtained, or whatever boiler may have been employed for the purpose.

A pound of coal or any other fuel has a definite heat producing capacity, and is capable of evaporating a definite quantity of water under given conditions. That is the limit beyond which even perfection cannot go, and yet I have known. and doubtless you have heard of. cases where inventors have claimed, and so-called engineers have certified to, much higher results.

The first step in generating steam is in burning the fuel to the best advantage. A pound of carbon will generate 14,500 B.t.u. during combustion into carbonic dioxide. and this will be the same. whatever the temperature or the rapidity at which the combustion may take place. If possible, we might oxidize it at as slow a rate as that with which iron rusts or wood rots in the open air, or we might burn it with the rapidity of gunpowder, a ton in a second, yet the total heat generated would be precisely the same. Again we may keep the temperature down to the lowest point at which combustion can take place, by bringing large bodies of air in contact with it, or otherwise, or we may supply it with just the right quantity of pure oxygen, and burn it at a temperature approaching that of dissociation, and still the heat units given off will be neither more nor less. It follows, therefore, that great latitude in the manner or rapidity of combustion may be taken without affecting the quantity of heat generated.

But in practice it is found that other considerations limit this latitude, and that there are certain conditions necessary in order to get the most available heat from a pound of coal. There are three ways, and only three, in which the heat developed by the combustion of coal in a steam boiler furnace may be expended.

1st, and principally. It should be conveyed to the water in the boiler, and be utilized in the production of steam. To be perfect, a boiler should so utilize all the heat of combustion, but there are no perfect boilers.

2nd. A portion of the heat of combustion is conveyed up the chimney in the waste gases. This is in proportion to the weight of the gases, and the difference between their temperature and that of the air and coal before they entered the fire.

3rd. Another portion is dissipated by radiation from the sides of the furnace. In a stove the heat is all used in these latter two ways, either it goes off through the chimney or is radiated into the surrounding space. It is one of the principal problems of boiler engineering to render the amount of heat thus lost as small as possible.

The loss from radiation is in proportion to the amount of surface, its nature, its temperature, and the time it is exposed. This loss can be almost entirely eliminated by thick walls and a smooth white or polished surface, but its amount is ordinarily so small that these extraordinary precautions do not pay in practice.

It is evident that the temperature of the escaping gases cannot be brought below that of the absorbing surfaces, while it may be much greater even than that of the fire. This is supposing that all of the escaping gases have passed through the fire. In case air is allowed to leak into the flues and mingle with the gases after they have left the heating surfaces, the temperature may be brought down to almost any point above that of the atmosphere, but without any reduction in the amount of heat wasted. It is in this way that those low chimney temperatures are sometimes attained which pass for proof of economy with the unobserving. All surplus air admitted to the fire, or to the gases before they leave the heating surfaces, increases the losses.

We are now prepared to see why and how the temperature and the rapidity of combustion in the boiler furnace affect the economy, and that though the amount of heat developed may be the same, the heat available for the generation of steam may be much less with one rate or temperature of combustion than another.

Assuming that there is no air passing up the chimney other than that which has passed through the fire, the higher the temperature of the fire and the lower that of the escaping gases the better the economy, for the losses by the chimney gases will bear the same proportion to the heat generated by the combustion as the temperature of those gases bears to the temperature of the fire. That is to say, if the temperature of the fire is 2500 degrees and that of the chimney gases 500 degrees above that of the atmosphere, the loss by the chimney will be $\frac{500}{2500} = 20$ per cent. Therefore as the escaping gases cannot be brought below the temperature of the absorbing surface, which is practically a fixed quantity the temperature of the fire must be high in order to secure good economy.

The losses by radiation being practically proportioned to the time occupied the more coal burned in a given furnace in a given time, the less will be the proportionate loss from that cause.

It therefore follows that we should burn our coal rapidly and at a high temperature, to secure the best available economy

REAR VIEW OF A 1166 HORSE-POWER STEEL-CASED BABCOCK & WILCOX CROSS-DRUM BOILER

PROPERTIES OF WATER

PURE water is a chemical compound of one volume of oxygen and two volumes of hydrogen, its chemical symbol being H_2O.

The density of water depends upon its temperature. Its density at four temperatures, much used in physical calculations, is given in Table 8.

TABLE 8

DENSITY OF WATER AT TEMPERATURES USED IN PHYSICAL CALCULATIONS

Temperature Degrees Fahrenheit	Density per Cubic Foot Pounds	Density per Cubic Inch Pound
At 32 degrees or freezing point at sea level . .	62.418	0.03612
At 39.2 degrees or point of maximum density	62.427	0.03613
At 62 degrees or standard temperature	62.355	0.03608
At 212 degrees or boiling point at sea level . .	59.846	0.03469

The density of water has been very carefully determined and the results of the more trustworthy determinations are in very good agreement. For temperatures up to 212 degrees Fahrenheit, the values are known to an accuracy of one or two parts in 100,000. The values in Table 8 are quoted from the Bureau of Standards Circular No. 19. "Standard Density and Volumetric Tables."

The United States standard gallon is a volume of 231 cubic inches. A gallon (231 cubic inches) of water at 62 degrees Fahrenheit weighs 8.3357 pounds.

In the above a pound of water means the quantity of water which has a mass of one pound; in other words, the quantity which has a weight, corrected for the effect of the buoyancy of the air, of one pound. If water is weighed in air in the

TABLE 9

VOLUME AND WEIGHT OF DISTILLED WATER AT VARIOUS TEMPERATURES*

Temperature Degrees Fahrenheit	Relative Volume Water at 39.2 Degrees=1	Weight per Cubic Foot Pounds	Temperature Degrees Fahrenheit	Relative Volume Water at 39.2 Degrees=1	Weight per Cubic Foot Pounds	Temperature Degrees Fahrenheit	Relative Volume Water at 39.2 Degrees=1	Weight per Cubic Foot Pounds	Temperature Degrees Fahrenheit	Relative Volume Water at 39.2 Degrees=1	Weight per Cubic Foot Pounds
32	1.000176	62.42	160	1.02337	61.00	290	1.0830	57.65	430	1.197	52.2
39.2	1.000000	62.43	170	1.02682	60.80	300	1.0890	57.33	440	1.208	51.7
40	1.000004	62.43	180	1.03047	60.58	310	1.0953	57.00	450	1.220	51.2
50	1.00027	62.42	190	1.03431	60.36	320	1.1019	56.66	460	1.232	50.7
60	1.00096	62.37	200	1.03835	60.12	330	1.1088	56.30	470	1.244	50.2
70	1.00201	62.30	210	1.04256	59.88	340	1.1160	55.94	480	1.256	49.7
80	1.00338	62.22	212	1.04343	59.83	350	1.1235	55.57	490	1.269	49.2
90	1.00504	62.11	220	1.0469	59.63	360	1.1313	55.18	500	1.283	48.7
100	1.00698	62.00	230	1.0515	59.37	370	1.1396	54.78	510	1.297	48.1
110	1.00915	61.86	240	1.0562	59.11	380	1.1483	54.36	520	1.312	47.6
120	1.01157	61.71	250	1.0611	58.83	390	1.1573	53.94	530	1.329	47.0
130	1.01420	61.55	260	1.0662	58.55	400	1.167	53.5	540	1.35	46.3
140	1.01705	61.38	270	1.0715	58.26	410	1.177	53.0	550	1.37	45.6
150	1.02011	61.20	280	1.0771	57.96	420	1.187	52.6	560	1.39	44.9

*Marks and Davis.

ordinary way, against brass or iron weights, the weight will be smaller than the mass by about one-tenth of one per cent, a difference which is negligible except where results are required to more than three significant figures. Table 20 gives the density of water at different temperatures. as stated by Marks and Davis.

The British Imperial gallon is defined as the volume of 10 Imperial pounds of distilled water weighed in air against brass weights, with the water and the air at a temperature of 62 degrees Fahrenheit, the barometer being at 30 inches. This volume equals 277.41 cubic inches.

Water is but slightly compressible and for all practical purposes may be considered non-compressible. The coefficient of compressibility ranges from 0.000051 per atmosphere at 35 degrees Fahrenheit to 0.000044 at 128 degrees, decreasing as the temperature increases.

Table 9 gives the weight in vacuo and the relative volume of a cubic foot of distilled water at various temperatures.

The density of water at the standard temperature being taken as 62.355 pounds per cubic foot, the pressure exerted by the column of water of any stated height, and conversely the height of any column required to produce a stated pressure, may be computed as follows:

The pressure in pounds per square foot $=62.355 \times$ height of column in feet.

The pressure in pounds per square inch $=0.433 \times$ height of column in feet.

Height of column in feet = pressure in pounds per square foot $\div 62.355$.

Height of column in feet = pressure in pounds per square inch $\div 0.433$.

Height of column in inches = pressure in pounds per square inch $\times 27.71$.

Height of column in inches = pressure in ounces per square inch $\times 1.73$.

By a change in the densities given above, the pressure exerted and height of column may be computed for temperatures other than 62 degrees. Table 20 gives the densities at temperatures from 32 to 340 degrees Fahrenheit.

Where an accuracy of more than about one-tenth of one per cent is required, the pressures in pounds per square foot computed by the relations just stated must be corrected for the displaced air column, the variation of g and the variation of density with temperature. The error due to neglecting the difference between the local and standard values of g may be greater than the error due to neglecting the difference between the standard temperature and the actual temperature of the water column. Similarly the error due to neglecting the effect of the displaced air in calculating the pressure in pounds per square foot due to a water column, by the above relations, may amount to as much as the change due to a variation of 10 to 15 degrees in the temperature of the water. The data generally adopted by American physicists and engineers are given in Circular 19 of the U. S. Bureau of Standards, "Standard Density and Volumetric Tables."

A pressure of one pound per square inch is exerted by a column of water 2.3093 feet or 27.71 inches high at 62 degrees Fahrenheit.

Water in its natural state is never found absolutely pure. It has a solvent action of wider range than any other liquid possesses, and this action is affected by temperature. The solubility of sodium chloride increases somewhat with rising temperature, the solubility of magnesium sulphate increases much more rapidly, while the solubility of sodium sulphate increases rapidly with the temperature up to about 90 degrees Fahrenheit, and then decreases as the temperature rises above that point.

Sea water contains about 3.5 per cent, by weight, of dissolved solids comprising sodium chloride, 2.72 per cent; magnesium chloride, 0.38 per cent; magnesium sulphate, 0.17 per cent; calcium sulphate, 0.13 per cent; potassium sulphate, 0.08 per cent; calcium carbonate, 0.01 per cent; magnesium bromide, 0.01 per cent (Encyclopædia Britannica, eleventh edition, XIX, 977).

TABLE 10

EFFECT OF ALTITUDE IN REDUCING THE BOILING POINT OF WATER

Altitude Feet	Barometer at 32 Degrees Fahrenheit Inches	Boiling Point Degrees Fahrenheit	Atmospheric Pressure Pounds per Square Inch	Altitude Feet	Barometer at 32 Degrees Fahrenheit Inches	Boiling Point Degrees Fahrenheit	Atmospheric Pressure Pounds per Square Inch
0	29.92	212	14.70	8,000	22.38	198	10.98
500	29.43	211	14.46	8,500	21.96	197	10.78
1,000	28.93	210	14.21	9,000	21.55	196	10.58
1,500	28.44	209	13.97	9,500	21.15	195	10.38
2,000	27.94	209	13.72	10,000	20.75	194	10.19
2,500	27.44	208	13.48	10,500	20.36	193	10.00
3,000	26.94	207	13.23	11,000	19.98	192	9.81
3,500	26.44	206	12.99	11,500	19.60	191	9.63
4,000	25.95	205	12.75	12,000	19.24	191	9.45
4,500	25.47	204	12.52	12,500	18.88	190	9.27
5,000	25.01	203	12.29	13,000	18.52	189	9.09
5,500	24.55	202	12.06	13,500	18.17	188	8.92
6,000	24.09	201	11.83	14,000	17.82	187	8.75
6,500	23.65	200	11.61	14,500	17.47	186	8.58
7,000	23.22	199	11.39	15,000	17.12	185	8.41
7,500	22.80	199	11.18	15,500	16.77	184	8.24

The boiling point of water decreases as the altitude above the sea level increases. Table 10 gives the reduction in barometric pressure with the altitude as reported by Gregg in the Monthly Weather Review, January, 1918, with the corresponding boiling points of water and atmospheric pressures in pounds per square inch, computed from data in the Smithsonian Physical Tables, 1921. Tables of this nature are only approximately correct and need revision as research establishes more closely the relation between altitude and barometric pressure.

The specific heat of a substance is defined as the quantity of heat, per degree, required to raise the temperature of unit mass of the substance under specified conditions. As measured under ordinary conditions, at constant (atmospheric) pressure the specific heat of water is found to be larger than that of all but a few other substances. The specific heat of water depends upon the temperature, and numerous careful measurements have been made to determine the manner of this variation, but the results are not in sufficiently good agreement to lead to conclusions that may be considered as final. The values that are used for the Marks and Davis Steam Tables and for the Callendar Steam Tables are given in Table 11. The values used by Callendar are based on his own measurements and those of Barnes (Philosophical Transactions, Series A, 1913, volume 212, page 30).

The British thermal unit is defined as 1/180th of the quantity of heat required to raise the temperature of one pound of water from 32 to 212 degrees Fahrenheit, and, according to this definition, the average specific heat of water between 32 and

212 degrees Fahrenheit is unity. It will be noted in Table 11 that the specific heat is arbitrarily chosen as equal to unity at 55 degrees Fahrenheit and 20 degrees Centigrade, respectively, in the two parts of the table.

TABLE 11

SPECIFIC HEAT OF WATER AT VARIOUS TEMPERATURES

MARKS AND DAVIS From Values of Barnes and Dieterici				CALLENDAR From Values of Barnes and Callendar					
Temperature Degrees Fahrenheit	Specific Heat	Temperature Degrees Fahrenheit	Specific Heat	Temperature		Specific Heat	Temperature		Specific Heat
				Degrees Centigrade	Degrees Fahrenheit		Degrees Centigrade	Degrees Fahrenheit	
30	1.0098	130	0.9979	0	32	1.0106	55	131	0.9994
40	1.0045	140	0.9986	5	41	1.0065	60	140	0.9999
50	1.0012	150	0.9994	10	50	1.0031	65	149	1.0005
55	1.0000	160	1.0002	15	59	1.0009	70	158	1.0012
60	0.9990	170	1.0010	20	68	1.0000	75	167	1.0020
70	0.9977	180	1.0019	25	77	0.9992	80	176	1.0029
80	0.9970	190	1.0029	30	86	0.9988	85	185	1.0038
90	0.9967	200	1.0039	35	95	0.9986	90	194	1.0048
100	0.9967	210	1.0052	40	104	0.9986	95	203	1.0059
110	0.9970	220	1.007	45	113	0.9987	100	212	1.0070
120	0.9974	230	1.009	50	122	0.9990	. .	. .	. .

In consequence of this variation in specific heat, the variation in the heat of the liquid of the water at different temperatures is not a constant. Table 20* gives the heat of the liquid in a pound of water at temperatures ranging from 32 to 340 degrees Fahrenheit.

The specific heat of ice at 32 degrees Fahrenheit is 0.506. The latent heat of fusion is 143.5 B.t.u. per pound (Bulletin, Bureau of Standards, 1915, volume 12, page 49). The specific heat of water vapor, the third form in which water may exist, depends very much on the manner in which the vapor is heated, whether at constant pressure, constant volume, or otherwise. The specific heat of saturated vapor requires special mention. To change saturated vapor at one temperature to saturated vapor at a higher temperature, the pressure must be increased. If the compression were effected adiabatically, that is, without adding or subtracting heat, the vapor would become superheated, consequently, during compression, some heat must be abstracted in order to keep the vapor in a saturated condition. In this process, the rise in temperature was accompanied by the removal of heat from the vapor, and the specific heat of saturated vapor is therefore negative. The specific heat of superheated steam is discussed later in this book.

*See page 116.

BOILER-FEED WATER

ALL natural waters contain some impurities which, when introduced into a boiler, may appear as solids. In view of the apparent present-day tendency toward large size boiler units and high overloads, the importance of the use of pure water for boiler-feed purposes cannot be overestimated.

Ordinarily, when water of sufficient purity for such use is not at hand, the supply available may be rendered suitable by some process of treatment. Against the cost of such treatment, there are many factors to be considered. With water in which there is a marked tendency toward scale formation, the interest and depreciation on the added boiler units necessary to allow for the systematic cleaning of certain units must be taken into consideration. Again there is a considerable loss in taking boilers off for cleaning and replacing them on the line. On the other hand, the decrease in capacity and efficiency accompanying an increased incrustation of boilers in use has been too generally discussed to need repetition here. Many experiments have been made and actual figures reported as to this decrease, but in general such figures apply only to the particular set of conditions found in the plant where the boiler in question was tested. So many factors enter into the effect of scale on capacity and economy that it is impossible to give any accurate figures on such decrease that will serve all cases, but that it is large has been thoroughly proven.

While it is almost invariably true that practically any cost of treatment will pay a return on the investment in the apparatus, the fact must not be overlooked that there are certain waters which should never be used for boiler-feed purposes

TABLE 12

APPROXIMATE CLASSIFICATION OF IMPURITIES FOUND IN FEED WATERS THEIR EFFECT AND ORDINARY METHODS OF RELIEF

Difficulty Resulting from Presence of	Nature of Difficulty	Ordinary Method of Overcoming or Relieving
Sediment, Mud, etc.	Incrustation	Settling tanks, filtration, blowing down.
Readily Soluble Salts	Incrustation and Priming	Blowing down.
Bicarbonates of Lime, Magnesia, etc.	Incrustation	Heating feed. Treatment by addition of lime or of lime and soda. Caustic soda and barium hydrate.
Sulphate of Lime	Incrustation	Treatment by addition of soda. Barium carbonate.
Chloride and Sulphate of Magnesium	Corrosion	Treatment by addition of carbonate of soda
Acid	Corrosion	Alkali.
Dissolved Carbonic Acid and Oxygen	Corrosion	Heating feed. Keeping air from feed. Addition of caustic soda or slacked lime.
Grease	Corrosion	Filter. Iron alum as coagulant. Neutralization by carbonate of soda Use of best hydrocarbon oils.
Organic Matter	Corrosion	Filter. Use of coagulant.
Organic Matter (Sewage)	Priming	Settling tanks. Filter in connection with coagulant.
Carbonate of Soda in large quantities	Priming	New feed supply. If from treatment, change

and which no treatment can render suitable for such purpose. In such cases, the only remedy is the securing of another feed supply or the employment of evaporators for distilling the feed water, as in marine service.

It is evident that the whole subject of boiler-feed waters and their treatment is one for the chemist rather than for the engineer. A brief outline of the difficulties that may be experienced from the use of poor feed water and a suggestion as to a method of overcoming certain of these difficulties is all that will be attempted here. Such a brief outline of the subject, however, will indicate the necessity for a chemical analysis of any water before a treatment is tried and the necessity of adapting the treatment in each case to the nature of the difficulties that may be experienced.

Table 12 gives a list of impurities which may be found in boiler-feed water, grouped according to their effect on boiler operation and stating the customary method used for overcoming the difficulty to which they lead.

SCALE—Scale is formed on boiler-heating surfaces by the depositing of impurities in the feed water in the form of a more or less hard adherent crust. Such deposits are due to the fact that water loses its soluble power at high temperatures or because the concentration becomes so high, due to evaporation, that the impurities crystallize and adhere to the boiler surfaces. The opportunity for formation of scale in a boiler will be apparent when it is realized that during a month's operation of a 100 horse-power boiler, 300 pounds of solid matter may be deposited from water containing only 7 grains per gallon, while some spring and well waters contain sufficient to cause a deposit of as high as 2000 pounds.

The salts usually responsible for such incrustation are the carbonates and sulphates of lime and magnesia, and boiler-feed treatment in general deals with the getting rid of these salts more or less completely.

Table 13 gives some data on the solubility of these mineral salts at various temperatures in grains per United States gallon (58,381 grains). Calcium and magnesium carbonates are rarely present as such in raw boiler-feed waters, but exist in the presence of CO_2, as bicarbonates. Carbon dioxide is insoluble at 212 degrees, and the solubility of the carbonates becomes zero. The variation of solubility at temperatures below 212 degrees is dependent upon the partial pressure of the carbon dioxide, the upper limits given at 60 degrees representing saturation of the water. Calcium sulphate, while somewhat insoluble above 212 degrees, becomes more so as the temperature increases.

TABLE 13

SOLUBILITY OF MINERAL SALTS IN WATER

IN GRAINS PER U. S. GALLON (58,381 GRAINS), EXCEPT AS NOTED

Temperature Degrees Fahrenheit	60 Degrees	212 Degrees
Calcium Carbonate*	2.5	1.5
Calcium Sulphate	140.0	125.0
Magnesium Carbonate*	1.0	1.8
Magnesium Sulphate	3.0 pounds	12.0 pounds
Sodium Chloride	3.5 pounds	4.0 pounds
Sodium Sulphate	1.1 pounds	5.0 pounds

CALCIUM SULPHATE AT TEMPERATURES ABOVE 212 DEGREES (CHRISTIE)

Temperature, degrees Fahrenheit	284	329	347–365	464	482
Corresponding gauge pressure	38	87	115–149	469	561
Grains per gallon	45.5	32.7	15.7	10.5	9.3

*In presence of CO_2.

Scale is also formed by the settling of mud and sediment carried in suspension in water. This may bake or be cemented into a hard scale when mixed with other scale-forming ingredients.

CORROSION—Corrosion, or a chemical action leading to the actual destruction of the boiler metal, is due to the solvent or oxidizing properties of the feed water. It results from the presence of acid, either free or developed,* in the feed, the admixture of air with the feed water, or as a result of galvanic action. In boilers it takes several forms:

1st. Pitting, which consists of isolated spots of active corrosion which does not attack the boiler as a whole.

2nd. General corrosion, produced by naturally acid waters and where the amount is so even and continuous that no accurate estimate of the metal eaten away may be made.

3rd. Grooving, which, while largely a mechanical action which may occur in neutral waters, is intensified by acidity.

FOAMING—This phenomenon, which ordinarily occurs with waters contaminated with sewage or organic growths, is due to the fact that the suspended particles collect on the surface of the water in the boiler and render difficult the liberation of steam bubbles rising to that surface. It sometimes occurs with water containing carbonates in solution in which a light flocculent precipitate will be formed on the surface of the water. Again, it is the result of an excess of sodium carbonate used in treatment for some other difficulty where animal or vegetable oil finds its way into the boiler.

PRIMING—Priming, or the passing off of steam from a boiler in belches, is caused by the concentration of sodium carbonate, sodium sulphate or sodium chloride in solution. Sodium sulphate is found in some quantity in the great majority of waters throughout the country, being more prevalent in southern waters, and in treated waters where calcium or magnesium sulphate is precipitated with soda ash.

TREATMENT OF FEED WATER—For scale formation. The treatment of feed water, carrying scale-forming ingredients, is along two main lines: 1st, by chemical means by which such impurities as are carried by the water are caused to precipitate; and 2nd, by means of heat, which results in the reduction of the power of water to hold certain salts in solution. The latter method alone is sufficient in the case of certain temporarily hard waters, but the heat treatment, in general, is used in connection with a chemical treatment to assist the latter.

Before going further into detail as to the treatment of water, it may be well to define certain terms used.

Hardness, which is the most widely known evidence of the presence in water of scale-forming matter, is that quality, the variation of which makes it more difficult to obtain a lather or suds from soap in one water than in another. This action is made use of in the soap test for hardness described later. Hardness is ordinarily classed as either temporary or permanent. Temporarily hard waters are those containing bicarbonates of lime and magnesium which may be precipitated by boiling at 212 degrees and which, if they contain no other scale-forming ingredients, become

*Some waters, not naturally acid, become so at high temperatures, as when chloride of magnesia decomposes with the formation of free hydrochloric acid; such phenomena become more serious with an increase in pressure and temperature.

" soft " under such treatment. Permanently hard waters are those containing mainly calcium sulphate, which is only precipitated at the high temperatures found in the boiler itself, 300 degrees Fahrenheit or more. The scale of hardness is an arbitrary one, based on the number of grains of solids per gallon, and waters may be classed on such a basis as follows: 1-10 grains per gallon, soft water; 10-20 grains per gallon, moderately hard water; above 25 grains per gallon, very hard water.

Alkalinity is a general term used for waters containing compounds with the power of neutralizing acids.

Causticity, as used in water treatment, is a term coined by A. McGill, indicating the presence of an excess of lime or caustic soda added during treatment. Though such presence would also indicate alkalinity, the term is arbitrarily used to apply to those hydrates whose presence is indicated by phenolphthalein.

Of the chemical methods of water treatment, there are five general processes:

1st. *Lime Process*—The lime process is used for waters containing bicarbonates of lime and magnesia. Slacked lime in solution, as lime water, is the reagent used. This combines with the carbonic acid which is present, either free or as bicarbonates, to form an insoluble carbonate of lime. The soluble bicarbonates of lime and magnesia, losing their carbonic acid, thereby become insoluble and precipitate.

2nd. *Soda Process*—The soda process is used for waters containing sulphates of lime and magnesia. Carbonate of soda and hydrate of soda (caustic soda) are used either alone or together as the reagents. Carbonate of soda, added to the water, decomposes the sulphates to form insoluble carbonates of lime or magnesium which precipitate, the neutral soda remaining in solution. If free carbonic acid (carbon dioxide) is present in the water, soluble bicarbonates of lime or magnesium are formed. Under the action of heat, the carbon dioxide will be driven off and the insoluble carbonates will be formed. Caustic soda, when used in this process, absorbs the free carbonic acid and also converts the bicarbonates of lime and magnesia into the insoluble carbonates, at the same time forming either sodium carbonate or bicarbonate, according to the amount present, which in turn decompose the sulphates present, converting them to either carbonates or bicarbonates.

3rd. *Lime and Soda Process*—This process, which is the combination of the first two, is by far the most generally used in water purification. Such a method is used where sulphates of lime and magnesia are contained in the water, together with such quantity of carbonic acid or bicarbonates as to impair the action of the soda. Sufficient soda is used to break down the sulphates of lime and magnesia and as much lime added as is required to absorb the carbonic acid not taken up in the soda reaction.

All of the apparatus for effecting such treatment of feed waters is approximately the same in its chemical action, the numerous systems differing in the methods of introduction and handling of the reagents.

The methods of testing water treated by an apparatus of this description follow:

There is considerable difference of opinion as to the correct ratio of alkalinity, hardness and causticity in properly treated water. According to some authorities, the causticity should be from two to three times the hardness, and the alkalinity from one and a half to two times the causticity. If too little lime is used, the causticity will be too low, the hardness and alkalinity too high, and scale will result. If too much is used, the causticity will be too high. Where insufficient soda ash is used, the hardness is too high and alkalinity too low, and scale is also formed. Where too great a quantity of soda ash is used, priming and other troubles result.

Alkalinity and causticity are tested with a standard solution of sulphuric acid. A standard soap solution is used for testing for hardness and a silver nitrate solution may also be used for determining whether an excess of lime has been used in the treatment.

Alkalinity: To 100 cubic centimeters of treated water, to which there has been added sufficient methylorange to color it, add the standard 10th normal acid solution, drop by drop, until the mixture is on the point of turning permanent red. As the standard 10th normal acid solution is first added, the red color, which shows quickly, disappears on shaking the mixture, and this color disappears more slowly as the critical point is approached. Number of cubic centimeters of the standard 10th normal acid solution used, multiplied by 50, equals parts per million of alkalinity as calcium carbonate.

Causticity: To 100 cubic centimeters of treated water, to which there has been added one drop of phenolphthalein dissolved in alcohol to give the water a pinkish color, add the standard 10th normal acid solution, drop by drop, shaking after each addition, until the color entirely disappears. Number of cubic centimeters of the standard 10th normal acid solution used, multiplied by 50, equals parts per million of causticity as calcium carbonate.

The alkalinity may be determined from the same sample tested for causticity by coloring it with methylorange and adding the acid until the sample is on the point of turning red. The total acid added in determining both causticity and alkalinity in this case is the measure of the alkalinity.

Hardness: 50 cubic centimeters of the treated water is used for this test. The total hardness, in parts per million of calcium carbonate for each 10th of a cubic centimeter of soap solution, when 50 cubic centimeters of the sample is titrated, is given in Table 14. The soap solution is added a little at a time and the whole violently shaken. Enough of the solution must be added to make a permanent foam or lather, that is, the soap bubbles must not disappear after the shaking is stopped.

TABLE 14

TOTAL HARDNESS, IN PARTS PER MILLION OF $CaCO_3$, FOR EACH TENTH OF A CUBIC CENTIMETER OF SOAP SOLUTION WHEN A 50 CUBIC CENTIMETER SAMPLE IS TITRATED

Soap Solution Cubic Centimeters	0.0	0.1	0.2	0.3	0.4	0.5	0.6	0.7	0.8	0.9
0.0	. . .	. . .	. . .	. . .	.	. .	. .	0.0	1.6	3.2
1.0	4.8	6.3	7.9	9.5	11.1	12.7	14.3	15.6	16.9	18.2
2.0	19.5	20.8	22.1	23.4	24.7	26.0	27.3	28.6	29.9	31.2
3.0	32.5	33.8	35.1	36.4	37.7	38.0	40.3	41.6	42.9	44.3
4.0	45.7	47.1	48.6	50.0	51.4	52.9	54.3	55.7	57.1	58.6
5.0	60.0	61.4	62.9	64.3	65.7	67.1	68.6	70.0	71.4	72.9
6.0	74.3	75.7	77.1	78.6	80.0	81.4	82.9	84.3	85.7	87.1
7.0	88.6	90.0	91.4	92.9	94.3	95.7	97.1	98.6	100.0	101.5

Four 509 Horse-power Babcock & Wilcox Boilers, Equipped with Babcock & Wilcox Superheaters, Operated by the Exposition Cotton Mills Atlanta Ga

Excess of Lime as Determined by Nitrate of Silver: If there is an excess of lime used in the treatment, a sample will become a dark brown by the addition of a small quantity of silver nitrate, otherwise a milky white solution will be formed.

Combined Heat and Chemical Treatment: Heat is used in many systems of feed-treatment apparatus as an adjunct to the chemical process. Heat alone will remove temporary hardness by the precipitation of carbonates of lime and magnesia and, when used in connection with the chemical process, leaves only the permanent hardness or the sulphates of lime to be taken care of by chemical treatment.

The chemicals used in the ordinary lime and soda process of feed-water treatment are common lime and soda. The efficiency of such apparatus will depend wholly upon the amount and character of the impurities in the water to be treated. Table 15 gives the amount of lime and soda required per 1000 gallons for each grain per gallon of the various impurities found in the water. This table is based on lime containing 90 per cent calcium oxide and soda containing 58 per cent sodium oxide, which correspond to the commercial quality ordinarily purchasable. From this table and the cost of the lime and soda, the cost of treating any water per 1000 gallons may readily be computed.

4th. *Zeolite Process*—Zeolite is a hydrated sodium aluminum silicate, and when water, containing calcium or magnesium, is passed through a bed of this material an exchange takes place whereby the sodium of the zeolite goes into solution and calcium and magnesium are removed. By a regenerative process with common salt the sodium is replaced in the zeolite.

Whenever waters high in calcium or magnesium carbonate or bicarbonate are treated by the zeolite process the treated water carries an equivalent amount of sodium carbonate or bicarbonate. Such waters, when concentrated in the boiler, cause priming, and are, apparently, chemically the same as certain waters that have frequently caused embrittlement of the boiler plates. Highly carbonated waters can be treated by the following process:

5th. *Lime Zeolite Process*—This process embodies the principles of the first and fourth processes just described. By treating with lime the carbonates are

TABLE 15

REAGENTS REQUIRED IN LIME AND SODA PROCESS FOR TREATING 1000 U S GALLONS OF WATER PER GRAIN PER GALLON OF CONTAINED IMPURITIES*

	Lime† Pound	Soda‡ Pound		Lime† Pound	Soda‡ Pound
Calcium Carbonate	0.098	...	Ferrous Carbonate	0.169	...
Calcium Sulphate	...	0.124	Ferrous Sulphate	0.070	0.110
Calcium Chloride	...	0.151	Ferric Sulphate	0.074	0.126
Calcium Nitrate	...	0.104	Aluminum Sulphate	0.087	0.147
Magnesium Carbonate	0.234	..	Free Sulphuric Acid	0.100	0.171
Magnesium Sulphate	0.079	0.141	Sodium Carbonate	0.093	...
Magnesium Chloride	0.103	0.177	Free Carbon Dioxide	0.223	..
Magnesium Nitrate	0.067	0.115	Hydrogen Sulphate	0.288	...

*L. M. Booth Company.

†Based on lime containing 90 per cent calcium oxide.

‡Based on soda containing 58 per cent sodium oxide.

removed and then the zeolite process is used to remove the permanent hardness. Priming dangers are materially reduced and the embrittling danger is removed.

A number of the chemical methods of testing and analyzing treated waters given in connection with the first three methods of treatment are also applicable to the zeolite and lime-zeolite processes.

LESS USUAL REAGENTS—Barium hydrate is sometimes used to reduce permanent hardness or the calcium sulphate component. Until recently, the high cost of barium hydrate has rendered its use prohibitive, but at present it is obtained as a by-product in cement manufacture and it may be purchased at a more reasonable figure than heretofore. It acts directly on the soluble sulphates to form barium sulphate which is insoluble and may be precipitated. Where this reagent is used, it is desirable that the reaction be allowed to take place outside of the boiler, though there are certain cases where its internal use is permissible.

Barium carbonate is sometimes used in removing calcium sulphate, the products of the reaction being barium sulphate and calcium carbonate, both of which are insoluble and may be precipitated. As barium carbonate in itself is insoluble, it cannot be added to water as a solution and its use should, therefore, be confined to treatment outside of the boiler.

Silicate of soda will precipitate calcium carbonate with the formation of a gelatinous silicate of lime and carbonate of soda. If calcium sulphate is also present, carbonate of soda is formed in the above reaction, which in turn will break down the sulphate.

Oxalate of soda is an expensive but efficient reagent which forms a precipitate of calcium oxalate of a particularly insoluble nature.

Alum and iron alum will act as efficient coagulants where organic matter is present in the water. Iron alum has not only this property but also that of reducing oil discharged from surface condensers to a condition in which it may be readily removed by filtration.

CORROSION—Where there is a corrosive action because of the presence of acid in the water or of oil containing fatty acids, which will decompose and cause pitting wherever the sludge can find a resting place, it may be overcome by the neutralization of the water by carbonate of soda. Such neutralization should be carried to the point where the water will just turn red litmus paper blue. As a preventive of such action arising from the presence of the oil, only the highest grades of hydrocarbon oils should be used.

Acidity will occur where sea water is present in a boiler. There is a possibility of such an occurrence in marine practice and in stationary plants using sea water for condensing, due to leaky condenser tubes, priming in the evaporators, etc. Such acidity is caused through the dissociation of magnesium chloride into hydrochloric acid and magnesia under high temperatures. The acid in contact with the metal forms an iron salt which immediately upon its formation is neutralized by the free magnesia in the water, thereby precipitating iron oxide and reforming magnesium chloride. The preventive for corrosion arising from such acidity is the keeping tight of the condenser. Where it is unavoidable that some sea water should find its way into a boiler, the acidity resulting should be neutralized by soda ash. This will convert the magnesium chloride into magnesium carbonate and sodium chloride, neither of which is corrosive, but both of which are scale-forming.

The presence of dissolved oxygen or air in feed water is well recognized as a cause of corrosion, but all phases of such corrosive action have not been definitely determined or thoroughly understood. The facts as to the effect of dissolved oxygen in the feed water, as demonstrated by experience, may be briefly summarized as follows:

1st. The action, with boilers in service, is slight, if any.

2nd. With boilers out of service and cold, there is a marked corrosive action at and just below the water line, particularly in the case of extremely soft and neutral waters.

3rd. Economizers, when the boiler or boilers so equipped are on the line, are especially subject to corrosive attack from this source, steel-tube economizers being much more susceptible to such attack than cast-iron economizers.

4th. Units using very soft or pure feed waters are more subject to corrosive trouble due to the presence of dissolved gases than are those using hard or alkaline waters. This is probably due to the fact that soft or neutral waters have a hydrogen ion value which is very close to the neutral point and which, in some cases, passes from the alkaline to the acid side.

The prevention of corrosive action resulting from the presence of dissolved oxygen or air is discussed in the chapter on "Economizers."

Galvanic action, resulting in the eating away of the boiler metal through electrolysis, was formerly considered practically the sole cause of corrosion. But little is known of such action aside from the fact that it does take place in certain instances. The means adopted as a remedy is usually the installation of zinc plates within the boiler, which must have positive metallic contact with the boiler metal. In this way, local electrolytic effects are overcome by a still greater electrolytic action at the expense of the more positive zinc. The positive contact necessary is difficult to maintain and it is questionable just what efficacy such plates have except for a short period after their installation when the contact is known to be positive. Aside from protection from such electrolytic action, however, the zinc plates have a distinct use where there is the liability of air in the feed, as they offer a substance much more readily oxidized by such air than the metal of the boiler.

FOAMING—Where foaming is caused by organic matter in suspension, it may be largely overcome by filtration or by the use of a coagulant in connection with filtration, the latter combination having come recently into considerable favor. Alum, or potash alum, and iron alum, which in reality contains no alumina and should rather be called potassia-ferric, are the coagulants generally used in connection with filtration. Such matter as is not removed by filtration may, under certain conditions, be handled by surface blowing. In some instances, settling tanks are used for the removal of matter in suspension, but where large quantities of water are required, filtration is ordinarily substituted on account of the time element and the large area necessary in settling tanks.

Where foaming occurs as the result of overtreatment of the feed water, the obvious remedy is a change in such treatment.

PRIMING—Where priming is caused by excessive concentration of salts within a boiler, it may be overcome largely by frequent blowing down. The degree of concentration allowable before priming will take place varies widely with conditions of operation and may be definitely determined only by experience with each individual set of conditions. It is the presence of the salts that cause priming that may result in the absolute unfitness of water for boiler-feed purposes. Where these salts exist in such quantities that the amount of blowing down necessary to keep the degree of concentration below the priming point results in excessive losses, the only remedy is the securing of another supply of feed, and the results will warrant the change almost regardless of the expense. In some few instances, the impurities may be taken care of by some method of water treatment, but such water should be submitted to an authority on the subject before any treatment apparatus is installed.

BOILER COMPOUNDS—The method of treatment of feed water by far the most generally used is by the use of some of the so-called boiler compounds. There are many reliable concerns handling such compounds who unquestionably secure the promised results, but there is a great tendency toward looking on the compound as a "cure all" for any water difficulties and care should be taken to deal only with reputable concerns.

The composition of these compounds is almost invariably based on soda with certain tannic substances and in some instances a gelatinous substance which is presumed to encircle scale particles and prevent their adhering to the boiler surfaces. The action of these compounds is ordinarily to reduce the calcium sulphate in the water by means of carbonate of soda and to precipitate it as a muddy form of calcium carbonate which may be blown off. The tannic compounds are used in connection with the soda with the idea of introducing organic matter into any scale already formed. When it has penetrated to the boiler metal, decomposition of the scale sets in, causing a disruptive effect which breaks the scale from the metal, sometimes in large slabs. It is this effect of boiler compounds that is to be most carefully guarded against or inevitable trouble will result from the presence of loose scale with the consequent danger of tube losses through burning.

When proper care is taken to suit the compound to the water in use, the results secured are fairly effective. In general, however, the use of compounds may only be recommended for the prevention of scale rather than with the view to removing scale which has already formed, that is, the compounds should be introduced with the feed water only when the boiler has been thoroughly cleaned.

BOILER-FEED WATER HEATERS AND PUMPS

BEFORE water fed to a boiler can be converted into steam, it must first be heated to the temperature corresponding to the pressure within the boiler. The temperature due to 200 pounds gauge pressure is approximately 388 degrees Fahrenheit, and if water is fed to the boiler at 60 degrees, each pound of water must have added to it approximately 333 B.t.u. to bring its temperature to the point at which it can be converted into steam. As it requires 1171.1 B.t.u. to raise the temperature of one pound of water from 60 to 388, and to convert it into steam at 200 pounds gauge pressure, the 333 B.t.u. required simply to raise the temperature of the water to that corresponding to the pressure will be approximately 28 per cent of the total. If, therefore, the temperature of the water can be increased from 60 to 388 degrees before it is introduced into the boiler, by the utilization of heat other than that resulting from the combustion of fuel under the boiler, there will be a saving in fuel burned under the boiler of $333 \div 1171.1 = 28$ per cent. Such a saving, if accomplished by the use of a heating medium that would otherwise be wasted and provided that the cost of operating and maintaining the apparatus used to bring about the saving is less than the value of the heat saved, is a direct net saving.

The saving in fuel due to the heating of the feed water by means of heat that would otherwise be wasted may be computed from the formula:

$$\text{Fuel saving, per cent} = \frac{100\,(h - h_1)}{H - h_1} \qquad (3)$$

where h = heat of the liquid at the temperature of the feed water after heating,

TABLE 16A

SAVING IN FUEL, IN PERCENTAGES, BY HEATING FEED WATER

GAUGE PRESSURE 180 POUNDS

Initial Temperature Degrees Fahrenheit	Final Temperature—Degrees Fahrenheit							Initial Temperature Degrees Fahrenheit	Final Temperature—Degrees Fahrenheit						
	120	140	160	180	200	250	300		120	140	160	180	200	250	300
32	7.35	9.02	10.69	12.36	14.04	18.20	22.38	95	2.20	3.97	5.73	7.49	9.25	13.66	18.07
35	7.12	8.79	10.46	12.14	13.82	18.00	22.18	100	1.77	3.54	5.31	7.08	8.85	13.28	17.70
40	6.72	8.41	10.09	11.77	13.45	17.65	21.86	110	.89	2.68	4.47	6.25	8.04	12.50	16.97
45	6.33	8.02	9.71	11.40	13.08	17.30	21.52	120	00	1.80	3.61	5.41	7.21	11.71	16.22
50	5.93	7.63	9.32	11.02	12.72	16.95	21.19	130	...	.91	2.73	4.55	6.37	10.91	15.46
55	5.53	7.24	8.94	10.64	12.34	16.60	20.86	140	...	00	1.84	3.67	5.51	10.09	14.68
60	5.13	6.84	8.55	10.27	11.97	16.24	20.52	150	..	...	.93	2.78	4.63	9.26	13.89
65	4.72	6.44	8.16	9.87	11.59	15.88	20.18	160	...	...	00	1.87	3.74	8.41	13.09
70	4.31	6.04	7.77	9.48	11.21	15.52	19.83	170	...	...	..	.94	2.83	7.55	12.27
75	3.90	5.64	7.36	9.09	10.82	15.16	19.48	180	...	..	...	.00	1.91	6.67	11.43
80	3.48	5.22	6.96	8.70	10.44	14.79	19.13	190	...	..	...	...	.96	5.77	10.58
85	3.06	4.80	6.55	8.30	10.05	14.41	18.78	200	...	...	..	...	.00	4.86	9.71
90	2.63	4.39	6.14	7.89	9.65	14.04	18.43	210	...	..	...	...	...	3.92	8.82

PORTION OF 15,000 HORSE-POWER INSTALLATION OF BABCOCK & WILCOX BOILERS EQUIPPED WITH BABCOCK & WILCOX CHAIN-GRATE STOKERS AT THE NORTHAMPTON, PA., PLANT OF THE ATLAS PORTLAND CEMENT CO. THIS COMPANY OPERATES A TOTAL OF 2[illegible],000 HORSE-POWER OF BABCOCK & WILCOX BOILERS IN ITS VARIOUS PLANTS

h_1=heat of the liquid at the temperature of the feed water before heating, and H= total heat above 32 degrees per pound of steam at the boiler pressure. Values of H may be found from Table 21.* The value of $(32—t_1)$ is not exact since the heat of the liquid is not always within exactly 32 B.t.u. of the temperature of the feed water. Any error arising from this source, however, is small enough to be neglected in computing net savings. Table 16A has been computed from this formula to show the fuel saving resulting from various increases in feed temperatures at a boiler pressure of 200 pounds gauge. The saving at other pressures will vary somewhat, being less for a given temperature rise as the pressure increases and greater as the pressure decreases.

Beside the saving in fuel effected by the use of feed water heaters, other advantages are secured. The time required for the conversion of the water into steam is diminished, and the steam capacity of the boiler correspondingly increased. Further, the feeding of cold water into a boiler has a tendency toward setting up temperature strains, which are diminished as the temperature of the feed approaches that due to the steam pressure. An important additional advantage of feed water heating in certain classes of apparatus is the precipitation in the heater of a large portion of the scale-forming ingredients before they enter the boiler, with a consequent saving in cleaning and losses through decreased efficiency and capacity.

FEED-WATER HEATERS—In general, feed-water heaters may be included in three main classes: closed heaters, open heaters, and economizers. The first two classes may use either live or exhaust steam as the heating medium, while the last utilizes the heat in the spent gases leaving the boiler. Because of the growing importance of the use of economizers in power-plant practice, this class of apparatus will be treated in a separate chapter.

In closed heaters the steam and water do not come into direct contact with each other, either the steam or the water passing through tubes. Because of this lack of actual contact between steam and water, the temperature of the feed can never be brought up to that of the steam in this type of heater, and except in the case of waters free from scale-forming matter, there is a tendency toward the formation on the heater surfaces of a scale that is difficult to remove. These two factors may lead to a lower efficiency than that obtainable with open heaters, whereas closed heaters are advantageous where there is oil in the steam, since the oil can be prevented from mingling with the feed water. In other cases, closed heaters may lead to a simplification of the plant.

Where closed heaters of the straight-tube type are used, care must be taken in their design to minimize the possibility of tube leakage, particularly where the service is intermittent.

In a closed type of heater, any dissolved gases contained in the feed water entering the heater ordinarily pass through the heater to the economizer or boiler. Corrosion of the piping or boiler and economizer may result unless care is taken to remove such gases from the feed leaving the heater.

Both closed and open heaters have been available for use for a sufficient length of time for their status in the art to have become pretty thoroughly determined, and in ordinary power-plant practice the closed heater is in much less general use than

* See page 118.

the open type. There is, however, a marked tendency in the larger power stations toward interstage heating of feed water, and for this practice closed heaters are ordinarily used.

Open heaters consist of a chamber in which feed water and exhaust steam come into direct contact and are mingled. They usually operate at atmospheric pressure or under a back pressure of from 10 to 15 pounds above the atmosphere, depending upon the back pressure of the auxiliaries furnishing the exhaust steam for the heater, and ordinarily have considerable storage space. The feed water in this class of heater is brought to the temperature of the exhaust steam used in heating, and where this temperature is sufficiently high, an appreciable percentage of the gases carried in solution by the feed may be driven off at the heater, with a resulting reduction in the possibility of economizer or boiler corrosion.

All open heaters should be provided with relief valves, and where the heaters are to operate against back pressure particular provisions should be made. Where the exhaust utilized for heating is from reciprocating engines and pumps, efficient oil separators should be provided.

Open heaters allow any matter that is precipitated by the action of heat to be readily removed, the method of removal varying in different designs. Some designs, in addition to providing special means for precipitation of scale-forming ingredients, are equipped with filters for the removal of other impurities, and frequently such heaters are used in connection with recording and indicating devices for measuring the amount of feed water.

Special so-called live-steam purifiers of the open-heater type are sometimes used under a pressure sufficient to bring the temperature of the feed up to 300 degrees. Such heaters, however, are used only under feed-water conditions in which it is advisable to precipitate in the heater the sulphates contained in the feed, though even with this temperature a certain percentage of sulphates remains in solution and is carried to the boiler, where scale will be formed if the concentration is allowed to build up to a sufficient extent. Live-steam purifiers have never come into general use.

Boiler-feed Pumps—The pumps used for boiler-feeding purposes are of two general classes, direct-acting steam pumps and centrifugal pumps.

While direct-acting pumps may be used with either simple, compound or triple-expansion steam ends, the first is ordinarily used. The steam consumption of the latter two designs will be appreciably less than that of the first, but this may be more than offset by cheaper first and maintenance costs, and in cases where all of the exhaust may be used for heating the feed there will be but little difference in plant efficiency. The simple direct-acting pump may have one or two water cylinders, being known as simplex or duplex. The former, because of fewer wearing parts, has a lower maintenance cost, but delivers the feed with a certain irregularity that may lead to strains in the feed piping. This irregularity of discharge is largely overcome with duplex pumps, but it may exist to some extent under certain conditions of service. Plungers in direct-acting pumps are preferable to pistons because of greater ease in adjusting packing to minimize slippage.

Direct-acting steam pumps are in general use in the smaller plants of from 1000 to 1500 boiler horse-power. The steam consumption is high and for compound steam ends will be approximately 100 pounds per indicated horse-power when run under the ordinary service conditions, while with simple steam ends in the smaller pumps, the steam consumption may be as high as 200 pounds. The power used in the feed pumps, even with such high steam consumption, represents only a small percentage of the steam generated by the total fuel burned.

Centrifugal pumps have not been developed in the smaller sizes to an extent where their advantages are sufficient to warrant their use in place of direct-acting pumps. In such pumps the limit of pressure obtainable in a single stage is not in excess of 100 pounds, and to meet the requirements of pressure it would be necessary to go to additional stages which would make the cost of the centrifugal pump with its drive such that it could not compete with the direct-drive pumps for the smaller plants.

In plants of over 1500 to 2000 boiler horse-power, the centrifugal feed pumps are in almost universal use. They are built in from two to five stages, depending upon the pressure at which the feed is to be delivered. This delivery pressure is ordinarily based upon from 60 to 100 pounds for each stage. Centrifugal pumps are continuous in their action and set up no pulsating strains in the feed piping. Because of relatively small clearances the slippage is less than in the direct-acting pump, particularly in the case of the larger centrifugals. In the event of a sudden shutting of the feed valves, this design of pump simply churns the water within it, without delivering any water, and no accident can occur. They are relatively small and require but little attention, the latter factor representing a saving in labor. While with its drive, a centrifugal pump costs appreciably more than a direct-acting pump, the lower maintenance cost resulting from the absence of pump valves probably more than offsets the difference in original expenditure.

Centrifugal feed pumps are either turbine or motor-driven, though until very recently the former method of drive was used almost entirely. This was largely because of its reliability and the ability to regulate the speed when so driven, which avoids the change in pressure characteristics of the centrifugal pump with varying discharge rates at constant speed which exists when driven by alternating current motors. With a motor drive, because of lower speeds, the pump efficiency is somewhat less than with a turbine drive, and to offset the lower efficiency it is ordinarily necessary to supply an extra stage in the pump to give the same pressure.

With the tendency in central station practice toward driving electrically as large a portion of the auxiliaries as possible, there will undoubtedly be developments in motor drives for boiler-feed pumps, and in several recent installations, such drives, both direct and alternating current, have been used. The use of a certain number of pumps, motor-driven at a constant speed, and an additional number, turbine-driven to handle variations in the load, may offer the proper solution for central station work.

A third method of boiler feeding is by the use of an injector, used either alone or supplementing the feed pump. Injectors are cheap, compact, have no wearing parts, no exhaust to handle and deliver feed to the boiler heated by the steam used in obtaining the injector action. They have the disadvantages of unreliability through wrong adjustment or variation in the conditions, trouble in starting after such stoppage, and inability to handle feed water at temperatures much, if any, over 100 degrees Fahrenheit.

Injectors use a greater amount of steam in feeding a given weight of water into a boiler than direct-acting pumps, and are rarely used with boilers of over 100 horse-power, except in the case of the locomotive boiler, where there is no exhaust steam available for heating the feed water. In this class of work the exhaust is used to produce draft, and an injector does not count against the efficiency.

A 331 Horse-power Babcock & Wilcox Boiler, Equipped with a Babcock & Wilcox Superheater and a Babcock & Wilcox Counterflow Economizer, at the Works of the Duratex Company, Newark, N. J.

ECONOMIZERS

ECONOMIZERS may be classed as closed heaters, utilizing the heat in the spent gases of combustion as they leave the boiler, for heating the feed water either after it has passed through some other form of heater or just as it comes from the source of supply.

The use of economizers in this country has passed through what might be considered a cycle, being in common use a number of years ago, then for a period falling rather into disuse, and at the present again coming into greater favor. While economizer installations have never been as general in this as in other countries, we appear to be approaching European practice more nearly, and it may be definitely stated that economizers are receiving today more favorable consideration than at any previous period. Nevertheless, engineers are not in entire agreement as to the value of this class of apparatus, as discussed later in the consideration of the advisability of economizer installations.

The present tendency toward the increased use of economizers is the result of three factors. First, and of greatest importance, is the increasing cost of fuel; obviously the higher the fuel cost, the greater is the expense in first cost, operating cost and maintenance cost warranted in endeavoring to obtain maximum fuel efficiency. The second factor is the higher boiler capacities now demanded and obtained in general practice. These higher capacities are accompanied, even under the best conditions of boiler design and of combustion, by an increase in gas temperatures leaving the boiler, and other things being equal, the higher the entering gas temperature the greater the return from a given economizer. The third factor is the tendency toward increased boiler pressures. As the pressure is increased there is a corresponding increase in the temperature of the boiler parts, and the temperature to which it is possible to cool the gases leaving the boiler is increased almost directly as the increase in temperature due to pressure.

Economizers in general may be classed as group and individual. In the former, a single economizer heats the feed water for and handles the gases from two or more boilers, while in the latter an individual economizer is installed with each boiler. In early practice, group economizers were used almost entirely, and were so arranged that they could be by-passed, in the case of necessity for shut down, without interfering with boiler operation. Present practice tends toward the use of individual economizers, ordinarily without provision for any by-passing of the gases. With properly designed and properly operated economizers, there should be no necessity for the shutting down of such a unit at periods other than those at which the boiler with which it is set is taken out of service. With no by-pass flues and the necessary by-pass dampers, there is a minimum of air or gas leakage. Such flues and dampers and, in some cases, the economizer settings, are difficult to keep tight. Where excessive air leakage exists, not only is the economizer return decreased through a lowering of gas temperatures, but the load on the fan is increased because of the increased volume of gas to be handled and the added draft resistance resulting from such increased volume. While group economizers will undoubtedly continue to be used in certain classes of power plants, the individual economizer represents better practice for central station work.

Individual economizers may be subdivided into two types, the integral and the independent. In the former, the economizer is within the same setting as the boiler and is connected directly to the circulation of the boiler with which it is installed, with no valves between it and the boiler. Independent economizers are separate from the boiler, with feed valves between the economizer and boiler. They are set either within the boiler setting, or overhead, or in the rear of the boiler in a separate setting or casing.

Materials of Construction—The material used in the construction of economizers has been almost universally cast iron until recently. Occasionally steel was used, but in general. because of excessive and rapid corrosion. the steel economizers gave unsatisfactory service and were removed. From the standpoint of corrosion, cast iron was looked upon as the only metal that would give a satisfactory length of life, and for the pressures in common use, cast-iron economizers proved satisfactory. With the coming of higher boiler pressures, however, doubts of the ability of cast-iron construction to withstand such pressures, and the possibility of adverse legislation against the use of this metal in economizers above certain working pressures, as in the case of boilers, led engineers to turn to other economizer material. Wrought-steel construction, because of safety and ability to withstand high pressures, naturally received first consideration. The increased tendency toward corrosion of wrought steel as compared with cast iron in economizer construction was appreciated, but it was felt that the increased safety of steel more than offset this tendency and warranted its use provided corrosion could be obviated or minimized through the use of any reasonable methods of corrosion prevention. Cast-iron economizers will, without question, continue to be used in many plants, but the steel-tube economizers are being generally adopted in high-pressure power-plant work.

The danger of corrosion is both external and internal. External corrosion, occurring principally at the cold end of the economizer, is the result of condensation of the moisture content of the gases on the comparatively cool surfaces of the economizer. Trouble through such corrosion occurs with cast-iron as well as with steel-tube economizers and is particularly rapid if the condensation occurs where coals high in sulphur are used as fuel, as the sulphur-bearing gases form sulphuric acid. External corrosion is almost entirely a function of the temperature of the economizer surface, and where the feed water is brought to the economizer at a temperature well above the dew point, there is but little danger of undue exterior corrosion except in the case of intermittent service. Where boilers and economizers are shut down or banked at frequent intervals, the temperature of the economizer surfaces will fall below the dew point, the soot adhering to the economizer becomes moist, and corrosion will be more or less rapid. Experience has shown that with a feed temperature to the economizer of 120 degrees Fahrenheit, but little danger from exterior corrosion should be encountered with coals having a small sulphur content, though the feed temperature should preferably be 140 degrees and, if possible, should be higher with coals having a high sulphur content.

Interior corrosion results from the presence of dissolved gases in the feed water entering the economizer, the most active corroding agent being oxygen. Water, reasonably pure, and after natural exposure to the air at ordinary temperatures, contains when fully saturated approximately 16 parts by volume per 1000 parts of water, the air being under standard conditions of pressure and temperature. The amount of gas that may be held in solution by water decreases as the temperature approaches its boiling point at a given pressure. When in contact with steel surfaces at temperatures as low as 100 degrees, water in such condition is actively corrosive, and between this temperature and the boiling point rapid and severe pitting will result. Such action, while severe in steel economizers, also occurs in cast-iron economizers, though to an appreciably less marked extent. It is this tendency toward rapid interior corrosion that caused steel to be considered impracticable for use in earlier economizer construction.

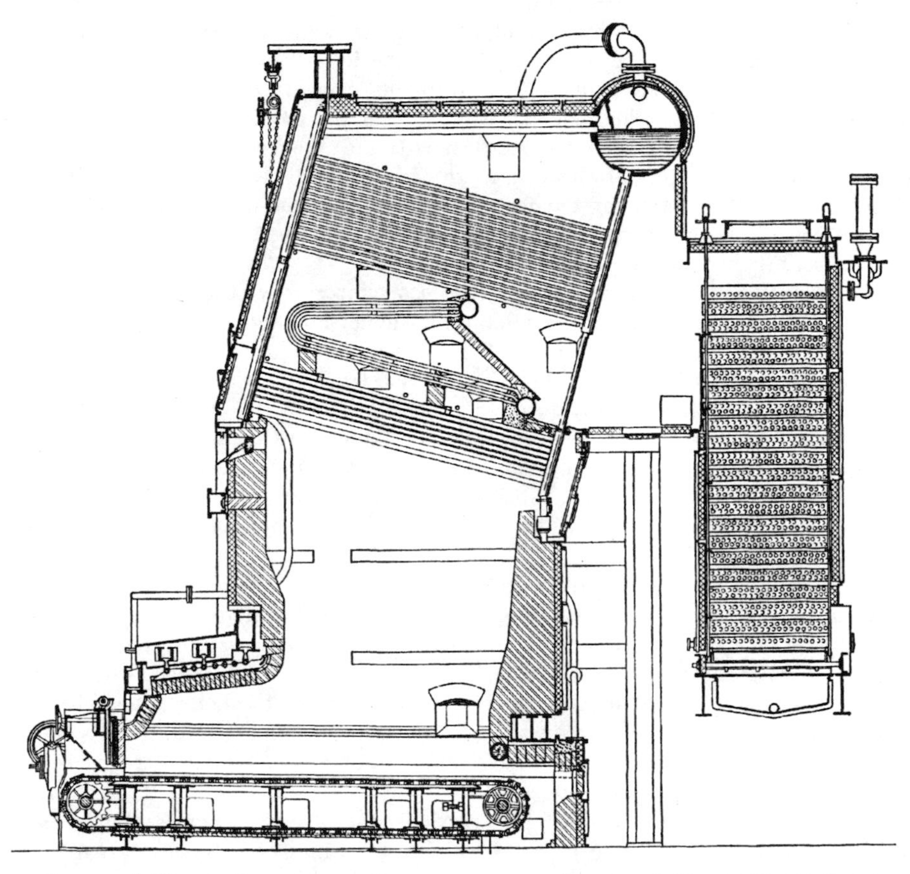

A Babcock & Wilcox Counterflow Economizer with a Babcock & Wilcox Boiler Built for 400 Pounds Working Pressure and Equipped with a Babcock & Wilcox Superheater and a Babcock & Wilcox Blast Chain Grate Stoker

Experience over a length of time sufficient for definite demonstration has shown that with air eliminated to the proper extent from feed water fed to a steel economizer, the danger of interior corrosion, if not entirely overcome, is, at least, reduced to a minimum.

There is a difference of opinion among engineers as to the best method for eliminating air from feed water. If the temperature of feed to the economizer is increased in open heaters, where provision is made for air expulsion, to a point where a proper degree of air elimination is assured, there will be no material interior corrosion. The heating of the feed to such a temperature, however, might result in the feed water entering the economizer at a temperature too high for the best economizer return, or in too large a loss of heat in the vapor passing from the heater with the discharged air. If 210 degrees is accepted as the temperature to which water must be heated at or about atmospheric pressure to obtain the necessary degree of air elimination, it means that the average temperature must be above 210 and that the temperature must never fall below this point. A feed temperature of 210 degrees ordinarily means an average of 210 degrees, and such temperature actually may range from 180 to 230 degrees. It is at the lower range of temperature that the dissolved oxygen would find its way into the economizer and lead to corrosion.

A second method of air removal, which, from the consideration of economizer return, because of lower feed temperature entering the economizer, is thermally more efficient than the primary heating of the feed water, is based on the principle of reducing the boiling point of the feed water and the elimination of dissolved air through boiling of the water at a temperature favorable to economizer performance, *i. e.*, the boiling of the feed under a vacuum.

A number of installations of steel-tube economizers have been made where low feed temperatures are used and air elimination is not attempted. In such cases the interior economizer surfaces have been given some non-corrosive protection. This method has been tried with greater or lesser success, but is not as reliable as the elimination of the air from the feed water.

To summarize, it may be safely stated that there are methods of air removal available which render the danger of interior corrosion in steel-tube economizers practically negligible and which still enable the temperature of the feed water entering the economizer to be such as will lead to the maximum economizer efficiency in heat absorption.

Economizer Design — Until recently, practically all cast-iron economizers were made up of vertical tubes, connecting headers at the top and bottom, the gases passing over the tubes horizontally. In some of the later designs, the tubes are arranged horizontally, with a horizontal gas flow, and in one design which leads to a compact arrangement of economizer set at the rear of a boiler, the tubes are horizontal and the gases pass downward over the economizer surface and are removed at or near the boiler-room floor line. The flow of water through all cast-iron economizers is through the tubes, in parallel, though in some of the larger group economizers the surface is divided into sections, the flow being in parallel in each section but counterflow to the extent that the water passes in series through the different sections.

The first steel-tube economizers of present practice consisted of boiler sections, placed above the boiler with the gases making three passes over the economizer surface in the same manner as in the boiler. This design, while reasonably satisfactory from the standpoint of heat transfer and efficiency of absorption, gave a relatively high draft resistance because of the turns. In later designs, the tubes were placed transverse to the boiler tubes, the gases making a single horizontal pass over the surface. Where proper gas passage areas were provided, the efficiency of heat absorption was as high as in the three-pass economizer and, because of the single gas pass, the draft resistance was appreciably less. In both of these designs, as in the case of cast-iron economizers, the water flow was in parallel.

Gas passage areas are relatively large in cast-iron and in both designs of steel economizers described, and the gas velocities relatively low. These factors, together with the parallel water flow, result in a rate of heat transfer considerably lower than is obtained in the latest designs of steel economizers.

The most recent steel-tube economizers are designed to give relatively high gas and water velocities, and by employing a strictly counterflow of the gas and water high rates of heat transfer and correspondingly high economizer efficiencies are obtained. The counterflow type of economizer developed by The Babcock & Wilcox Company and illustrated on page 103 is representative of the modern steel-tube type. This economizer is made up of forged-steel boxes and straight tubes. The water is fed to the lowermost box at one side of the economizer, from which it passes through a single row of tubes to the lowermost box on the other side; from this it passes to the second box on the inlet side through another single row of tubes, and successively, in each instance through a single row of tubes, from side to side of the economizer from bottom to top, from which point it is fed to the boiler. Space is left between adjacent boxes on the same side of the economizer so that tubes may be readily removed and replaced almost regardless of the length of the tube or the width of the aisle between the units. Economizers of this design may be placed above the boiler or in the rear. The latter arrangement leads to a compact boiler and economizer setting and permits a fan location at approximately the boiler-room floor line.

With the water flowing upward through successive rows of tubes and the gas passing downward, a strictly counterflow effect is obtained, with a corresponding maximum rate of heat transfer.

Reference has been made to the relatively high gas velocities in the modern design of steel economizers. As indicated in the chapter on "Waste Heat," high gas velocities tend toward high rates of heat transfer, but any increase in gas velocity is also accompanied by an increase in draft resistance. There is obviously a point above which velocity should not be increased, as any increase in transfer rate resulting from increased velocity above this point would be offset by additional power required for fan drive to take care of the added draft resistance. One amount and arrangement of economizer surface may give a transfer rate greatly in excess of that obtainable with the same surface differently arranged. The higher rate, however, can only be accomplished by the use of higher gas velocity, which, in turn, leads to higher draft resistance. The difference in power required for fan drive to give the necessary draft with the two arrangements of surface may, when capitalized, warrant the expenditure of an appreciably greater sum for the greater amount of surface that would be necessary with the arrangement giving the lower draft and lower transfer rate. Further, in capitalizing the difference in power required, the selling price rather than the cost of production of power may be, in some cases, the proper basis for evaluation, as any power used for the fan drive cannot be sold and represents a net deduction from the maximum plant output.

Next in importance to the arrangement of economizer surface is the amount of surface to be installed. With the arrangement determined to give the desired gas velocity and heat transfer rate, any additional surface installed that does not change the velocity over the economizer as a whole increases the economizer return.

Successive additions of units of economizer surface, however, give an increase in economizer return at diminishing rates, and a point is reached in the addition of units of surface where the increased return does not warrant the cost of added surface. This point is dependent upon fuel and economizer surface costs.

An important point in determining the limit of economizer surface that should be installed is the pressure at which the boiler served by the economizer is operated, this pressure determining the upper limit of water temperature allowable from the economizer. No exact figure for this maximum water temperature from the economizer can be given, but ordinarily this may be taken as approximately 50 degrees below the temperature due to the pressure within the boiler under the most unsteady conditions of operation at maximum boiler load. If the amount of economizer surface is such that this temperature, under the worst conditions, will be exceeded, difficulties may result. If, with too great an amount of economizer surface, the boiler is overfed and the feed is shut off, the water within the economizer may reach the steaming temperature before the feed is again turned on. Opening the feed valves with the water within the economizer at or near the steam temperature would cause a belching of the water from the economizer. which would either partially or wholly drive the water from it, the amount varying with the form of the economizer. Under such conditions a considerable period might elapse before the economizer could be refilled and an irregular action of feed would result. The refilling with comparatively cool water after the steam had been formed in the economizer might lead to temperature strains and leakage or a water-hammer action that might be serious in its results. These actions, again, vary greatly with the form of the economizer and need not be feared with certain designs.

Another factor of primary importance in determining the amount of economizer surface to be installed, and a factor which is not ordinarily appreciated, is the class of fuel burned and the methods used in burning different fuels. This factor resolves itself into a question of the ratio of gas to water weights. and the variation in such ratios with different fuels, and with different combustion conditions even with the same fuel, is great. As an example that by no means represents the extreme case, consider oil burned with mechanical burners and blast-furnace gas burned with the most efficient design of pressure burners. In the case of the former fuel, it is readily possible to complete combustion with a weight of products of less than 40 pounds per horse-power developed, while with the latter, even under the best combustion conditions, such weight may be approximately 80 pounds per horse-power developed. For a given boiler output, then, we would have for the same weight of water passing through the economizer a weight of gas passing over the economizer twice as great in the case of blast-furnace gas as with oil fuel, and in all probability the gas temperature entering the economizer would be higher with the former than with the latter fuel. It is evident that for a given feed temperature rise, the amount of economizer surface to be installed with the blast-furnace gas-fired boiler will be appreciably less than with the oil-fired unit. Table 16B, which is based on the ordinary efficiencies obtained with the different fuels considered, indicates the wide variation in the ratio of gas to water weights. This table does not take into consideration the gas temperatures entering the economizer, which, in general, would be higher as the ratio is higher and would accentuate the difference in economizer performance with different fuels.

TABLE 16B

RATIO OF GAS WEIGHTS TO WATER WEIGHTS WITH VARIOUS FUELS

RATE OF EVAPORATION, 60,000 POUNDS PER HOUR

Fuel	CO_2 Per Cent	Total Gas Pounds per Hour	Gas Weight / Water Weight
Coal	13.5	94,000	1.566 : 1
Wood	14.3	116,500	1.942 : 1
Oil	14.0	72,500	1.208 : 1
Natural Gas	10.0	77,800	1.297 : 1
By-product Coke Oven Gas	8.0	78,000	1.300 : 1
Blast-furnace Gas	20.0	161,000	2.683 : 1

Present economizer practice with coal in individual separate economizers is to install the equivalent from 55 to 70 per cent of the boiler surface in the economizer, although in some cases it is made as great as the boiler surface. The greater the ratio of gas to water weight, the less the amount of economizer surface that can be installed without having trouble with steaming in the economizer. Integral economizers are, to an extent, limited in size by the design of boiler with which they are installed. Ordinarily this type contains approximately 30 per cent of the heating surface of the boiler.

Economizer Operation—Theoretically, economizers function so automatically that the only features of operation that must be given attention are the cleaning, exterior and interior, and keeping the air leakage at a minimum.

Cast-iron economizers of the earlier installations were almost universally equipped with scrapers designed to remove mechanically any soot and deposit forming on the exterior of the tubes. Such scrapers are still used and give satisfactory service with certain classes of fuels. The tendency of scrapers, where the deposit is of a pasty nature, is to form a coating on the outside of the tubes over which the scrapers readily slide but which reduces the heat-absorbing ability of the economizer surface. Further, the necessity for openings into the economizer setting through which the scraper chains pass leads to an amount of air leakage that may or may not be excessive. As in the case of by-pass dampers and flues, such leakage, if great, leads to a reduction in gas temperature and an increased load on the fan. A low exit gas temperature from an economizer where the amount of air infiltration is high is no measure of economizer efficiency.

With steel-tube economizers and with cast-iron economizers not equipped with scrapers, it has been found that surfaces may be kept externally as clean as boiler surfaces with the proper installation of mechanical soot blowers. Such blowers must, of course, be properly located, properly "stopped" and properly operated. With economizers, even to a greater extent than with the boilers, care must be taken to make sure that any condensation in the blowing system is removed before actually blowing the tubes. Certain designs of economizers can be cleaned by washing with a stream or streams of water. Provided the external surfaces can be rapidly dried after washing, such a method of cleaning will give satisfactory results; otherwise this method will lead to more rapid corrosion than would result from other cleaning methods.

In all economizer practice, both cast iron and steel tube, where feed-water conditions are not good, there will be a certain amount of sludge deposited in the interior of the economizer, though by the use of proper blow-down methods such a deposit may be handled satisfactorily. With the growing care given to feed-water treatment, interior deposits on economizer surfaces are causing less and less trouble.

ADVISABILITY OF ECONOMIZER INSTALLATION—The advisability of an economizer installation is dependent upon a great number of factors, all of which must be given due consideration, and engineers are not in entire agreement regarding economizer practice in general.

Within recent years it has been more thoroughly appreciated that the greater the amount of boiler-heating surface installed per foot of furnace width, in boilers of a given type having the same system of baffling, provided the furnace width is within the limits of efficient combustion rates, the greater the efficiency of heat absorption at a given combustion rate or at a given rate of steam output. The appreciation of this fact leads to the question of the limits in boiler-heating surface per foot of furnace width to which it is desirable or advisable to go, and also to the question whether, in maintaining a given width of furnace, additional surface should be added as boiler or economizer surface.

Those engineers favoring economizer installation base their judgment on the better temperature gradient that exists with economizers. The temperature of the heating surface of a boiler is ordinarily assumed to be that due to the pressure within the boiler, while that of the economizer surface will be the mean of the temperature of the feed water flowing through the economizer. Under such conditions, due to the greater temperature difference between gases and absorbing surfaces, the total absorption by a given amount of economizer surface would be greater than the total absorption by an equal amount of surface added as boiler-heating surface. While the temperature difference would have but a slight effect on the rate of heat transfer itself, it might have an appreciable effect on the total heat absorption.

Those engineers who favor an increase in boiler-heating surface per foot of furnace width by an amount equal to what would be installed as economizer surface, while recognizing the theoretical and actual advantage of the economizer surface over the same amount as added boiler surface, claim that such advantage is not sufficient to offset other factors entering into the problem. They claim for the increase in boiler surface, lower first cost, lower maintenance cost, lower draft loss, reduced possibility of air leakage and, in general, greater simplicity of installation and operation of the unit as a whole.

If economizers are installed, it is of interest to note that for a given total heating surface in boiler and economizer, the percentage of total surface installed as economizer surface can usually be varied over a considerable range without affecting to any appreciable extent the efficiency of the combined unit.

Obviously, the cost of fuel is the primary factor in determining the expense warranted in increased investment in a boiler and economizer installation over that for the same amount of surface in a boiler alone. The return due to one or two per cent net additive efficiency in the case of the economizer unit may or may not, when capitalized, justify the increased first cost as the fuel cost increases or decreases.

It is to be remembered that economizer return, as represented by gas temperature drop or feed-water temperature rise, is gross and not net. With any of the modern designs of economizer, at the boiler capacities commonly developed, the installation of an induced draft fan is necessary, and such fans, except in the case of very high stacks, must be operated a major portion of the time, since from the standpoint of commercial efficiency the operation of a boiler and economizer unit at a rate of output that could be handled with natural draft would not be warranted. The power consumption of the fan operating over the greater proportion of the time must be deducted from the gross increased efficiency of the economizer unit. It is true that where the amount of surface corresponding to that which would be installed as economizer surface is added as boiler surface, an induced draft fan would be necessary for the high or peak-load conditions; it is probable that with the boiler unit of the greater amount of heating surface it would not be necessary, however, with a reasonable height of stack, to operate the fan for as great a percentage of the time as with the economizer unit.

In the consideration of the advisability of economizer installation, the class of operation available is an all-important factor. While engineers to a greater and greater extent appreciate the necessity of improving the class of operation and labor in the boiler room, there is, unfortunately, a great way to go before this class is brought to a par with that now employed in the average engine room. The boiler and economizer installation leads, without question, to a somewhat greater complication in the boiler-room operation and also, without question, the class of operation and labor ordinarily available limits the amount of such complication that is warranted.

From a consideration of these features, it is evidently impossible to make any general statement as to the advisability of recommending a boiler and economizer installation as against an installation of the same amount of heating surface in a boiler unit alone. Each class of installation has its field and every individual installation must be considered as an entirely separate engineering problem, proper weight being given not only to the features that have been outlined but also to the special features that affect any individual power-plant problem.

Air Heaters—Air heaters or, as they are called abroad, air economizers, utilize the heat in the gases coming either directly from the boiler or from the economizer for heating the air used for combustion. From the standpoint of maximum efficiency, it would appear as logical to reduce gas temperatures to a minimum with this class of apparatus as with water economizers. Any quantity of sensible heat introduced into the furnace with the air for combustion represents heat available for absorption to as great an extent as does heat evolved in the actual burning of the fuel, and the use of heated air may result in an increase in efficiency of combustion greater than that actually corresponding to the amount of heat imparted to the air.

Air heaters have been in common use in both marine and stationary practice in Europe for a number of years, though they have as yet not been used to any extent in stationary boiler practice in this country. Increasing fuel costs of the past few years, however, have awakened considerable interest in this class of apparatus, and as higher efficiencies are sought, air heaters will undoubtedly be used to a greater extent. In spite of the large number of installations of air heaters abroad, there is a surprising lack of definite information as to their performance.

Air heaters are of two general types, tube and plate heaters. In the former, the air for combustion passes through the tubes and the gases over them, while in the latter the air and gas pass through adjacent channels formed by plates.

The transfer rates obtained with air heaters are low, requiring for adequate return large surfaces with gas and air passages more or less restricted. Mr. W. H. Patchell gives air-heater transfer rates varying from 1.28 to 3.08 B.t.u. per hour per square foot per degree difference in temperature, though the average reported is less than 2.0 with an added efficiency due to the heater installation of from 3.1 to 5.4 per cent. These efficiencies, however, are with air heaters directly connected to boilers, and where economizers are installed between the boiler and the heater, the efficiency would probably be not in excess of 3 per cent as a maximum. There appear to be no data available on the maximum air temperature allowable from the standpoint of stoker maintenance cost. The highest air temperature given by Mr. Patchell is 303 degrees, obtained with a heater containing approximately 90 per cent of the boiler surface.

Air heaters are of special advantage from the standpoint of efficiency of combustion with low-grade fuels, where any increase in furnace temperature is of assistance. Unfortunately, it is this class of fuel, particularly where the sulphur content is high, with which air heaters would operate to the greatest disadvantage. for with such fuels there would be the greatest tendency toward fouling of the surfaces and, because of restricted areas, cleaning would be difficult. With the temperatures found, sweating would occur at the cold end which would lead to rapid corrosion, though with air heaters the expense of replacing corroded parts would be much less than in the case of economizers.

With certain classes of fuel anything done toward increasing furnace temperatures would result in clinkering difficulties, and with any class of fuel increased furnace temperature is accompanied by increased brickwork maintenance.

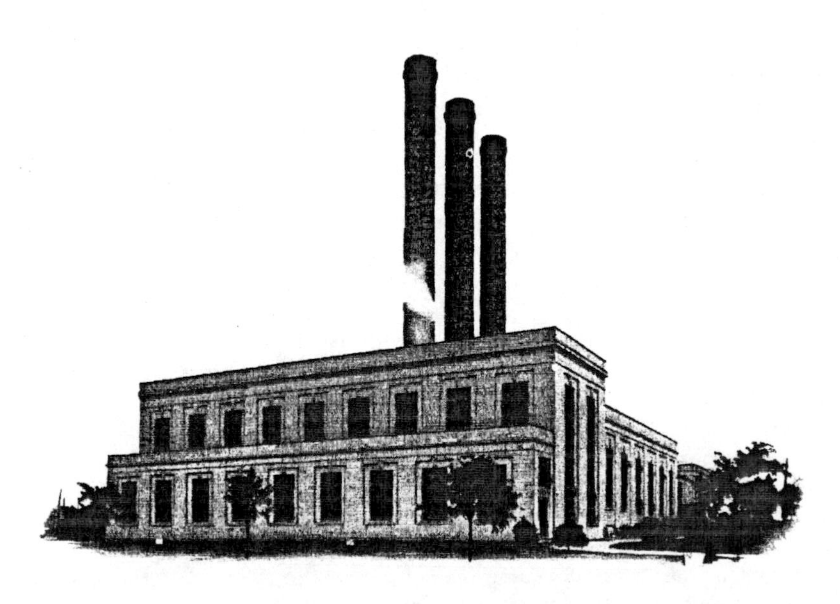

Bennings Station, Potomac Electric Power Co., Where 12,700 Horse-power of Babcock & Wilcox Boilers Are Operated

STEAM

WHEN a given weight of a perfect gas is compressed or expanded at a constant temperature, the product of the pressure and volume is a constant. Vapors, which are liquids in aeriform condition, on the other hand, can exist only at a definite pressure corresponding to each temperature if in the saturated state, that is, the pressure is a function of the temperature only. Steam is water vapor, and at a pressure of, say, 150 pounds absolute per square inch, saturated steam can exist only at a temperature of 358 degrees Fahrenheit. Hence, if the pressure of saturated steam be fixed, its temperature is also fixed, and *vice versa.*

Saturated steam is water vapor in the condition in which it is generated from water with which it is in contact. Or it is steam which is at the maximum pressure and density possible at its temperature. If any change be made in the temperature or pressure of steam, there will be a corresponding change in its condition. If the pressure be increased or the temperature decreased, a portion of the steam will be condensed. If the temperature be increased or the pressure decreased, a portion of the water with which the steam is in contact will be evaporated into steam. Steam will remain saturated just so long as it is of the same pressure and temperature as the water with which it can remain in contact without a gain or loss of heat. Moreover. saturated steam cannot have its temperature lowered without a lowering of its pressure, any loss of heat being made up by the latent heat of such portion as will be condensed. Nor can the temperature of saturated steam be increased except when accompanied by a corresponding increase in pressure, any added heat being expended in the evaporation into steam of a portion of the water with which it is in contact.

Dry saturated steam contains no water. In some cases, saturated steam is accompanied by water which is carried along with it, either in the form of a spray or is blown along the surface of the piping, and the steam is then said to be wet. The percentage weight of the steam in a mixture of steam and water is called the quality of the steam. Thus, if in a mixture of 100 pounds of steam and water there is three-quarters of a pound of water, the quality of the steam will be 99.25.

Heat may be added to steam not in contact with water, such an addition of heat resulting in an increase of temperature and pressure if the volume be kept constant. or an increase in temperature and volume if the pressure remain constant. Steam whose temperature thus exceeds that of saturated steam at a corresponding pressure is said to be superheated and its properties approximate those of a perfect gas.

As pointed out in the chapter on heat, the heat necessary to raise one pound of water from 32 degrees Fahrenheit to the point of ebullition is called the *heat of the liquid.* The heat absorbed during ebullition consists of that necessary to dissociate the molecules, or the *inner latent heat,* and that necessary to overcome the resistance to the increase in volume, or the *outer latent heat.* These two make up the *latent heat of evaporation* and the sum of this latent heat of evaporation and the heat of the liquid make the *total heat* of the steam. These values for various pressures are given in the steam tables, pages 118 to 122.

The specific volume of saturated steam at any pressure is the volume in cubic feet of one pound of steam at that pressure.

The density of saturated steam, that is, its weight per cubic foot, is obviously the reciprocal of the specific volume. This density varies as the $\frac{16}{17}$ power over the ordinary range of pressures used in steam-boiler work and may be found by the formula, $D = .003027\ p^{.941}$, which is correct within 0.15 per cent up to 250 pounds pressure.

The relative volume of steam is the ratio of the volume of a given weight to the volume of the same weight of water at 39.2 degrees Fahrenheit and is equal to the specific volume times 62.427.

As vapors are liquids in their gaseous form and the boiling point is the point of change in this condition, it is clear that this point is dependent upon the pressure under which the liquid exists. This fact is of great practical importance in steam condenser work and in many operations involving boiling in an open vessel, since in the latter case its altitude will have considerable influence. The relation between altitude and boiling point of water is shown in Table 10.

The conditions of feed temperature and steam pressure in boiler tests, fuel performances and the like will be found to vary widely in different trials. In order to secure a means for comparison of different trials, it is necessary to reduce all results to some common basis. The method which has been adopted for the reduction to a comparable basis is to transform the evaporation under actual conditions of steam pressure and feed temperature which exist in the trial to an equivalent evaporation under a set of standard conditions. These standard conditions presuppose a feed-water temperature of 212 degrees Fahrenheit and a steam pressure equal to the normal atmospheric pressure at sea level, 14.7 pounds absolute. Under such conditions, steam would be generated *at* a temperature of 212 degrees, the temperature corresponding to atmospheric pressure at sea level, *from* water at 212 degrees. The weight of water which *would* be evaporated under the assumed standard conditions by exactly the amount of heat absorbed by the boiler under actual conditions existing in the trial, is, therefore, called the equivalent evaporation "from and at 212 degrees."

The factor for reducing the weight of water actually converted into steam from the temperature of the feed, at the steam pressure existing in the trial, to the equivalent evaporation under standard conditions is called the *factor of evaporation* This factor is the ratio of the heat added to each pound of steam in heating the water from the temperature of the feed in the trial to the temperature corresponding to the pressure existing in the trial to the total heat added to evaporate one pound of water from and at 212 degrees. The heat added during the trial is obviously the difference between the total heat of evaporation of the steam at the pressure existing in the trial and the heat of the liquid in the water at the temperature at which it was fed in the trial. To illustrate by an example:

In a boiler trial the temperature of the feed water is 60 degrees Fahrenheit and the pressure under which steam is delivered is 160.3 pounds gauge pressure or 175 pounds absolute pressure. The total heat of one pound of steam at 175 pounds pressure is 1195.9 B.t.u. measured above the standard temperature of 32 degrees Fahrenheit. But the water fed to the boiler contained 28.08 B.t.u. as the heat of the liquid measured above 32 degrees Fahrenheit. Therefore, to each pound of steam there has been added 1167.82 B.t.u. To evaporate one pound of water under standard conditions would, on the other hand, have required but 970.4 B.t.u., which, as described, is the latent heat of evaporation at 212 degrees Fahrenheit. Expressed differently, the total heat of one pound of steam at the pressure corresponding to a temperature of 212 degrees is 1150.4 B.t.u. One pound of water at 212 degrees contains 180 B.t.u. of sensible heat above 32 degrees Fahrenheit. Hence, under standard conditions, 1150.4—180=970.4 B.t.u. is added in the changing of one pound of water into steam at atmospheric pressure and a temperature of 212 degrees. This is in effect the definition of the latent heat of evaporation.

Hence, if conditions of the trial had been standard, only 970.4 B.t.u. would be required and the ratio of 1167.82 to 970.4 B.t.u. is the ratio determining the factor of evaporation. The factor in the assumed case is 1167.82 ÷ 970.4=1.2034 and if the same amount of heat had been absorbed under standard conditions as was absorbed in the trial condition, 1.2034 times the amount of steam would have been generated. Expressed as a formula for use with any set of conditions, the factor is.

$$F = \frac{H - h}{970.4} \qquad (4)$$

Where H=the total heat of steam above 32 degrees Fahrenheit from steam tables,

h=sensible heat of feed water above 32 degrees Fahrenheit from Table 20.

In the form above, the factor may be determined with either saturated or superheated steam, provided that in the latter case values of H are available for varying degrees of superheat and pressures.

In the case of saturated steam containing moisture, the factor may be computed from the formula

$$F = \frac{H - \frac{L \times m}{100} - h}{970.4} \qquad (5)$$

Where L = latent heat of steam at observed pressure, and

m = percentage of moisture in steam.

Table 17 gives factors of evaporation for saturated steam boiler trials to cover a large range of conditions. Except for the most refined work, intermediate values may be determined by interpolation.

Steam gauges indicate the pressure above the atmosphere. As has been pointed out, the atmospheric pressure changes according to the altitude and the variation in the barometer. Hence, calculations involving the properties of steam are based on *absolute* pressures, which are equal to the gauge pressure plus the atmospheric pressure in pounds to the square inch. This latter is generally assumed to be 14.7 pounds per square inch at sea level, but for other levels it must be determined from the barometric reading at that place.

Vacuum gauges indicate the difference, expressed in inches of mercury, between atmospheric pressure and the pressure within the vessel to which the gauge is attached. For approximate purposes, 2.04 inches height of mercury may be considered equal to a pressure of one pound per square inch at the ordinary temperatures at which mercury gauges are used. Hence for any reading of the vacuum gauge in inches, G, the absolute pressure for any barometer reading in inches, B, will be $(B - G) \div 2.04$. If the barometer is 30 inches measured at ordinary temperatures and not corrected to 32 degrees Fahrenheit and the vacuum gauge 24 inches, the absolute pressure will be $(30 - 24) \div 2.04 = 2.9$ pounds per square inch.

TABLE 17

FACTORS OF EVAPORATION

CALCULATED FROM MARKS AND DAVIS TABLES

Feed Temperature Degrees Fahrenheit	Steam Pressure by Gauge																				
	50	60	70	80	90	100	110	120	130	140	150	160	170	180	190	200	210	220	230	240	250
32	1.2143	1.2170	1.2194	1.2215	1.2233	1.2251	1.2265	1.2280	1.2292	1.2304	1.2314	1.2323	1.2333	1.2342	1.2350	1.2357	1.2364	1.2372	1.2378	1.2384	1.2390
40	1.2060	1.2087	1.2111	1.2131	1.2150	1.2168	1.2181	1.2196	1.2209	1.2221	1.2231	1.2241	1.2250	1.2259	1.2267	1.2274	1.2282	1.2289	1.2295	1.2301	1.2307
50	1.1957	1.1984	1.2008	1.2028	1.2047	1.2065	1.2079	1.2093	1.2106	1.2117	1.2128	1.2137	1.2147	1.2156	1.2164	1.2171	1.2178	1.2186	1.2192	1.2198	1.2204
60	1.1854	1.1881	1.1905	1.1925	1.1944	1.1961	1.1976	1.1990	1.2003	1.2014	1.2025	1.2034	1.2044	1.2053	1.2061	1.2068	1.2075	1.2083	1.2089	1.2095	1.2101
70	1.1751	1.1778	1.1802	1.1822	1.1841	1.1859	1.1873	1.1887	1.1900	1.1911	1.1922	1.1931	1.1941	1.1950	1.1958	1.1965	1.1972	1.1980	1.1986	1.1992	1.1998
80	1.1649	1.1675	1.1699	1.1720	1.1738	1.1756	1.1770	1.1785	1.1797	1.1809	1.1819	1.1828	1.1838	1.1847	1.1855	1.1863	1.1869	1.1877	1.1883	1.1889	1.1895
90	1.1545	1.1572	1.1596	1.1617	1.1636	1.1653	1.1668	1.1682	1.1695	1.1706	1.1717	1.1725	1.1735	1.1744	1.1752	1.1760	1.1766	1.1774	1.1780	1.1786	1.1792
100	1.1443	1.1470	1.1493	1.1514	1.1533	1.1550	1.1565	1.1579	1.1592	1.1603	1.1614	1.1623	1.1633	1.1642	1.1650	1.1657	1.1664	1.1671	1.1678	1.1684	1.1690
110	1.1340	1.1367	1.1391	1.1411	1.1430	1.1448	1.1462	1.1477	1.1489	1.1500	1.1511	1.1520	1.1530	1.1539	1.1547	1.1554	1.1562	1.1569	1.1575	1.1581	1.1587
120	1.1237	1.1264	1.1288	1.1309	1.1327	1.1345	1.1359	1.1374	1.1386	1.1398	1.1408	1.1418	1.1427	1.1436	1.1444	1.1452	1.1459	1.1466	1.1472	1.1478	1.1484
130	1.1134	1.1161	1.1185	1.1206	1.1225	1.1242	1.1257	1.1271	1.1284	1.1295	1.1305	1.1315	1.1324	1.1333	1.1341	1.1349	1.1356	1.1363	1.1369	1.1375	1.1381
140	1.1031	1.1058	1.1082	1.1103	1.1122	1.1139	1.1154	1.1168	1.1181	1.1192	1.1203	1.1212	1.1221	1.1230	1.1239	1.1246	1.1253	1.1260	1.1266	1.1272	1.1278
150	1.0928	1.0955	1.0979	1.1000	1.1019	1.1036	1.1051	1.1065	1.1078	1.1089	1.1099	1.1109	1.1118	1.1127	1.1136	1.1143	1.1150	1.1157	1.1163	1.1169	1.1176
160	1.0825	1.0852	1.0876	1.0897	1.0916	1.0933	1.0948	1.0962	1.0975	1.0986	1.0997	1.1006	1.1015	1.1024	1.1033	1.1040	1.1047	1.1054	1.1060	1.1066	1.1073
170	1.0722	1.0749	1.0773	1.0794	1.0813	1.0830	1.0845	1.0859	1.0872	1.0883	1.0893	1.0903	1.0912	1.0921	1.0930	1.0937	1.0944	1.0951	1.0957	1.0963	1.0969
180	1.0619	1.0646	1.0670	1.0691	1.0709	1.0727	1.0741	1.0756	1.0768	1.0780	1.0790	1.0800	1.0809	1.0818	1.0826	1.0834	1.0841	1.0848	1.0854	1.0860	1.0866
190	1.0516	1.0543	1.0567	1.0587	1.0606	1.0624	1.0638	1.0653	1.0665	1.0676	1.0687	1.0696	1.0706	1.0715	1.0723	1.0730	1.0737	1.0745	1.0751	1.0757	1.0763
200	1.0412	1.0439	1.0463	1.0484	1.0503	1.0520	1.0535	1.0549	1.0562	1.0573	1.0584	1.0593	1.0602	1.0611	1.0620	1.0627	1.0634	1.0641	1.0647	1.0653	1.0660
210	1.0309	1.0336	1.0360	1.0380	1.0399	1.0417	1.0432	1.0446	1.0458	1.0469	1.0480	1.0489	1.0499	1.0508	1.0516	1.0523	1.0531	1.0538	1.0544	1.0550	1.0556

The temperature, pressure and other properties of steam for varying amounts of vacuum and the pressure above vacuum corresponding to each inch of reading of the vacuum gauge are given in Table 18.

TABLE 18

PROPERTIES OF SATURATED STEAM FOR VARYING AMOUNTS OF VACUUM

CALCULATED FROM MARKS AND DAVIS TABLES

Vacuum Height Inches	Absolute Pressure Pounds	Temperature Degrees Fahrenheit	Heat of the Liquid Above 32 Degrees B.t.u.	Latent Heat Above 32 Degrees B.t.u.	Total Heat Above 32 Degrees B.t.u.	Density or Weight per Cubic Foot Pound
29.5	.207	54.1	22.18	1061.0	1083.2	0.000678
29	.452	76.6	44.64	1048.7	1093.3	0.001415
28.5	.698	90.1	58.09	1041.1	1099.2	0.002137
28	.944	99.9	67.87	1035.6	1103.5	0.002843
27	1.44	112.5	80.4	1028.6	1109.0	0.00421
26	1.93	124.5	92.3	1022.0	1114.3	0.00577
25	2.42	132.6	100.5	1017.3	1117.8	0.00689
24	2.91	140.1	108.0	1013.1	1121.1	0.00821
22	3.89	151.7	119.6	1006.4	1126.0	0.01078
20	4.87	161 1	128.9	1001.0	1129.9	0.01331
18	5.86	168.9	136.8	996.4	1133.2	0.01581
16	6.84	175.8	143.6	992.4	1136.0	0.01827
14	7.82	181.8	149.7	988.8	1138.5	0.02070
12	8.80	187.2	155.1	985.6	1140.7	0.02312
10	9.79	192.2	160.1	982.6	1142.7	0.02554
5	12 24	202.9	170.8	976.0	1146.8	0.03148

From the steam tables, the condensed Table 19 of the properties of steam at different pressures may be constructed. From such a table there may be drawn the following conclusions:

TABLE 19

VARIATION IN PROPERTIES OF SATURATED STEAM WITH PRESSURE

Pressure Pounds Absolute	Temperature Degrees Fahrenheit	Heat of Liquid B.t.u	Latent Heat B.t.u.	Total Heat B.t.u.
14.7	212.0	180.0	970.4	1150.4
20.0	228.0	196.1	960.0	1156.2
100.0	327.8	298.3	888.0	1186.3
300.0	417.5	392.7	811.3	1204.1

As the pressure and temperature increase, the latent heat decreases. This decrease, however, is less rapid than the corresponding increase in the heat of the liquid and hence the total heat increases with an increase in the pressure and temperature. The percentage increase in the total heat is small, being 0.5, 3.1 and 4.7 per cent for 20, 100 and 300 pounds absolute pressure, respectively, above the total heat in one pound of steam at 14.7 pounds absolute. The temperatures, on the other hand, increase at the rates of 7.5, 54.6 and 96.9 per cent. The efficiency of a perfect steam engine is proportional to the expression $(t-t_1)\div t$, in which t and t_1 are the absolute temperatures of the saturated steam at admission and exhaust, respectively. While actual engines only approximate the ideal engine in efficiency, yet they follow the same general law. Since the exhaust temperature cannot be lowered beyond present practice, it follows that the only available method of increasing the efficiency is by an increase in the temperature of the steam at admission. How this may be accomplished by an increase of pressure is clearly shown, for the increase of fuel necessary to increase the pressure is negligible, as shown by the total heat, while the increase in economy, due to the higher pressure, will result directly from the rapid increase of the corresponding temperature.

TABLE 20

HEAT UNITS PER POUND AND WEIGHT PER CUBIC FOOT OF WATER BETWEEN 32 DEGREES FAHRENHEIT AND 340 DEGREES FAHRENHEIT

Temperature Degrees Fahrenheit	Heat Units per Pound	Weight per Cubic Foot	Temperature Degrees Fahrenheit	Heat Units per Pound	Weight per Cubic Foot	Temperature Degrees Fahrenheit	Heat Units per Pound	Weight per Cubic Foot	Temperature Degrees Fahrenheit	Heat Units per Pound	Weight per Cubic Foot	Temperature Degrees Fahrenheit	Heat Units per Pound	Weight per Cubic Foot	Temperature Degrees Fahrenheit	Heat Units per Pound	Weight per Cubic Foot
32	0.00	62.42	70	38.06	62.30	108	75.95	61.90	146	113.86	61.27	184	151.89	60.49	222	190.1	59.58
33	1.01	62.42	71	39.06	62.30	109	76.94	61.88	147	114.86	61.25	185	152.89	60.47	223	191.1	59.55
34	2.01	62.42	72	40.05	62.29	110	77.94	61.86	148	115.86	61.24	186	153.89	60.45	224	192.1	59.53
35	3.02	62.43	73	41.05	62.28	111	78.94	61.85	149	116.86	61.22	187	154.90	60.42	225	193.1	59.50
36	4.03	62.43	74	42.05	62.27	112	79.93	61.83	150	117.86	61.20	188	155.90	60.40	226	194.1	59.48
37	5.04	62.43	75	43.05	62.26	113	80.93	61.82	151	118.86	61.18	189	156.90	60.38	227	195.2	59.45
38	6.04	62.43	76	44.04	62.26	114	81.93	61.80	152	119.86	61.16	190	157.91	60.36	228	196.2	59.42
39	7.05	62.43	77	45.04	62.25	115	82.92	61.79	153	120.86	61.14	191	158.91	60.33	229	197.2	59.40
40	8.05	62.43	78	46.04	62.24	116	83.92	61.77	154	121.86	61.12	192	159.91	60.31	230	198.2	59.37
41	9.05	62.43	79	47.04	62.23	117	84.92	61.75	155	122.86	61.10	193	160.91	60.29	231	199.2	59.34
42	10.06	62.43	80	48.03	62.22	118	85.92	61.74	156	123.86	61.08	194	161.92	60.27	232	200.2	59.32
43	11.06	62.43	81	49.03	62.21	119	86.91	61.72	157	124.86	61.06	195	162.92	60.24	233	201.2	59.29
44	12.06	62.43	82	50.03	62.20	120	87.91	61.71	158	125.86	61.04	196	163.92	60.22	234	202.2	59.27
45	13.07	62.42	83	51.02	62.19	121	88.91	61.69	159	126.86	61.02	197	164.93	60.19	235	203.2	59.24
46	14.07	62.42	84	52.02	62.18	122	89.91	61.68	160	127.86	61.00	198	165.93	60.17	236	204.2	59.21
47	15.07	62.42	85	53.02	62.17	123	90.90	61.66	161	128.86	60.98	199	166.94	60.15	237	205.3	59.19
48	16.07	62.42	86	54.01	62.16	124	91.90	61.65	162	129.86	60.96	200	167.94	60.12	238	206.3	59.16
49	17.08	62.42	87	55.01	62.15	125	92.90	61.63	163	130.86	60.94	201	168.94	60.10	239	207.3	59.14
50	18.08	62.42	88	56.01	62.14	126	93.90	61.61	164	131.86	60.92	202	169.95	60.07	240	208.3	59.11
51	19.08	62.41	89	57.00	62.13	127	94.89	61.59	165	132.86	60.90	203	170.95	60.05	241	209.3	59.08
52	20.08	62.41	90	58.00	62.12	128	95.89	61.58	166	133.86	60.88	204	171.96	60.02	242	210.3	59.05
53	21.08	62.41	91	59.00	62.11	129	96.89	61.56	167	134.86	60.86	205	172.96	60.00	243	211.4	59.03
54	22.08	62.40	92	60.00	62.09	130	97.89	61.55	168	135.86	60.84	206	173.97	59.98	244	212.4	59.00
55	23.08	62.40	93	60.99	62.08	131	98.89	61.53	169	136.86	60.82	207	174.97	59.95	245	213.4	58.97
56	24.08	62.39	94	61.99	62.07	132	99.88	61.52	170	137.87	60.80	208	175.98	59.93	246	214.4	58.94
57	25.08	62.39	95	62.99	62.06	133	100.88	61.50	171	138.87	60.78	209	176.98	59.90	247	215.4	58.91
58	26.08	62.38	96	63.98	62.05	134	101.88	61.49	172	139.87	60.76	210	177.99	59.88	248	216.4	58.89
59	27.08	62.37	97	64.98	62.04	135	102.88	61.47	173	140.87	60.73	211	178.99	59.85	249	217.4	58.86
60	28.08	62.37	98	65.98	62.03	136	103.88	61.45	174	141.87	60.71	212	180.00	59.83	250	218.5	58.83
61	29.08	62.36	99	66.97	62.02	137	104.87	61.43	175	142.87	60.69	213	181.0	59.80	260	228.6	58.55
62	30.08	62.36	100	67.97	62.00	138	105.87	61.41	176	143.87	60.67	214	182.0	59.78	270	238.8	58.26
63	31.07	62.35	101	68.97	61.99	139	106.87	61.40	177	144.88	60.65	215	183.0	59.75	280	249.0	57.96
64	32.07	62.35	102	69.96	61.98	140	107.87	61.38	178	145.88	60.62	216	184.0	59.73	290	259.3	57.65
65	33.07	62.34	103	70.96	61.97	141	108.87	61.36	179	146.88	60.60	217	185.0	59.70	300	269.6	57.33
66	34.07	62.33	104	71.96	61.95	142	109.87	61.34	180	147.88	60.58	218	186.1	59.68	310	279.9	57.00
67	35.07	62.33	105	72.95	61.94	143	110.87	61.33	181	148.88	60.56	219	187.1	59.65	320	290.2	56.66
68	36.07	62.32	106	73.95	61.93	144	111.87	61.31	182	149.89	60.53	220	188.1	59.63	330	300.6	56.30
69	37.06	62.31	107	74.95	61.91	145	112.86	61.29	183	150.89	60.51	221	189.1	59.60	340	311.0	55.94

The gain due to superheat cannot be predicted from the formula for the efficiency of a perfect steam engine. This formula is not applicable in cases where superheat is present, since only a relatively small amount of the heat in the steam is imparted at the maximum or superheated temperature.

The advantage of the use of high-pressure steam may be also indicated by considering the question from the aspect of volume. With an increase of pressure comes a decrease in volume, thus one pound of saturated steam at 100 pounds absolute pressure occupies 4.43 cubic feet, while at 200 pounds pressure it occupies 2.29 cubic feet. If then, in separate cylinders of the same dimensions, one pound of steam at 100 pounds absolute pressure and one pound at 200 pounds absolute pressure enter and are allowed to expand to the full volume of each cylinder, the high-pressure steam, having more room and a greater range for expansion than the low-pressure steam, will thus do more work. This increase in the amount of work, as was the increase in temperature is large relative to the additional fuel required as indicated by the total heat. In general it may be stated that the fuel required to impart a given amount of heat to a boiler is practically independent of the steam pressure, since the temperature of the fire is so high as compared with the steam temperature that a variation in the steam temperature does not produce an appreciable effect.

The formulæ for the algebraic expression of the relation between saturated steam pressures, temperatures and steam volumes have been up to the present time empirical. These relations have, however, been determined by experiment and, from the experimental data, tables have been computed which render unnecessary the use of empirical formulæ. Such formulæ may be found in any standard work of thermodynamics. The following tables cover all practical cases.

Table 20 gives the heat units contained in water above 32 degrees Fahrenheit at different temperatures.

Table 21 gives the properties of saturated steam for various pressures.

Table 22 gives the properties of superheated steam at various pressures and temperatures.

These tables are based on those computed by Lionel S. Marks and Harvey N. Davis, these being generally accepted as being the most correct.

TABLE 21

PROPERTIES OF SATURATED STEAM

REPRODUCED BY PERMISSION FROM MARKS AND DAVIS "STEAM TABLES AND DIAGRAMS"

Pressure Pounds Absolute	Temperature Degrees Fahrenheit	Specific Volume Cubic Feet per Pound	Heat of the Liquid, B.t.u.	Latent Heat of Evaporation B.t.u.	Total Heat of Steam, B.t.u.	Pressure Pounds Absolute
1	101.83	333.0	69.8	1034.6	1104.4	1
2	126.15	173.5	94.0	1021.0	1115.0	2
3	141.52	118.5	109.4	1012.3	1121.6	3
4	153.01	90.5	120.9	1005.7	1126.5	4
5	162.28	73.33	130.1	1000.3	1130.5	5
6	170.06	61.89	137.9	995.8	1133.7	6
7	176.85	53.56	144.7	991.8	1136.5	7
8	182.86	47.27	150.8	988.2	1139.0	8
9	188.27	42.36	156.2	985.0	1141.1	9
10	193.22	38.38	161.1	982.0	1143.1	10
11	197.75	35.10	165.7	979.2	1144.9	11
12	201.96	32.36	169.9	976.6	1146.5	12
13	205.87	30.03	173.8	974.2	1148.0	13
14	209.55	28.02	177.5	971.9	1149.4	14
15	213.0	26.27	181.0	969.7	1150.7	15
16	216.3	24.79	184.4	967.6	1152.0	16
17	219.4	23.38	187.5	965.6	1153.1	17
18	222.4	22.16	190.5	963.7	1154.2	18
19	225.2	21.07	193.4	961.8	1155.2	19
20	228.0	20.08	196.1	960.0	1156.2	20
22	233.1	18.37	201.3	956.7	1158.0	22
24	237.8	16.93	206.1	953.5	1159.6	24
26	242.2	15.72	210.6	950.6	1161.2	26
28	246.4	14.67	214.8	947.8	1162.6	28
30	250.3	13.74	218.8	945.1	1163.9	30
32	254.1	12.93	222.6	942.5	1165.1	32
34	257.6	12.22	226.2	940.1	1166.3	34
36	261.0	11.58	229.6	937.7	1167.3	36
38	264.2	11.01	232.9	935.5	1168.4	38
40	267.3	10.49	236.1	933.3	1169.4	40
42	270.2	10.02	239.1	931.2	1170.3	42
44	273.1	9.59	242.0	929.2	1171.2	44
46	275.8	9.20	244.8	927.2	1172.0	46
48	278.5	8.84	247.5	925.3	1172.8	48
50	281.0	8.51	250.1	923.5	1173.6	50
52	283.5	8.20	252.6	921.7	1174.3	52
54	285.9	7.91	255.1	919.9	1175.0	54
56	288.2	7.65	257.5	918.2	1175.7	56
58	290.5	7.40	259.8	916.5	1176.4	58
60	292.7	7.17	262.1	914.9	1177.0	60
62	294.9	6.95	264.3	913.3	1177.6	62
64	297.0	6.75	266.4	911.8	1178.2	64
66	299.0	6.56	268.5	910.2	1178.8	66
68	301.0	6.38	270.6	908.7	1179.3	68

PROPERTIES OF SATURATED STEAM—CONTINUED

Pressure Pounds Absolute	Temperature Degrees Fahrenheit	Specific Volume Cubic Feet per Pound	Heat of the Liquid, B.t.u.	Latent Heat of Evaporation B.t.u.	Total Heat of Steam, B.t.u	Pressure Pounds Absolute
70	302.9	6.20	272.6	907.2	1179.8	70
72	304.8	6.04	274.5	905.8	1180.4	72
74	306.7	5.89	276.5	904.4	1180.9	74
76	308.5	5.74	278.3	903.0	1181.4	76
78	310.3	5.60	280.2	901.7	1181.8	78
80	312.0	5.47	282.0	900.3	1182.3	80
82	313.8	5.34	283.8	899.0	1182.8	82
84	315.4	5.22	285.5	897.7	1183.2	84
86	317.1	5.10	287.2	896.4	1183.6	86
88	318.7	5.00	288.9	895.2	1184.0	88
90	320.3	4.89	290.5	893.9	1184.4	90
92	321.8	4.79	292.1	892.7	1184.8	92
94	323.4	4.69	293.7	891.5	1185.2	94
96	324.9	4.60	295.3	890.3	1185.6	96
98	326.4	4.51	296.8	889.2	1186.0	98
100	327.8	4.429	298.3	888.0	1186.3	100
105	331.4	4.230	302.0	885.2	1187.2	105
110	334.8	4.047	305.5	882.5	1188.0	110
115	338.1	3.880	309.0	879.8	1188.8	115
120	341.3	3.726	312.3	877.2	1189.6	120
125	344.4	3.583	315.5	874.7	1190.3	125
130	347.4	3.452	318.6	872.3	1191.0	130
135	350.3	3.331	321.7	869.9	1191.6	135
140	353.1	3.219	324.6	867.6	1192.2	140
145	355.8	3.112	327.4	865.4	1192.8	145
150	358.5	3.012	330.2	863.2	1193.4	150
155	361.0	2.920	332.9	861.0	1194.0	155
160	363.6	2.834	335.6	858.8	1194.5	160
165	366.0	2.753	338.2	856.8	1195.0	165
170	368.5	2.675	340.7	854.7	1195.4	170
175	370.8	2.602	343.2	852.7	1195.9	175
180	373.1	2.533	345.6	850.8	1196.4	180
185	375.4	2.468	348.0	848.8	1196.8	185
190	377.6	2.406	350.4	846.9	1197.3	190
195	379.8	2.346	352.7	845.0	1197.7	195
200	381.9	2.290	354.9	843.2	1198.1	200
205	384.0	2.237	357.1	841.4	1198.5	205
210	386.0	2.187	359.2	839.6	1198.8	210
215	388.0	2.138	361.4	837.9	1199.2	215
220	389.9	2.091	363.4	836.2	1199.6	220
225	391.9	2.046	365.5	834.4	1199.9	225
230	393.8	2.004	367.5	832.8	1200.2	230
235	395.6	1.964	369.4	831.1	1200.6	235
240	397.4	1.924	371.4	829.5	1200.9	240
245	399.3	1.887	373.3	827.9	1201.2	245
250	401.1	1.850	375.2	826.3	1201.5	250

TABLE 22

PROPERTIES OF SUPERHEATED STEAM

REPRODUCED BY PERMISSION FROM MARKS AND DAVIS "STEAM TABLES AND DIAGRAMS"

(Copyright, 1909, by Longmans, Green & Co.)

Pressure Pounds Absolute		Saturated Steam	Degrees of Superheat							Pressure Pounds Absolute
			50	100	150	200	250	300		
	t	162.3	212.3	262.3	312.3	362.3	412.3	462.3	t	
5	v	73.3	79.7	85.7	91.8	97.8	103.8	109.8	v	5
	h	1130.5	1153.5	1176.4	1199.5	1222.5	1245.6	1268.7	h	
	t	193.2	243.2	293.2	343.2	393.2	443.2	493.2	t	
10	v	38.4	41.5	44.6	47.7	50.7	53.7	56.7	v	10
	h	1143.1	1166.3	1189.5	1212.7	1236.0	1259.3	1282.5	h	
	t	213.0	263.0	313.0	363.0	413.0	463.0	513.0	t	
15	v	26.27	28.40	30.46	32.50	34.53	36.56	38.58	v	15
	h	1150.7	1174.2	1197.6	1221.0	1244.4	1267.7	1291.1	h	
	t	228.0	278.0	328.0	378.0	428.0	478.0	528.0	t	
20	v	20.08	21.69	23.25	24.80	26.33	27.85	29.37	v	20
	h	1156.2	1179.9	1203.5	1227.1	1250.6	1274.1	1297.6	h	
	t	240.1	290.1	340.1	390.1	440.1	490.1	540.1	t	
25	v	16.30	17.60	18.86	20.10	21.32	22.55	23.77	v	25
	h	1160.4	1184.4	1208.2	1231.9	1255.6	1279.2	1302.8	h	
	t	250.4	300.4	350.4	400.4	450.4	500.4	550.4	t	
30	v	13.74	14.83	15.89	16.93	17.97	18.99	20.00	v	30
	h	1163.9	1188.1	1212.1	1236.0	1259.7	1283.4	1307.1	h	
	t	259.3	309.3	359.3	409.3	459.3	509.3	559.3	t	
35	v	11.89	12.85	13.75	14.65	15.54	16.42	17.30	v	35
	h	1166.8	1191.3	1215.4	1239.4	1263.3	1287.1	1310.8	h	
	t	267.3	317.3	367.3	417.3	467.3	517.3	567.3	t	
40	v	10.49	11.33	12.13	12.93	13.70	14.48	15.25	v	40
	h	1169.4	1194.0	1218.4	1242.4	1266.4	1290.3	1314.1	h	
	t	274.5	324.5	374.5	424.5	474.5	524.5	574.5	t	
45	v	9.39	10.14	10.86	11.57	12.27	12.96	13.65	v	45
	h	1171.6	1196.6	1221.0	1245.2	1269.3	1293.2	1317.0	h	
	t	281.0	331.0	381.0	431.0	481.0	531.0	581.0	t	
50	v	8.51	9.19	9.84	10.48	11.11	11.74	12.36	v	50
	h	1173.6	1198.8	1223.4	1247.7	1271.8	1295.8	1319.7	h	
	t	287.1	337.1	387.1	437.1	487.1	537.1	587.1	t	
55	v	7.78	8.40	9.00	9.59	10.16	10.73	11.30	v	55
	h	1175.4	1200.8	1225.6	1250.0	1274.2	1298.1	1322.0	h	
	t	292.7	342.7	392.7	442.7	492.7	542.7	592.7	t	
60	v	7.17	7.75	8.30	8.84	9.36	9.89	10.41	v	60
	h	1177.0	1202.6	1227.6	1252.1	1276.4	1300.4	1324.3	h	
	t	298.0	348.0	398.0	448.0	498.0	548.0	598.0	t	
65	v	6.65	7.20	7.70	8.20	8.69	9.17	9.65	v	65
	h	1178.5	1204.4	1229.5	1254.0	1278.4	1302.4	1326.4	h	
	t	302.9	352.9	402.9	452.9	502.9	552.9	602.9	t	
70	v	6.20	6.71	7.18	7.65	8.11	8.56	9.01	v	70
	h	1179.8	1205.9	1231.2	1255.8	1280.2	1304.3	1328.3	h	
	t	307.6	357.6	407.6	457.6	507.6	557.6	607.6	t	
75	v	5.81	6.28	6.73	7.17	7.60	8.02	8.44	v	75
	h	1181.1	1207.5	1232.8	1257.5	1282.0	1306.1	1330.1	h	
	t	312.0	362.0	412.0	462.0	512.0	562.0	612.0	t	
80	v	5.47	5.92	6.34	6.75	7.17	7.56	7.95	v	80
	h	1182.3	1208.8	1234.3	1259.0	1283.6	1307.8	1331.9	h	
	t	316.3	366.3	416.3	466.3	516.3	566.3	616.3	t	
85	v	5.16	5.59	5.99	6.38	6.76	7.14	7.51	v	85
	h	1183.4	1210.2	1235.8	1260.6	1285.2	1309.4	1333.5	h	

t = Temperature, degrees Fahrenheit.

v = Specific volume, in cubic feet, per pound.

h = Total heat from water at 32 degrees, B.t.u.

Pressure Pounds Absolute		Saturated Steam	Degrees of Superheat							Pressure Pounds Absolute
			50	100	150	200	250	300		
	t	320.3	370.3	420.3	470.3	520.3	570.3	620.3	t	
90	v	4.89	5.29	5.67	6.04	6.40	6.76	7.11	v	90
	h	1184.4	1211.4	1237.2	1262.0	1286.6	1310.8	1334.9	h	
	t	324.1	374.1	424.1	474.1	524.1	574.1	624.1	t	
95	v	4.65	5.03	5.39	5.74	6.09	6.43	6.76	v	95
	h	1185.4	1212.6	1238.4	1263.4	1288.1	1312.3	1336.4	h	
	t	327.8	377.8	427.8	477.8	527.8	577.8	627.8	t	
100	v	4.43	4.79	5.14	5.47	5.80	6.12	6.44	v	100
	h	1186.3	1213.8	1239.7	1264.7	1289.4	1313.6	1337.8	h	
	t	331.4	381.4	431.4	481.4	531.4	581.4	631.4	t	
105	v	4.23	4.58	4.91	5.23	5.54	5.85	6.15	v	105
	h	1187.2	1214.9	1240.8	1265.9	1290.6	1314.9	1339.1	h	
	t	334.8	384.8	434.8	484.8	534.8	584.8	634.8	t	
110	v	4.05	4.38	4.70	5.01	5.31	5.61	5.90	v	110
	h	1188.0	1215.9	1242.0	1267.1	1291.9	1316.2	1340.4	h	
	t	338.1	388.1	438.1	488.1	538.1	588.1	638.1	t	
115	v	3.88	4.20	4.51	4.81	5.09	5.38	5.66	v	115
	h	1188.8	1216.9	1243.1	1268.2	1293.0	1317.3	1341.5	h	
	t	341.3	391.3	441.3	491.3	541.3	591.3	641.3	t	
120	v	3.73	4.04	4.33	4.62	4.89	5.17	5.44	v	120
	h	1189.6	1217.9	1244.1	1269.3	1294.1	1318.4	1342.7	h	
	t	344.4	394.4	444.4	494.4	544.4	594.4	644.4	t	
125	v	3.58	3.88	4.17	4.45	4.71	4.97	5.23	v	125
	h	1190.3	1218.8	1245.1	1270.4	1295.2	1319.5	1343.8	h	
	t	347.4	397.4	447.4	497.4	547.4	597.4	647.4	t	
130	v	3.45	3.74	4.02	4.28	4.54	4.80	5.05	v	130
	h	1191.0	1219.7	1246.1	1271.4	1296.2	1320.6	1344.9	h	
	t	350.3	400.3	450.3	500.3	550.3	600.3	650.3	t	
135	v	3.33	3.61	3.88	4.14	4.38	4.63	4.87	v	135
	h	1191.6	1220.6	1247.0	1272.3	1297.2	1321.6	1345.9	h	
	t	353.1	403.1	453.1	503.1	553.1	603.1	653.1	t	
140	v	3.22	3.49	3.75	4.00	4.24	4.48	4.71	v	140
	h	1192.2	1221.4	1248.0	1273.3	1298.2	1322.6	1346.9	h	
	t	355.8	405.8	455.8	505.8	555.8	605.8	655.8	t	
145	v	3.12	3.38	3.63	3.87	4.10	4.33	4.56	v	145
	h	1192.8	1222.2	1248.8	1274.2	1299.1	1323.6	1347.9	h	
	t	358.5	408.5	458.5	508.5	558.5	608.5	658.5	t	
150	v	3.01	3.27	3.51	3.75	3.97	4.19	4.41	v	150
	h	1193.4	1223.0	1249.6	1275.1	1300.0	1324.5	1348.8	h	
	t	361.0	411.0	461.0	511.0	561.0	611.0	661.0	t	
155	v	2.92	3.17	3.41	3.63	3.85	4.06	4.28	v	155
	h	1194.0	1223.6	1250.5	1276.0	1300.8	1325.3	1349.7	h	
	t	363.6	413.6	463.6	513.6	563.6	613.6	663.6	t	
160	v	2.83	3.07	3.30	3.53	3.74	3.95	4.15	v	160
	h	1194.5	1224.5	1251.3	1276.8	1301.7	1326.2	1350.6	h	
	t	366.0	416.0	466.0	516.0	566.0	616.0	666.0	t	
165	v	2.75	2.99	3.21	3.43	3.64	3.84	4.04	v	165
	h	1195.0	1225.2	1252.0	1277.6	1302.5	1327.1	1351.5	h	
	t	368.5	418.5	468.5	518.5	568.5	618.5	668.5	t	
170	v	2.68	2.91	3.12	3.34	3.54	3.73	3.92	v	170
	h	1195.4	1225.9	1252.8	1278.4	1303.3	1327.9	1352.3	h	

t = Temperature, degrees Fahrenheit.
v = Specific volume, in cubic feet, per pound.
h = Total heat from water at 32 degrees, B.t.u.

PROPERTIES OF SUPERHEATED STEAM—Continued

Pressure Pounds Absolute		Saturated Steam	Degrees of Superheat							Pressure Pounds Absolute
			50	100	150	200	250	300		
	t	370.8	420.8	470.8	520.8	570.8	620.8	670.8	t	
175	v	2.60	2.83	3.04	3.24	3.44	3.63	3.82	v	175
	h	1195.9	1226.6	1253.6	1279.1	1304.1	1328.7	1353.2	h	
	t	373.1	423.1	473.1	523.1	573.1	623.1	673.1	t	
180	v	2.53	2.75	2.96	3.16	3.35	3.54	3.72	v	180
	h	1196.4	1227.2	1254.3	1279.9	1304.8	1329.5	1353.9	h	
	t	375.4	425.4	475.4	525.4	575.4	625.4	675.4	t	
185	v	2.47	2.68	2.89	3.08	3.27	3.45	3.63	v	185
	h	1196.8	1227.9	1255.0	1280.6	1305.6	1330.2	1354.7	h	
	t	377.6	427.6	477.6	527.6	577.6	627.6	677.6	t	
190	v	2.41	2.62	2.81	3.00	3.19	3.37	3.55	v	190
	h	1197.3	1228.6	1255.7	1281.3	1306.3	1330.9	1355.5	h	
	t	379.8	429.8	479.8	529.8	579.8	629.8	679.8	t	
195	v	2.35	2.55	2.75	2.93	3.11	3.29	3.46	v	195
	h	1197.7	1229.2	1256.4	1282.0	1307.0	1331.6	1356.2	h	
	t	381.9	431.9	481.9	531.9	581.9	631.9	681.9	t	
200	v	2.29	2.49	2.68	2.86	3.04	3.21	3.38	v	200
	h	1198.1	1229.8	1257.1	1282.6	1307.7	1332.4	1357.0	h	
	t	384.0	434.0	484.0	534.0	584.0	634.0	684.0	t	
205	v	2.24	2.44	2.62	2.80	2.97	3.14	3.30	v	205
	h	1198.5	1230.4	1257.7	1283.3	1308.3	1333.0	1357.7	h	
	t	386.0	436.0	486.0	536.0	586.0	636.0	686.0	t	
210	v	2.19	2.38	2.56	2.74	2.91	3.07	3.23	v	210
	h	1198.8	1231.0	1258.4	1284.0	1309.0	1333.7	1358.4	h	
	t	388.0	438.0	488.0	538.0	588.0	638.0	688.0	t	
215	v	2.14	2.33	2.51	2.68	2.84	3.00	3.16	v	215
	h	1199.2	1231.6	1259.0	1284.6	1309.7	1334.4	1359.1	h	
	t	389.9	439.9	489.9	539.9	589.9	639.9	689.9	t	
220	v	2.09	2.28	2.45	2.62	2.78	2.94	3.10	v	220
	h	1199.6	1232.2	1259.6	1285.2	1310.3	1335.1	1359.8	h	
	t	391.9	441.9	491.9	541.9	591.9	641.9	691.9	t	
225	v	2.05	2.23	2.40	2.57	2.72	2.88	3.03	v	225
	h	1199.9	1232.7	1260.2	1285.9	1310.9	1335.7	1360.3	h	
	t	393.8	443.8	493.8	543.8	593.8	643.8	693.8	t	
230	v	2.00	2.18	2.35	2.51	2.67	2.82	2.97	v	230
	h	1200.2	1233.2	1260.7	1286.5	1311.6	1336.3	1361.0	h	
	t	395.6	445.6	495.6	545.6	595.6	645.6	695.6	t	
235	v	1.96	2.14	2.30	2.46	2.62	2.77	2.91	v	235
	h	1200.6	1233.8	1261.4	1287.1	1312.2	1337.0	1361.7	h	
	t	397.4	447.4	497.4	547.4	597.4	647.4	697.4	t	
240	v	1.92	2.09	2.26	2.42	2.57	2.71	2.85	v	240
	h	1200.9	1234.3	1261.9	1287.6	1312.8	1337.6	1362.3	h	
	t	399.3	449.3	499.3	549.3	599.3	649.3	699.3	t	
245	v	1.89	2.05	2.22	2.37	2.52	2.66	2.80	v	245
	h	1201.2	1234.8	1262.5	1288.2	1313.3	1338.2	1362.9	h	
	t	401.0	451.0	501.0	551.0	601.0	651.0	701.0	t	
250	v	1.85	2.02	2.17	2.33	2.47	2.61	2.75	v	250
	h	1201.5	1235.4	1263.0	1288.8	1313.9	1338.8	1363.5	h	
	t	402.8	452.8	502.8	552.8	602.8	652.8	702.8	t	
255	v	1.81	1.98	2.14	2.28	2.43	2.56	2.70	v	255
	h	1201.8	1235.9	1263.6	1289.3	1314.5	1339.3	1364.1	h	

t = Temperature, degrees Fahrenheit.
v = Specific volume, in cubic feet, per pound.
h = Total heat from water at 32 degrees, B.t.u.

MOISTURE IN STEAM

THE presence of moisture in steam causes a loss, not only in the practical waste of the heat utilized to raise this moisture from the temperature of the feed water to the temperature of the steam, but also through the increased initial condensation in an engine cylinder and through friction and other actions in a steam turbine. The presence of such moisture also interferes with proper cylinder lubrication, causes a knocking in the engine and a water hammer in the steam pipes. In steam turbines it will cause erosion of the blades.

The percentage by weight of steam in a mixture of steam and water is called the *quality of the steam.*

The apparatus used to determine the moisture content of steam is called a calorimeter, though since it may not measure the heat in the steam, the name is not descriptive of the function of the apparatus. The first form used was the "barrel calorimeter," but the liability of error was so great that its use was abandoned. Modern calorimeters are in general of either the throttling or separator type.

THROTTLING CALORIMETER—Fig. 4 shows a typical form of throttling calorimeter. Steam is drawn from a vertical main through the sampling nipple, passes around the first thermometer cup, then through a one-eighth-inch orifice in a disk between two flanges, and lastly, around the second thermometer cup and to the atmosphere. Thermometers are inserted in the wells, which should be filled with mercury or heavy cylinder oil.

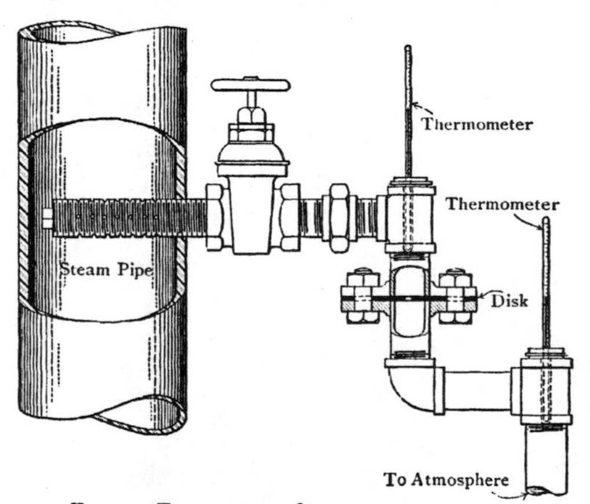

FIG. 4. THROTTLING CALORIMETER AND SAMPLING NOZZLE

The instrument and all pipes and fittings leading to it should be thoroughly insulated to diminish radiation losses. Care must be taken to prevent the orifice from becoming choked with dirt and to see that no leaks occur. The exhaust pipe should be short to prevent back pressure below the disk.

When steam passes through an orifice from a higher to a lower pressure, as is the case with the throttling calorimeter, no external work has to be done in overcoming a resistance. Hence, if there is no loss from radiation, the quantity of heat in the steam will be exactly the same after passing the orifice as before passing. If the higher steam pressure is 160 pounds gauge and the lower pressure that of the atmosphere, the total heat in a pound of dry steam at the former pressure is 1195.9 B.t.u. and at the latter pressure 1150.4 B.t.u., a difference of 45.4 B.t.u. As this heat will still exist in the steam at the lower pressure, since there is no external work done, its effect must be to superheat the steam. Assuming the specific heat of superheated steam to be 0.47, each pound passing through will be superheated $45.4 \div 0.47 = 96.6$ degrees. If, however, the steam had contained one per cent of moisture, it would have contained less heat units per pound than if it were dry. Since the latent heat of

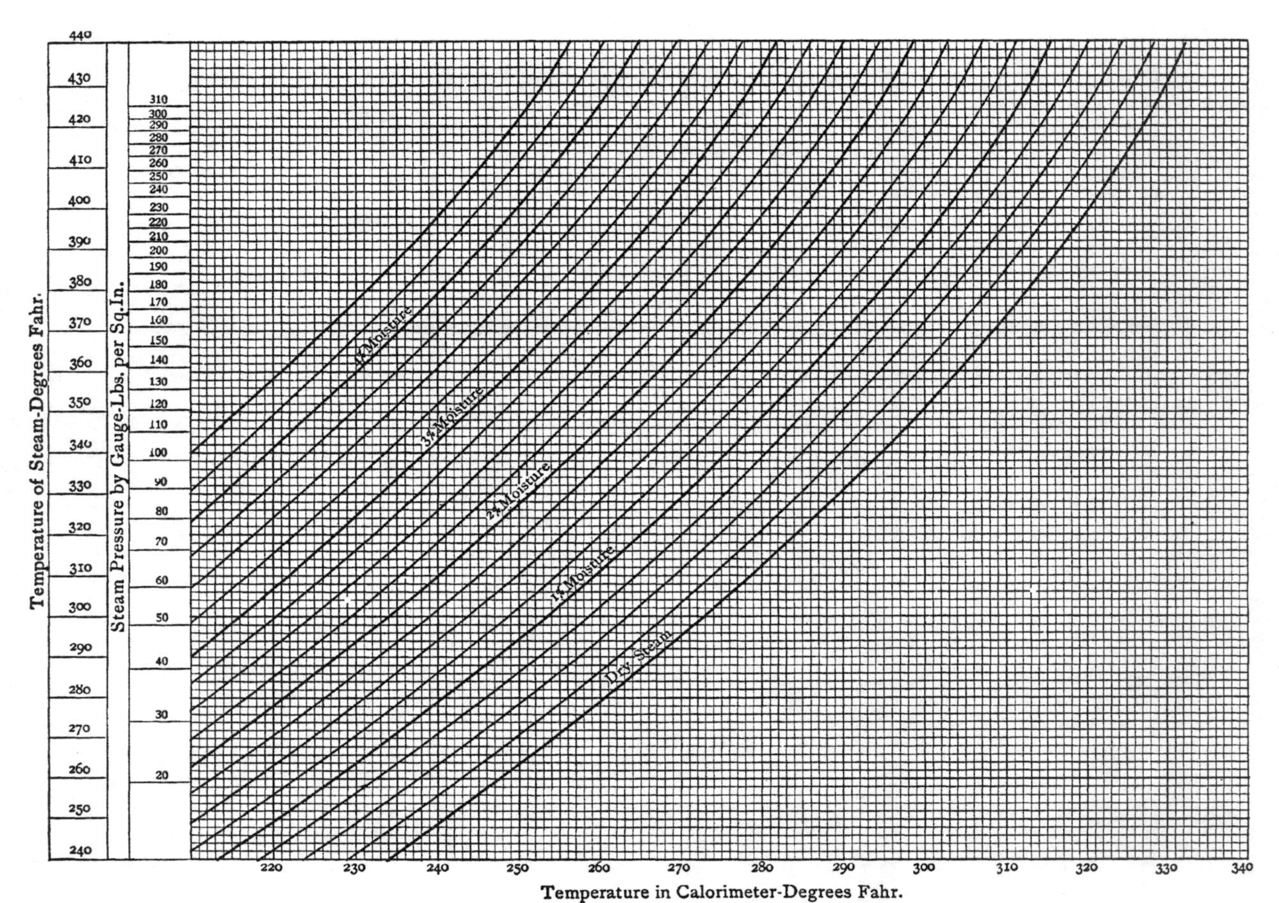

Fig. 5. Graphic Method of Determining Moisture Contained in Steam from Calorimeter Readings

steam at 160 pounds gauge pressure is 852.8 B.t.u., it follows that the one per cent of moisture would have required 8.5 B.t.u. to evaporate it, leaving only 45.4—8.5=36.9 B.t.u. available for superheating; hence, the superheat would be 36.9÷0.47=78.5 degrees, as against 96.6 degrees for dry steam. In a similar manner, the degree of superheat for other percentages of moisture may be determined. The action of the throttling calorimeter is based upon the foregoing facts, as shown below:

Let H = total heat of one pound of steam at boiler pressure,

L = latent heat of steam at boiler pressure,

h = total heat of steam at the reduced pressure after passing the orifice,

t_1 = temperature of saturated steam at the reduced pressure,

t_2 = temperature of steam after expanding through the orifice in the disk,

0.47 = the specific heat of saturated steam at atmospheric pressure,

x = proportion by weight of moisture in steam.

The difference in B.t.u. in a pound of steam at the boiler pressure and after passing the orifice is the heat available for evaporating the moisture content and superheating the steam. Therefore,

$$H-h=xL+0.47\,(t_2-t_1)$$

$$\text{or } x=\frac{H-h-0.47\,(t_2-t_1)}{L} \qquad (6)$$

Almost invariably the lower pressure is taken as that of the atmosphere. Under such conditions, h = 1150.4 and t_1 = 212 degrees. The formula thus becomes:

$$x=\frac{H-1150.4-0.47\,(t_2-212)}{L} \qquad (7)$$

For practical work it is more convenient to dispense with the upper thermometer in the calorimeter and to measure the pressure in the steam main by an accurate steam pressure gauge.

A chart may be used for determining the value of x for approximate work without the necessity for computation. Such a chart is shown in Fig. 5 and its use is as follows: Assume a gauge pressure of 180 pounds and a thermometer reading of 295 degrees. The intersection of the vertical line from the scale of temperatures as shown by the calorimeter thermometer and the horizontal line from the scale of gauge pressures will indicate directly the percentage of moisture in the steam as read from the diagonal scale. In the present instance, this percentage is 1.0.

Sources of Error in the Apparatus—A slight error may arise from the value, 0.47, used as the specific heat of superheated steam at atmospheric pressure. This value, however, is very nearly correct and any error resulting from its use will be negligible.

There is ordinarily a larger source of error due to the fact that the stem of the thermometer is not heated to its full length, to an initial error in the thermometer and to radiation losses.

With an ordinary thermometer immersed in the well to the 100 degrees mark, the error when registering 300 degrees would be about 3 degrees and the true temperature be 303 degrees.*

The steam is evidently losing heat through radiation from the moment it enters the sampling nipple. The heat available for evaporating moisture and superheating

* See Stem Correction, page 64.

steam after it has passed through the orifice into the lower pressure will be diminished by just the amount lost through radiation, and the value of t_2, as shown by the calorimeter thermometer, will, therefore, be lower than if there were no such loss. The method of correcting for the thermometer and radiation error recommended by the Power Test Committee of the American Society of Mechanical Engineers is by referring the readings as found on the boiler trial to a "normal" reading of the thermometer. This normal reading is the reading of the lower calorimeter thermometer for dry saturated steam, and should be determined by attaching the instrument to a horizontal steam pipe in such a way that the sampling nozzle projects upward to near the top of the pipe. there being no perforations in the nozzle and the steam taken only through its open upper end. The test should be made with the steam in a quiescent state and with the steam pressure maintained as nearly as possible at the pressure observed in the main trial, the calorimeter thermometer to be the same as was used on the trial or one exactly similar.

With a normal reading thus obtained for a pressure approximately the same as existed in the trial, the true percentage of moisture in the steam, that is, with the proper correction made for radiation, may be calculated as follows:

Let T denote the normal reading for the conditions existing in the trial. The effect of radiation from the instrument as pointed out will be to lower the temperature of the steam at the lower pressure. Let x_1 represent the proportion of water in the steam which will lower its temperature an amount equal to the loss by radiation. Then,

$$x_1 = \frac{H - h - 0.47\,(T - t_1)}{L}$$

This amount of moisture, x_1, was not in the steam originally but is the result of condensation in the instrument through radiation. Hence, the true amount of moisture in the steam represented by X is the difference between the amount as determined in the trial and that resulting from condensation, or,

$$X = x - x_1$$

$$= \frac{H - h - 0.47\,(t_2 - t_1)}{L} - \frac{H - h - 0.47\,(T - t_1)}{L}$$

$$= \frac{0.47\,(T - t_2)}{L} \qquad (8)$$

As T and t_2 are taken with the same thermometer under the same set of conditions, any error in the reading of the thermometers will be approximately the same for the temperatures T and t_2 and the above method therefore corrects for both the radiation and thermometer errors. The theoretical readings for dry steam, where there are no losses due to radiation, are obtainable from formula (7) by letting $x=0$ and solving for t_2. The difference between the theoretical reading and the normal reading for no moisture will be the thermometer and radiation correction to be applied in order that the correct reading of t_2 may be obtained.

For any calorimeter within the range of its ordinary use, such a thermometer and radiation correction taken from one normal reading is approximately correct for any conditions with the same or a duplicate thermometer.

The percentage of moisture in the steam, corrected for thermometer error and radiation and the correction to be applied to the particular calorimeter used, would be determined as follows: Assume a gauge pressure in the trial to be 180 pounds and the thermometer reading to be 295 degrees. A normal reading, taken in the manner described, gives a value of $T=303$ degrees; then, the percentage of moisture corrected for thermometer error and radiation is,

$$x=\frac{0.47\ (303-295)}{845.0}$$

$$=0.45 \text{ per cent.}$$

The theoretical reading for dry steam will be,

$$0=\frac{1197.7-1150.4-0.47\ (t_2-212)}{845.0}$$

$$t_2=313 \text{ degrees.}$$

The thermometer and radiation correction to be applied to the instrument used, therefore, over the ordinary range of pressure is

Correction = 313 — 303 = 10 degrees

The chart may be used in the determination of the correct reading of moisture percentage and the permanent radiation correction for the instrument used, without computation, as follows: Assume the same trial pressure, feed temperature and normal reading as above. If the normal reading is found to be 303 degrees, the correction for thermometer and radiation will be the theoretical reading for dry steam as found from the chart, less this normal reading, or 10 degrees correction. The correct temperature for the trial in question is, therefore, 305 degrees. The moisture corresponding to this temperature and 180 pounds gauge pressure will be found from the chart to be 0.45 per cent.

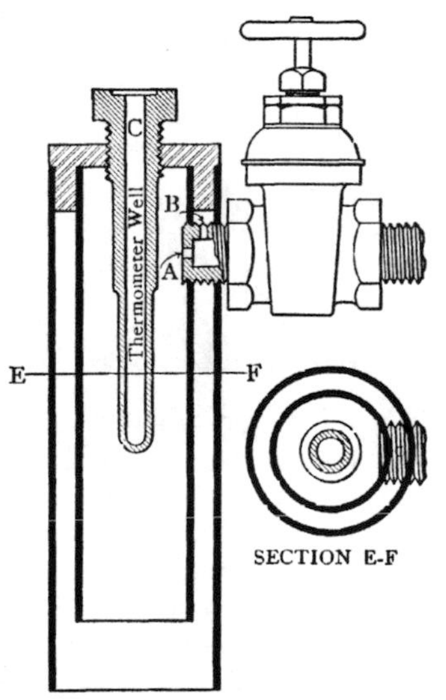

Fig. 6. Compact Throttling Calorimeter

There are many forms of throttling calorimeter, all of which work upon the same principle. The simplest one is probably that shown in Fig. 4. An extremely convenient and compact design is shown in Fig. 6. This calorimeter consists of two concentric metal cylinders screwed to a cap containing a thermometer well. The steam pressure is measured by a gauge placed in the supply pipe or other convenient location. Steam passes through the orifice *A* and expands to atmospheric pressure, its temperature at this pressure being measured by a thermometer placed in the cup *C*. To prevent radiation losses, as far as possible, the annular space between the two cylinders is used as a jacket, steam being supplied to this space through the hole *B*.

The limits of moisture within which the throttling calorimeter will work are, at sea level, from 2.88 per cent at 50 pounds gauge pressure and 7.17 per cent moisture at 250 pounds pressure.

Separating Calorimeter—The separating calorimeter mechanically separates the entrained water from the steam and collects it in a reservoir, where its amount is either indicated by a gauge glass or is drained off and weighed. Fig. 7 shows a calorimeter of this type. The steam passes out of the calorimeter through an orifice of known size, so that its total amount can be calculated or it can be weighed. A gauge is ordinarily provided with this type of calorimeter, which shows the pressure in its inner chamber and the flow of steam for a given period, this latter scale being graduated by trial.

The instrument, like a throttling calorimeter, should be well insulated to prevent losses from radiation.

While theoretically the separating calorimeter is not limited in capacity, it is well in cases where the percentage of moisture present in the steam is known to be high, to attach a throttling calorimeter to its exhaust. This, in effect, is the using of the separating calorimeter as a small separator between the sampling nozzle and the throttling instrument, and is necessary to insure the determination of the full percentage of moisture in the steam. The sum of the percentages shown by the two instruments is the moisture content of the steam.

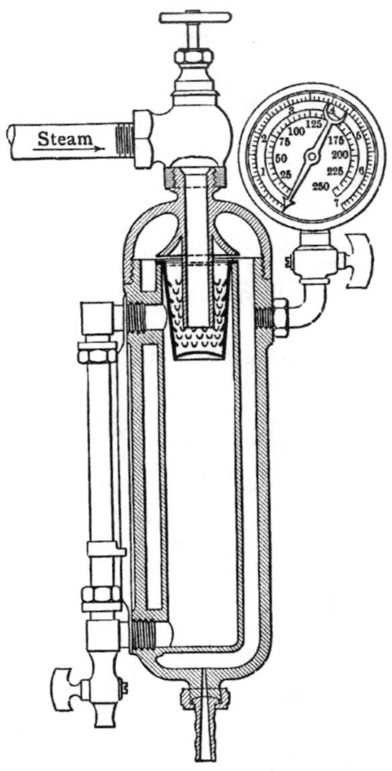

Fig. 7 Separating Calorimeter

The steam passing through a separating calorimeter may be calculated by Napier's formula, the size of the orifice being known. There are objections to such a calculation, however, in that it is difficult to determine accurately the areas of such small orifices. Further, small orifices have a tendency to become partly closed by sediment that may be carried by the steam. The more accurate method of determining the amount of steam passing through the instrument is as follows:

A hose should be attached to the separator outlet leading to a vessel of water on a platform scale graduated to $\frac{1}{100}$ of a pound. The steam outlet should be connected to another vessel of water resting on a second scale. In each case, the weight of each vessel and its contents should be noted. When ready for an observation, the instrument should be blown out thoroughly so that there will be no water within the separator. The separator drip should then be closed and the steam hose inserted into the vessel of water at the same instant. When the separator has accumulated a sufficient quantity of water, the valve of the instrument should be closed and the hose removed from the vessel of water. The separator should be emptied into the vessel on its scale. The final weight of each vessel and its contents is to be noted and the differences between the final and original weights will represent the weight of moisture collected by the separator and the weight of steam from which the moisture has been taken. The proportion of moisture can then be calculated from the following formula:

$$x = \frac{100\,w}{W + w} \qquad (9)$$

Where x = percentage of moisture in steam,

W = weight of steam condensed,

w = weight of moisture as taken out by the separating calorimeter.

Sampling Nozzle—The principal source of error in steam calorimeter determinations is the failure to obtain an average sample of the steam delivered by the boiler, and it is extremely doubtful whether such a sample is ever obtained. The two governing features in obtaining such a sample are the type of sampling nozzle used and its location.

The American Society of Mechanical Engineers recommends a sampling nozzle made of one-half-inch iron pipe closed at the inner end and the interior portion perforated with not less than twenty one-eighth-inch holes equally distributed from end to end and preferably drilled in irregular or spiral rows, with the first hole not less than one-half inch from the wall of the pipe. Many engineers object to the use of a perforated sampling nozzle because it ordinarily indicates a higher percentage of moisture than is actually present in the steam. This is due to the fact that if the perforations come close to the inner surface of the pipe, the moisture, which in many instances clings to this surface, will flow into the calorimeter and cause a large error. Where a perforated nozzle is used, in general it may be said that the perforations should be at least one inch from the inner pipe surface.

A sampling nozzle, open at the inner end and unperforated, undoubtedly gives as accurate a measure as can be obtained of the moisture in the steam passing that end. It would appear that a satisfactory method of obtaining an average sample of the steam would result from the use of an open end unperforated nozzle passing through a stuffing box which would allow the end to be placed at any point across the diameter of the steam pipe.

Incidental to a test of a 15,000-kilowatt steam engine turbine unit, Mr. H. G. Stott and Mr. R. G. S. Pigott, finding no experimental data bearing on the subject of low-pressure steam quality determinations, made a special investigation of the subject and the sampling nozzle illustrated in Fig. 8 was developed. In speaking of sampling nozzles in the determination of the moisture content of low pressure steam, Mr. Pigott says, "the ordinary standard perforated pipe sampler is absolutely worthless in giving a true sample and it is vital that the sample be abstracted from the main without changing its direction or velocity until it is safely within the sample pipe and entirely isolated from the rest of the steam."

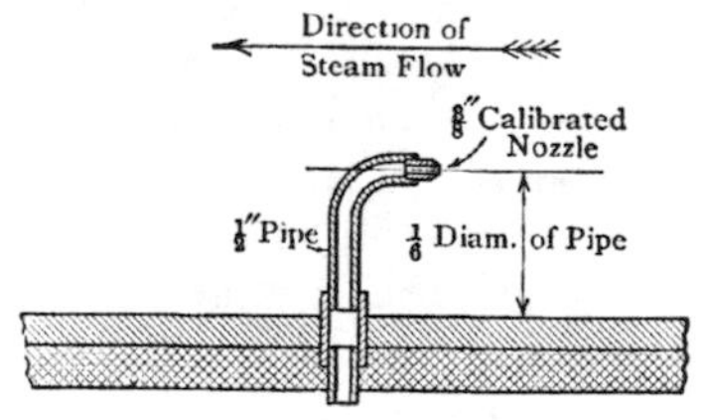

Fig. 8. Stott and Pigott Sampling Nozzle

It would appear that the nozzle illustrated is undoubtedly the best that has been developed for use in the determination of the moisture content of steam, not only in the case of low, but also in high-pressure steam.

Location of Sampling Nozzle—The calorimeter should be located as near as possible to the point from which the steam is taken and the sampling nozzle should be placed in a section of the main pipe near the boiler and where there is no chance of moisture pocketing in the pipe. The American Society of Mechanical Engineers recommends that a sampling nozzle of which a description has been given, should be located in a vertical main, rising from the boiler with its closed end extending nearly

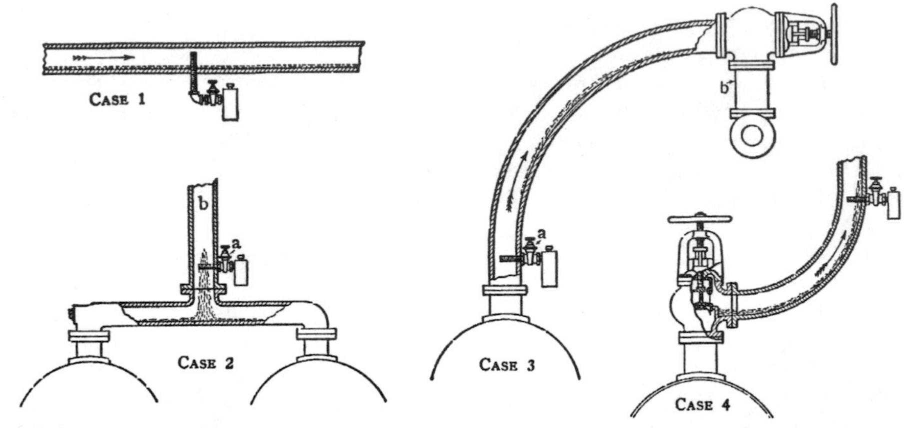

FIG. 9. ILLUSTRATING THE MANNER IN WHICH ERRONEOUS CALORIMETER READINGS MAY BE OBTAINED DUE TO IMPROPER LOCATION OF SAMPLING NOZZLE

Case 1—Horizontal pipe. Water flows at bottom. If perforations in nozzle are too near bottom of pipe, water piles against nozzle, flows into calorimeter and gives false reading. Case 2—If nozzle is located too near junction of two horizontal runs, as at *a*, condensation from vertical pipe, which collects at this point, will be thrown against the nozzle by the velocity of the steam, resulting in a false reading. Nozzle should be located far enough above junction to be removed from water kept in motion by the steam velocity, as at *b*. Case 3—Condensation in bend will be held by velocity of the steam as shown. When velocity is diminished during firing intervals and the like, moisture flows back against nozzle, *a*, and false reading is obtained. A true reading will be obtained at *b* provided condensation is not blown over on nozzle. Case 4—Where non-return valve is placed before a bend, condensation will collect on steam line side and water will be swept by steam velocity against nozzle and false readings result.

across the pipe. Where non-return valves are used, or where there are horizontal connections leading from the boiler to a vertical outlet, water may collect at the lower end of the uptake pipe and be blown upward in a spray which will not be carried away by the steam owing to a lack of velocity. A sample taken from the lower part of this pipe will show a greater amount of moisture than a true sample. With goose-neck connections a small amount of water may collect on the bottom of the pipe near the upper end where the inclination is such that the tendency to flow backward is ordinarily counterbalanced by the flow of steam forward over its surface; but when the velocity momentarily decreases the water flows back to the lower end of the goose-neck and increases the moisture at that point, making it an undesirable location for sampling. In any case, it should be borne in mind that with low velocities the tendency is for drops of entrained water to settle to the bottom of the pipe, and to be temporarily broken up into spray whenever an abrupt bend or other disturbance is met.

Fig. 9 indicates certain locations of sampling nozzles from which erroneous results will be obtained, the reasons being obvious from a study of the cuts.

Before taking any calorimeter reading, steam should be allowed to flow through the instrument freely until it is thoroughly heated. The method of using a throttling calorimeter is evident from the description of the instrument given and the principle upon which it works.

SUPERHEATED STEAM

WHILE, as has been seen, the use of superheated steam is old, but little was known until recently of its properties.

In the preparation and computation of a table giving the properties of superheated steam the important factor is its specific heat. Without this value it is impossible to determine the total heat of superheated steam, that is. the amount of heat that must be applied to bring saturated steam to any given superheated condition.

Regnault, in 1862, determined the specific heat of superheated steam, basing his value, which he gives as 0.48, upon four series of experiments, all of which covered approximately the same temperature range and all at atmospheric pressure. While Regnault's experiments in no way proved that the value of specific heat as given by him was independent of either pressure or temperature, this value was accepted for some forty years and applied to higher pressures and temperatures as well as to those within the range of his experiments.

With the revival of the use of superheated steam in the nineties, the assumption that the specific heat was constant regardless of pressure and temperature was found to be incorrect, and a number of investigators turned their attention to the subject. Grindley and Greissmann's experiments, in 1900, appeared to indicate that the specific heat increased with the temperature and was independent of the pressure. Other investigators. however. proved this untrue. Among those giving their services in this investigation were Lorenz, Linde, Holborn and Henning, Callendar, Carpenter, Thomas, and Knoblauch and Jakob. The experiments of the last two investigators were probably the most laborious and comprehensive and they made special efforts to eliminate the presence of moisture in the steam in observations near the saturation point, an error which unavoidably crept into other investigations.

Table 22* gives, in condensed form, the properties of superheated steam as calculated by Lionel S. Marks and Harvey N. Davis. The values as given by these authorities are considered reliable and are generally accepted in engineering practice.

The continually increasing use of steam at pressures and temperatures above the range covered by these tables and the differences between the steam tables of well-known authors led The American Society of Mechanical Engineers in 1921 to become sponsor for a program of research by engineers and physicists to determine primarily the properties of steam over a wider range of temperatures than had been done previously. The results of these and earlier investigations are to be used in preparing new steam tables which. it is hoped. will be adopted as standard.

In determining the mean specific heat of superheated steam for various temperatures and pressures and from these values the total heat, Messrs. Marks and Davis use the values as determined by Knoblauch and Jakob, modifying their C_p (or specific heat at constant pressure) curves in two respects; the first consists in modifications of the C_p curves at low pressures near the point of saturation because of thermodynamic evidence and because of Regnault's experiments at atmospheric pressure; the second modification is in the C_p curves for high degrees of superheat to follow Holborn and Henning's curve, which they consider authentic.

* See pages 120 to 122.

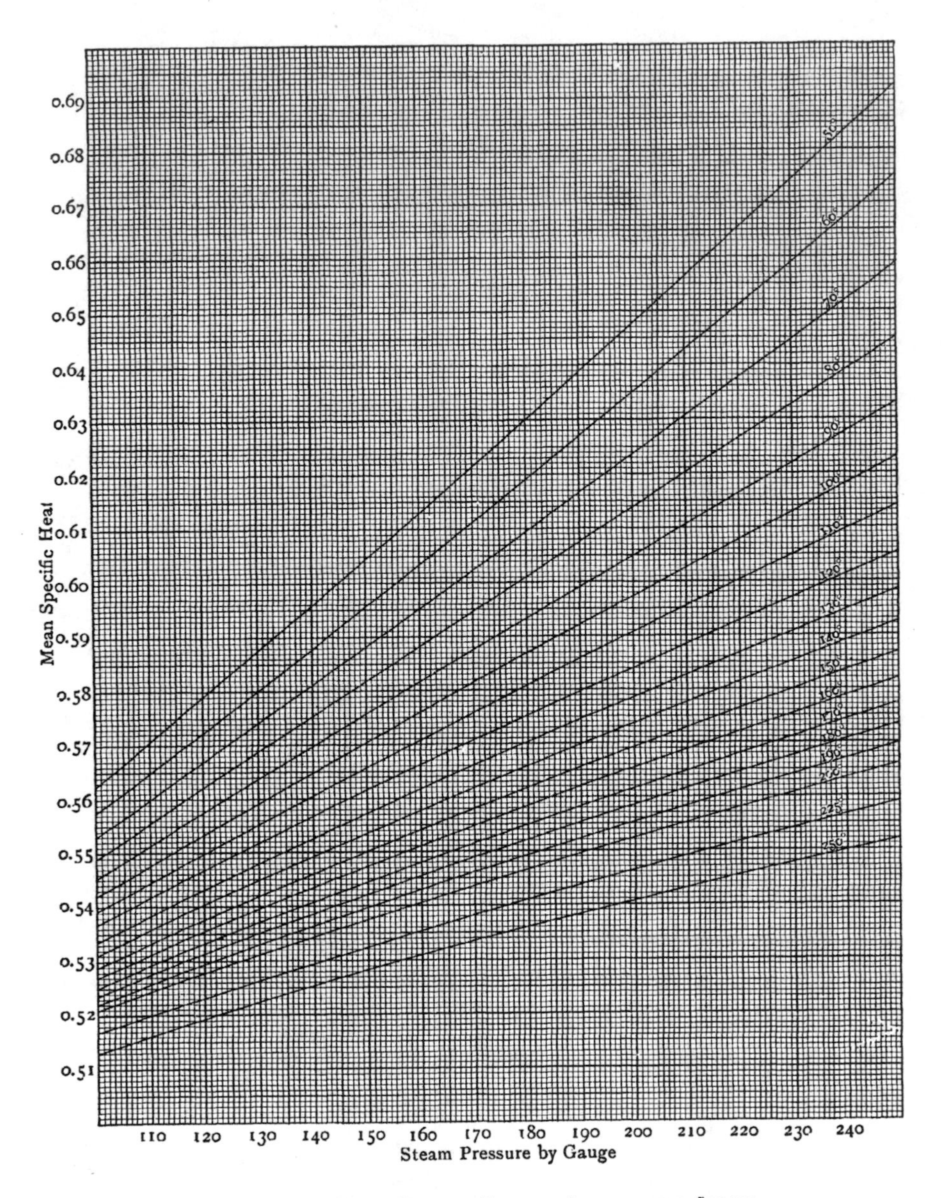

FIG. 10 MEAN SPECIFIC HEAT OF SUPERHEATED STEAM
CURVES REPRESENT DEGREES OF SUPERHEAT
Calculated from Marks and Davis

The values given for total heat at various temperatures may be represented by:

$$H = h + r + C_p (T_{sup} - T_{sat})$$

Where h = heat of liquid at existing pressure,

r = latent heat of evaporation of saturated steam at this pressure,

C_p = mean specific heat at existing pressure and ultimate temperature,

T_{sup} and T_{sat} = temperatures of superheated and of saturated steam.

The mean specific heat must be distinguished from the actual specific heat at any temperature and pressure. The actual specific heat at a given pressure and temperature is that corresponding to a change in temperature of one degree, while the mean specific heat is the average for all temperatures from that of saturated steam up to the ultimate temperature of the superheated steam at the existing pressure.

The temperatures given in the table are the ultimate temperatures of the superheated steam, the degree of superheat being represented by these ultimate temperatures less that of saturated steam at a corresponding pressure.

The specific volumes are based on values as given by Knoblauch, Linde and Klebe. Of all the corrective characteristic equations for superheated steam, that advanced by these authorities is probably the most satisfactory. This is:

$$pv = BT - p(1 + ap)\left(C\left\{\frac{373}{T}\right\}^3 - D\right)$$

Where p = pressure in kilograms per square meter,

v = volume in cubic meters,

T = absolute temperature, degrees Centigrade,

a, B, C and D = constants.

Reduced to English units, the pressures being in pounds per square inch, the volume in cubic feet per pound and the temperature in degrees Fahrenheit, this becomes:

$$pv = 0.5962\,T - p(1 + 0.0014p)\left(\frac{150{,}300{,}000}{T^3} - 0.0833\right)$$

In engineering work it is frequently necessary to know the mean specific heat of superheated steam at a given pressure and temperature. Fig. 10 shows graphically values of this mean specific heat for various temperatures and pressures and has been computed from values of total heat as given in Marks and Davis' steam tables. The calculation involved, expressed as a formula, is:

$$Sp.\,Ht. = \frac{H - h}{T_{sup} - T_{sat}} \qquad (10)$$

Where $Sp.\,Ht.$ = mean specific heat at a given pressure and temperature,

H = total heat of superheated steam at a given temperature and pressure,

h = total heat of saturated steam at the corresponding pressure,

T_{sup} and T_{sat} = temperatures of superheated and saturated steam.

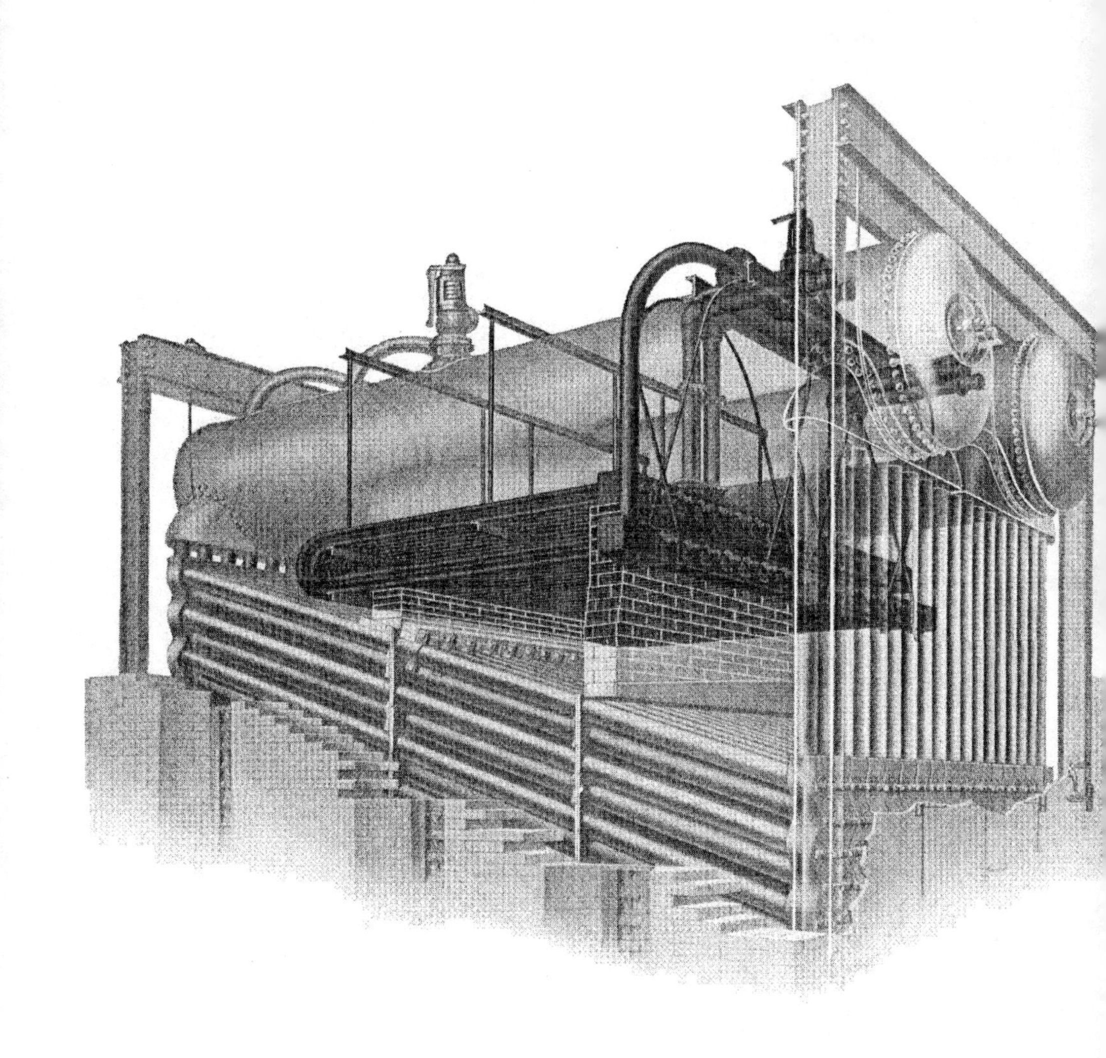

Phantom View of Babcock & Wilcox Superheater as Ordinarily Installed in Babcock & Wilcox Boilers, Showing Location, Method of Support and Method of Making Inlet and Outlet Connections

FACTOR OF EVAPORATION WITH SUPERHEATED STEAM—When superheat is present in steam during a boiler trial and superheated steam tables are available, the method of calculating the factor of evaporation is the same as in the case of saturated steam, namely:

$$Factor = \frac{H-h}{L} \qquad (4)$$

Where H = total heat in superheated steam,
h = sensible heat in the feed water above 32 degrees Fahrenheit,
L = 970.4, the latent heat of evaporation at atmospheric pressure.

ADVANTAGES OF SUPERHEATED STEAM—That there are advantages to be secured through the use of superheated steam is probably most conclusively shown by the fact that superheaters are installed, almost without exception, in the largest and most economical power plants throughout the world. Regardless of any such evidence, however, there is a deep-rooted conviction in the minds of certain engineers that the use of superheated steam will involve operating difficulties which, taken in connection with the added first cost, will more than offset any fuel saving. There are, of course, conditions under which the installation of superheaters would be in no way advisable. While such instances are perhaps rare, nevertheless, when a superheater installation is contemplated, it must be considered in all of its phases. The actual saving possible by the use of superheated steam must be balanced against such factors as the initial cost and upkeep of the superheater, the efficiency of the design of superheater to be installed, the nature of the service of the plant in question, the design of the prime movers, a consideration of pipe and fittings, and the like.

No general statement of the saving possible may be made since this may vary widely with a number of the factors above enumerated.

The logical method of approaching the subject would appear to be, first, a consideration of the saving in fuel possible through the use of superheated steam regardless of other factors; and, second, a consideration of the bearing of such factors on the advisability of a superheater installation.

In considering the saving possible by the use of superheated steam, it is too often assumed that there is only a saving in the prime movers, a saving which is at least partially offset by an increase in the fuel consumption of the boilers generating steam. This misconception is due to the fact that the fuel consumption of the boiler is only considered in connection with a definite weight of steam. It is true that where such a definite weight is to be superheated, an added amount of fuel must be burned. With a properly designed superheater, where the combined efficiency of the boiler and superheater will be at least as high as of a boiler alone, the approximate increase in coal consumption for producing a given weight of steam will be as follows:

Superheat Degrees	*Added Fuel Per Cent*	*Superheat Degrees*	*Added Fuel Per Cent*
25	1.59	100	5.69
50	3.07	150	8.19
75	4.38	200	10.58

These figures represent the added fuel necessary for superheating a definite weight of steam to the number of degrees given. A heat unit basis is standard in the consideration of boiler evaporation and from this standpoint, again providing the efficiency of the boiler and superheater is as high as of the boiler alone, there is no additional fuel required to generate steam containing a definite number of heat units whether such units be due to superheat or saturation. That is, if 6 per cent more fuel is required to generate and superheat to 100 degrees a definite weight of steam, over what would be required to produce the same weight of saturated steam, that steam, when superheated, will contain 6 per cent more heat units above the feed-water temperature than if saturated.

The statement that for an additional number of heat units generated in the furnace of a boiler equipped with a superheater, there will be an equivalent increase in the number of heat units appearing in the superheated steam over saturated, is based, as indicated, upon an equal efficiency for the boiler and superheater as for the boiler in which no superheater is installed. As a matter of fact, the efficiency of a boiler and superheater, where the latter is properly designed and located, will be slightly higher for the same set of furnace conditions than will be the efficiency of a boiler in which no superheater is installed. A superheater, properly placed within the boiler setting in such a way that the products of combustion for generating saturated steam are utilized as well for superheating that steam, will not in any way alter furnace conditions. With a given set of such furnace conditions, for a given amount of coal burned, the fact that additional surface, whether as boiler heating or superheating surface, is placed in such a manner that the gases must sweep over it, will tend to lower the temperature of the exit gases. It is such a lowering of exit-gas temperature that is the ultimate indication of added efficiency. Though the amount of this added efficiency is difficult to determine by test, that there is an increase is unquestionable.

Where a properly designed superheater is installed in a boiler, the boiler heating surface, in the generation of a definite number of heat units, is relieved of a portion of the work which would be required were these heat units delivered in saturated steam. This results either in a reduction in the capacity at which it is necessary to operate the boiler itself, apart from the superheater, in the developing of a definite number of boiler horse-power, with a consequent saving in the apparatus due to a decreased load, or it enables the same number of horse-power to be developed from a smaller number of boilers, with the boiler heating surface doing the same amount of work as if no superheaters were installed. Such a superheater needs practically no attention, is not subject to a large upkeep cost or depreciation, and performs its function without in any way interfering with the operation of the boiler.

Following the course of the steam in the plant, the advantage of the use of superheated steam is next seen in the general absence of water in the pipes. While it is possible for a pipe through which superheated steam is flowing to carry water also, there is usually an entire absence of such water in the piping system, especially where the piping is well covered. The thermal conductivity of superheated steam, that is, its power to receive from or to give out heat to surrounding bodies, is much lower than that of saturated steam and its heat, therefore, will not be transmitted so rapidly to the walls of the pipe as will the heat from saturated steam. When a pipe is carrying saturated steam, assuming no loss in pressure, the amount of heat radiated usually represents an equivalent condensation. Where such a pipe carries superheated steam, again assuming no loss in pressure, the amount of radiation represents only a decrease in the amount of superheat, for condensation cannot take place until the temperature of the steam is lowered to that of the saturated steam at the existing pressure and the temperature of the walls is higher than the temperature of saturated steam where the pipes are well covered and the steam is superheated an ordinary amount. Obviously, therefore, where the degree of superheat is sufficiently high at the boiler, an amount of heat could be radiated far in excess of what is found in well-covered steam lines, and the steam as delivered to the prime movers and auxiliaries would still be dry or superheated.

The loss through drips resulting from such line condensation in the average plant using saturated steam is one which is ordinarily largely underestimated. Such a loss. which is frequently in excess of 5 per cent, can be greatly reduced. if not entirely eliminated, where superheated steam is used.

It is in the prime movers that the advantages of the use of superheated steam are most clearly seen.

In an engine, steam is admitted into a space that has been cooled by the steam exhausted during the previous stroke. The heat necessary to warm the cylinder walls from the temperature to which they have been reduced by the exhaust can be supplied. in the absence of jackets, only by the entering steam. and even where jackets are used a large amount of heat must be supplied in this way. If this steam is saturated, such an adding of heat to the walls at the expense of the heat of the entering steam results in the condensation of a portion. This initial condensation is seldom less than from 20 to 30 per cent of the total weight of steam entering the cylinder. It is obvious that if the steam entering be superheated, it must be reduced to the temperature of saturated steam at a corresponding pressure before any condensation can take place. If the steam be superheated sufficiently to allow a reduction in temperature equivalent to the quantity of heat that must be imparted to the cylinder walls and still remain superheated, it is clear that initial condensation is avoided. In the case of a simple engine, where the range of temperature change is a maximum. the degree of superheat necessary to offset a cylinder condensation of. say, 20 per cent in the case of saturated steam, would be excessive, notwithstanding the lower conductivity of superheated steam. As cylinders are added, however, the range of temperature change between the entering steam and the cylinder walls is decreased and proportionately the degree of superheat necessary to prevent initial condensation.

With saturated steam, the heat utilized in warming the cylinder walls to the temperature of the entering steam is mainly lost, in so far as its ability to perform work in the cylinder is concerned. It is true that as expansion progresses a portion of the steam that has been so condensed will re-evaporate, though the greater portion of such re-evaporation takes place at the point of exhaust. The latent heat of the water given up in re-evaporation is utilized in changing the condition of the working fluid and is not available for work. Furthermore, a portion of the saturated steam condenses during adiabatic expansion, this condensation increasing as expansion progresses. In high-speed engines using saturated steam the condensation due to adiabatic expansion is just about offset by the re-evaporation of the initial condensation. Since superheated steam cannot condense until its temperature has been reduced to that of saturated steam at a corresponding pressure, not only is initial condensation prevented by its use but also such

The Webster Station of the Galveston Houston Electric Railway, Where Steam Is Provided by Three 520 Horse-power Babcock & Wilcox Boilers Equipped with Babcock & Wilcox Superheaters

condensation as would occur during expansion. When superheated sufficiently, the steam delivered by the exhaust will be dry. The number of heat units lost in overcoming condensation effects will be the same whether superheated or saturated steam is the working fluid, but in the case of saturated steam the water of condensation has no power to do work, while superheated steam, even after having given up the equivalent number of heat units to correspond to the condensation in the case of saturated steam, will still have the power of expansion and the ability to do work.

With superheated steam, therefore, a larger proportion of the heat is utilized in the developing of power than with saturated steam, where a large amount is lost in changing the condition of the working fluid. This results in a lower heat consumption in an engine using superheated steam, that is, the expenditure of a less number of heat units in the developing of one indicated horse-power. The "heat consumption" furnishes the true basis for the comparison of efficiencies of different engines, just as a comparison of boiler results is based on an evaporation from and at 212 degrees.

The water consumption of an engine in pounds per indicated horse-power is in no sense a true indication of its efficiency. The initial pressures and corresponding temperatures in two different cases may vary widely and thus through the resulting difference in the temperature of the exhaust affect the temperature of condensed steam returned to the boiler.

The lower the heat consumption of an engine per indicated horse-power, the higher its economy. Since the use of superheated steam decreases this heat consumption, as has been shown, the number of heat units to be imparted in generating steam is reduced, this in turn leading to the lowering of the amount of fuel which must be burned.

No accurate statement can be made as to the saving possible through the use of superheated steam with reciprocating engines. In highly economical plants, where the water consumption per indicated horse-power is low, the gain would be less than would result from the use of superheated steam in less economical plants where the water consumption is higher. Broadly speaking, it may vary from 3 to 5 per cent for 100 degrees of superheat in large and economical plants using engines in which there is a high ratio of expansion, to from 10 to 25 per cent for 100 degrees of superheat where less economical steam motors are in service.

Experience has unquestionably shown that the use of superheated steam with turbines leads to an appreciable gain in economy. This fact is so well established that engineering practice does not countenance the installation of turbines for use with saturated steam.

Where saturated steam is used with turbines, even when it is dry upon entering the first stage, the work done in expanding the steam through progressive stages causes the condensation of a sufficient amount of steam to give trouble through the presence of water in the low-pressure stages. When the entering steam is superheated, the amount of water in low-pressure stages of the turbine is reduced to a point where no trouble will be caused.

If the saturated steam entering a turbine contains moisture, the effect of such moisture is an appreciable lowering of the turbine's economy. It is stated on good authority that one per cent of moisture contained in the steam will reduce the economy approximately 2 per cent.

The water rate of a large economical steam turbine with superheated steam is reduced about one per cent for every 12 degrees of superheat up to 200 degrees of superheat. To superheat one pound of steam 12 degrees requires about 7 B.t.u. and if 1050 B.t.u. are required at the boiler to evaporate one pound of water into saturated steam from the temperature of the feed water, the heat required for the superheated steam will be 1057 B.t.u. One per cent of saving, therefore, in the water consumption corresponds to a net saving of about one-third of one per cent in the coal consumption. On this basis 100 degrees of superheat with an economical steam turbine will result in somewhat over 3 per cent of saving in the coal for equal boiler efficiencies. As a boiler with a properly designed superheater placed within the setting is more economical for a given capacity than a boiler without a superheater, the minimum gain in the coal consumption will be, say, 4 or 5 per cent as compared to a plant with the same boilers without superheaters.

The above estimates are on the basis of a thoroughly dry saturated steam or steam just at the point of being superheated or containing a few degrees of superheat. If the saturated steam is moist, the saving due to superheat is more, and ordinarily the gain in economy due to superheated steam for equal boiler efficiencies, as compared with commercially dry steam, is, say, 5 per cent for each 100 degrees of superheat.

Aside from any thermodynamic gain through the use of superheated steam with turbines, there is an objection to the use of saturated steam from the standpoint of turbine construction. If saturated steam is used, the erosion of the turbine buckets by water carried in the steam may become a serious factor, while with superheated steam this is negligible.

For any type of steam motor it may be broadly stated that the lower its economy, the greater the saving that will be effected through the use of superheat.

The figures that have been given refer to the possible saving through the use of superheated steam in the prime movers alone. Where the auxiliaries are of a design that can properly handle superheated steam there will be a larger percentage of gain than in the prime movers. This is due to the fact that it is the auxiliaries in a plant that ordinarily show the lowest economy and the lower the efficiency of the steam apparatus, the greater the saving possible.

An example from actual practice will perhaps best illustrate and emphasize the foregoing facts. In October, 1909, a series of comparable tests was conducted by The Babcock & Wilcox Company on the steam yacht "Idalia" to determine the steam consumption both with saturated and superheated steam of the main engine on that yacht, including as well the feed pump, circulating pump and air pump. These tests are more representative than are most tests of like character in that the saving in the steam consumption of the auxiliaries, which were much more wasteful than the main engine, formed an important factor. A résumé of these tests was published in the Journal of the Society of Naval Engineers, November, 1909.

The main engine of the "Idalia" is four cylinder, triple expansion, $11\frac{1}{2} \times 19 \times (2)\ 22\frac{11}{16} \times 18$ inches stroke. Steam is supplied by a Babcock & Wilcox marine boiler having 2500 square feet of boiler-heating surface, 340 square feet of superheating surface and 65 square feet of grate surface.

The auxiliaries consist of a feed pump $6 \times 4 \times 6$ inches, an independent air pump $6 \times 12 \times 8$ inches, and a centrifugal pump driven by a reciprocating engine

$5\frac{7}{16} \times 5$ inches. Under ordinary operating conditions the superheat existing is about 100 degrees Fahrenheit.

Tests were made with various degrees of superheat, the amount being varied by by-passing the gases, and in the tests with the lower amounts of superheat by passing a portion of the steam from the boiler to the steam main without passing it through the superheater. Steam temperature readings were taken at the engine throttle. In the tests with saturated steam, the superheater was completely cut out of the system. Careful calorimeter measurements were taken, showing that the saturated steam delivered to the superheater was dry.

The weight of steam used was determined from the weight of the condensed steam discharged from the surface condenser, the water being pumped from the hot well into a tank mounted on platform scales. The same indicators, thermometers and gauges were used in all the tests so that the results are directly comparable. The indicators used were of the outside spring type so that there was no effect of the temperature of the steam. All tests were of sufficient duration to show a uniformity of results by hours. A summary of the results secured is given in Table 23, which shows the water rate per indicated horse-power and the heat consumption. The latter figures are computed on the basis of the heat imparted to the steam above the actual temperature of the feed water, and. as stated, these are the figures that are directly comparable.

The table shows that the saving in steam consumption with 105 degrees of superheat was 15.3 per cent and in heat consumption about 10 per cent. This may be safely stated to be a conservative representation of the saving that may be accomplished by the use of superheated steam in a plant as a whole, where superheated

TABLE 23

RESULTS OF "IDALIA" TESTS

Date 1909	Oct. 11	Oct. 14	Oct. 14	Oct. 12	Oct. 13
Degrees of Superheat, Fahrenheit	0	57	88	96	105
Pressures, pounds per square inch above Atmospheric Pressure — Throttle	190	196	201	198	203
Pressures, pounds per square inch above Atmospheric Pressure — First Receiver	68.4	66.0	64.3	61.9	63.0
Pressures, pounds per square inch above Atmospheric Pressure — Second Receiver	9.7	9.2	8.7	7.8	8.4
Vacuum, inches	25.5	25.9	25.9	25.4	25.2
Temperature, Degrees Fahrenheit — Feed	201	206	205	202	200
Temperature, Degrees Fahrenheit — Hot Well	116	109.5	115	111.5	111
Revolutions per minute — Air Pump	57	56	53	54	45
Revolutions per minute — Circulating Pump	196	198	196	198	197
Revolutions per minute — Main Engine	194.3	191.5	195.1	191.5	193.1
Indicated Horse-power, Main Engine	512.3	495.2	521.1	498.3	502.2
Water per hour, total pounds	9397	8430	8234	7902	7790
Water per indicated Horse-power, pounds	18.3	17.0	15.8	15.8	15.5
B.t.u. per minute per indicated Horse-power	314	300	284	286	283
Percentage Saving of Steam	..	7.1	13.7	13.7	15.3
Percentage Saving of Fuel (computed)	..	4.4	9.5	8.9	9.9

steam is furnished not only to the main engine but also to the auxiliaries. The figures may be taken as conservative for the reason that in addition to the saving shown in the table there would be, in an ordinary plant, a saving much greater than is generally realized in the drips, where the loss with saturated steam is greatly in excess of that with superheated steam.

THE WINDSOR STATION OF THE BEACH BOTTOM POWER COMPANY, OWNED JOINTLY BY THE WEST PENN POWER COMPANY, OPERATING 11,616 HORSE-POWER OF BABCOCK & WILCOX BOILERS EQUIPPED WITH BABCOCK & WILCOX ECONOMIZERS AND SUPERHEATERS, AND BY THE OHIO POWER COMPANY, OPERATING 20,208 HORSE-POWER OF BABCOCK & WILCOX BOILERS EQUIPPED WITH BABCOCK & WILCOX SUPERHEATERS: THIS VIEW SHOWS PART OF THE WEST PENN INSTALLATION

THE PROPERTIES OF AIR AND GASES

ATMOSPHERIC air is a mechanical mixture and not a chemical compound. Its constituents are oxygen, nitrogen, and small amounts of carbon dioxide, water vapor, argon and other inert gases. In combustion work, the gases other than carbon dioxide are included in the nitrogen content, and while there is some slight difference in the proportion of oxygen and nitrogen as given by different authorities, the generally accepted values are:

By volume: oxygen 20.91 per cent. nitrogen 79.09 per cent.

By weight: oxygen 23.15 per cent, nitrogen 76.85 per cent.

All gases have a characteristic equation based upon that of a perfect gas, from which, with three values known, the fourth may be computed. This equation is:

$$PV = RT \qquad (11)$$

Where P = absolute pressure in pounds per square foot,

V = volume per pound in cubic feet,

T = absolute temperature, degrees Fahrenheit,

R = a constant, varying with the gas and derived from the relations existing between the volume, pressure and temperature of the gas in question.

Air is ordinarily considered the standard to which other gases are referred, and for air the value of R is 53.33. The values of R for gases other than air are given in Table 25.

The density of a gas other than air is the weight of unit volume of the gas divided by the weight of an equal volume of pure dry air. The weight per cubic

TABLE 24

VOLUME AND WEIGHT OF AIR

AT ATMOSPHERIC PRESSURE

Temperature Degrees Fahrenheit	Volume One Pound Cubic Foot	Weight per Cubic Foot Pound	Temperature Degrees Fahrenheit	Volume One Pound Cubic Foot	Weight per Cubic Foot Pound	Temperature Degrees Fahrenheit	Volume One Pound Cubic Foot	Weight per Cubic Foot Pound
32	12.390	.080710	160	15.615	.064041	340	20.151	.049625
50	12.843	.077863	170	15.867	.063024	360	20.655	.048414
55	12.969	.077107	180	16.119	.062039	380	21.159	.047261
60	13.095	.076365	190	16.371	.061084	400	21.663	.046162
65	13.221	.075637	200	16.623	060158	425	22.293	.044857
70	13.347	.074923	210	16.875	.059259	450	22.923	.043624
75	13.473	.074223	212	16.925	.059084	475	23.554	.042456
80	13.599	.073535	220	17.127	058388	500	24.184	.041350
85	13.725	.072860	230	17.379	.057541	525	24.814	.040300
90	13.851	.072197	240	17.631	.056718	550	25.444	.039302
95	13.977	.071546	250	17.883	.055919	575	26.074	.038352
100	14.103	.070907	260	18.135	055142	600	26.704	.037448
110	14.355	.069662	270	18.387	.054386	650	27.964	.035760
120	14.607	.068460	280	18.639	053651	700	29.224	.034219
130	14.859	.067299	290	18.891	.052935	750	30.484	.032804
140	15.111	.066177	300	19.143	.052238	800	31.744	.031502
150	15.363	.065092	320	19.647	050898	850	33.004	.030299

The Cannon Street Station of the New Bedford Gas & Edison Light Company Where 20,800 Horse-power of Babcock & Wilcox Boilers Are Operated

foot of a gas is thus the weight per cubic foot of air times the density, air and gas being under the same conditions of pressure and temperature. The volume of a gas is the reciprocal of its weight. It is perhaps more convenient to compute weight and volumetric data of gases from their relative densities and a table of weights and volumes of air than from the characteristic equation.

Since the volume, and hence the weight, of a gas is a function of both temperature and pressure, it is necessary, in order that there may be a suitable basis for comparison, that all volumes be referred to some standard set of conditions. Such conditions, for gases now ordinarily accepted as standard, are a pressure of 14.6963 pounds per square inch (2116.27 pounds per square foot), and a temperature of 32 degrees Fahrenheit.

Table 24 gives the volume and weight of air at atmospheric pressure and varying temperatures.

Table 25 gives the relative densities of the gases other than air encountered in combustion work, together with the weights and volumes under standard conditions. With these values and those of Table 24, the volume and weight of these gases at temperatures other than 32 degrees may be readily computed.

TABLE 25

DENSITY, WEIGHT AND VOLUME OF GASES

AT ATMOSPHERIC PRESSURE AND 32 DEGREES FAHRENHEIT

Substance	Molecular Symbol	Relative Density		Weight per Cubic Foot Pound	Cubic Feet per Pound	Value of Constant R in PV=RT
		Air=1	Hydrogen=1*			
Air		1.0000	....	.08071	12.390	53.33
Oxygen	O_2	1.1053	16	.08921	11.209	48.24
Hydrogen	H_2	0.0696	1	.00562	177.936	765.8
Nitrogen	N_2	0.9673	14	.07807	12.809	55.13
Carbon Monoxide	CO	0.9672	14	.07806	12.811	55.15
Carbon Dioxide	CO_2	1.5291	22	.12341	8.103	34.88
Methane	CH_4	0.5576	8	.04500	22.222	95.64
Acetylene	C_2H_2	0.9200	13	.07425	13.468	57.97
Ethylene	C_2H_4	0.9674	14	.07808	12.807	55.12
Ethane	C_2H_6	1.0494	15	.08470	11.806	50.81
Sulphur Dioxide	SO_2	2.2639	32	.18272	5.473	23.56
Sulphur	S_2	....	....	145	.0069	....
Carbon † ‡	..	.	.	125	0080	

* Based on approximate molecular weights.

† Solid.

‡ If carbon can be conceived to exist as a gas under standard conditions its relative density would be 0.820, its weight per cubic foot .0668 pound, and its volume 14.97 cubic feet per pound.

The specific heat of air at constant pressure varies with its temperature. A number of determinations of this value have been made, certain of those accepted as most authentic being given in Table 26.

Because of the discrepancy in the values as given by different authorities, it is customary to use a constant value of 0.24 for the specific heat of air at constant pressure regardless of temperature. This value finds its most common use in combustion work in the computation of the amount of heat lost in the dry chimney gases* leaving a boiler or economizer. Since such loss is based on dry gases, the error arising from the arbitrarily assumed constant value of the specific heat of air, particularly at moderate temperatures, is probably within the limits of error of boiler testing as a whole. On the other hand, in problems involving the total weight of gas, including moisture, the variation in specific heat with the composition of the gas as well as with its temperature may be appreciable, and where accurate results are desired the specific heat must be computed for each fuel and set of combustion

TABLE 26

SPECIFIC HEAT OF AIR AT CONSTANT PRESSURE AND VARIOUS TEMPERATURES

Temperature Range		Specific Heat	Authority
Degrees Centigrade	Degrees Fahrenheit		
-30- 10	-22- 50	0.2377	Regnault
0-100	32- 212	0.2374	Regnault
0-200	32- 392	0.2375	Regnault
20-440	68- 824	0.2366	Holborn and Curtis
20-630	68-1166	0.2429	Holborn and Curtis
20-800	68-1472	0.2430	Holborn and Curtis
0-200	32- 392	0.2389	Wiedemann

conditions. Problems in which the variation in specific heat with the composition and temperature of the gas are of primary importance include those of theoretical furnace temperature, superheater and economizer performance, and waste heat boiler performance.

Over a given temperature range, the specific heat of any gas is primarily affected by the moisture and hydrogen content of the fuel burned to produce the gas. These constituents appear in the products of combustion as superheated steam, the specific heat of which is high. As an example of this effect on the specific heat of the products of combustion with different fuels, assume two industrial furnaces with which waste heat boilers are to be set, in one of which coal is burned and in the other oil is the fuel. Assume, further, that the gases leave the furnace or enter the boilers at 1200 degrees in each instance, and that the boilers are so designed as to cool these gases to 450 degrees. In computing the heat available for absorption, the mean specific heat of the gases then must be taken for the temperature range 1200-450 degrees, and we have:

Fuel	*Temperature Range*	*H_2 Per Cent*	*CO_2 Per Cent*	*Excess Air Per Cent*	*Mean Specific Heat*
Coal	1200 — 450	5.0	13.06	40	25.736
Oil	1200 — 450	12.7	13.80	11	26.738

From these figures it is clear that for a given weight of gas cooled from 1200 to 450 degrees, the amount of heat given up in the case of oil fuel burned in the

*See page 172.

primary furnace would be approximately 4 per cent greater than in the case of coal, and this difference would be still greater if a gaseous fuel high in hydrocarbons, such as by-product coke oven gas, were assumed instead of oil.

The general formula for the specific heat of any gas at constant pressure may be expressed:

$$C_p = A + Bt + Ct^2 + Dt^3 \qquad (12)$$

The mean specific heat over a temperature range $t_1 - t_2$ is then

$$C_p = \int_{t_1}^{t_2} \frac{A + Bt + Ct^2 + Dt^3}{(t_2 - t_1)} dt \qquad (13)$$

or by integration

$$C_{p_{t_1 - t_2}} = A + \frac{B}{2}(t_2 + t_1) + \frac{C}{3}\left[(t_2 + t_1)^2 - t_2 t_1\right]$$

$$+ \frac{D}{4}(t_2 + t_1)\left[(t_2 + t_1)^2 - 2 t_2 t_1\right] \qquad (14)$$

Where the lower limit of the temperature range, t_1, is zero, the mean specific heat between zero and t_2, from formula *(14)*, becomes

$$C_{p_{0 - t_2}} = A + \frac{B}{2} t_2 + \frac{C}{3} t_2^2 + \frac{D}{4} t_2^3 \qquad (15)$$

Distinction was made in the chapter on " Heat and Its Measurement " between specific heat at constant volume and specific heat at constant pressure. Except in the case of water vapor, the variation with pressure in the specific heat of the gases ordinarily encountered in combustion work is negligible. In the case of water vapor, where it is necessary to deal with any considerable range of pressures, this variation would be an appreciable factor, but in the commercial gases involved in combustion, the partial pressure exerted by water vapor, either in gases before combustion or in the exhaust gases, is rarely over one pound absolute. With such a limited pressure range and in view of the fact that the water vapor content of the gases is small, the effect of such variation in pressure on the specific heat of the gas as a whole may be neglected.

The range of pressure in the gases encountered in boiler work is so limited, varying from that at which the commercial gases are introduced into the furnace for combustion to the suction under which they are drawn over the boiler heating surfaces, that in the computation of combustion data the gases may be safely assumed to be at a constant pressure. The specific heat at constant pressure is the specific heat which should be used, and any results based on the assumption of a constant pressure of the gases as a whole, and in which the variation in the specific heat of the water vapor content with change of pressure is neglected, will be well within the limits of accuracy of practically all combustion data computation.

The values of the constants A, B and C of formula *(12)* for the instantaneous specific heat of the gases commonly encountered in combustion work are given in Table 27. The values of $B/2$, $C/3$ and $D/4$ for use in the integrated formula for determining the mean specific heat, may readily be computed from the values of this table.

The values of the constants for carbon dioxide below 2200 degrees are those of Holborn and Henning, while above this temperature a modification of their formula is used. The values for carbon monoxide, nitrogen and hydrogen are also from their investigations, while those for oxygen are a modification of the values as determined by Langen and Pier to make them come into agreement with other determinations. The values for water vapor are based on Marks and Davis' steam tables. The specific heat of methane and ethylene has not been as definitely determined as that of the other gases, but the values of the constants in the formula for these gases as given in Table 27 are those ordinarily accepted as being authentic.

TABLE 27

SPECIFIC HEAT CONSTANTS

Gas	A	B	C	D
Oxygen	0.2154	.000019	. . . .	. . . . .
Nitrogen	0.2343	.000021	. . . .	. . . . .
Carbon Monoxide	0.2343	.000021	. . . . .	. . . . .
Carbon Dioxide	0.1983	835×10^{-7}	-16.7×10^{-9}	. . . . .
Carbon Dioxide*	0.1991	873×10^{-7}	-23.4×10^{-9}	0.22×10^{-11}
Water Vapor	0.4541	32×10^{-7}	2825×10^{-11}	. . . . .
Methane	0.481	.00056	. . . . .	. . . . .
Ethylene	0.335	.00042	. . . .	. . . . .

* For temperatures above 2200 degrees Fahrenheit.

The mean specific heat at constant pressure and at ordinary temperatures of the commoner gases is given in Table 28.

TABLE 28

MEAN SPECIFIC HEATS AT CONSTANT PRESSURE AND ORDINARY TEMPERATURES

	Molecular Symbol	Mean Specific Heat	
		0—60 Degrees Fahrenheit	0—600 Degrees Fahrenheit
Air	. . .	.2381	.2484
Oxygen	O_2	.2160	.2211
Hydrogen	H_2	3.3850	3.4750
Nitrogen	N_2	.2349	.2406
Carbon Monoxide	CO	.2349	.2406
Carbon Dioxide	CO_2	.2008	.2214
Water Vapor	H_2O	.4542	.4586
Methane*	CH_4	.498	.649
Ethylene†	C_2H_4	.348	.461
Sulphur Dioxide	SO_2	.1544	. . .

* Methane—$Cp_{0\text{-}t} = 0.481 + 0.00028\,t$.

† Ethylene—$Cp_{0\text{-}t} = 0.335 + 0.00021\,t$.

The specific heat of a gaseous mixture is found by multiplying the percentages by weight of the constituent gases by the specific heat of the individual constituents and dividing the sum of the products by 100. To illustrate, assume oil so burned as to give a weight of products of combustion per pound of oil, and the corresponding percentages by weight, as follows:

	Weight of Products per Pound of Oil Burned Pounds	*Percentage by Weight of Products*
CO_2	3.088	18.239
O_2	0.357	2.108
N_2	12.343	72.902
H_2O	1.143	6.751

Assume that this gas is cooled in passing through an economizer from 600 to 300 degrees Fahrenheit. The mean specific heat over this temperature range is computed in the following manner:

$$CO_2 \quad C_{300-600} = 0.1983 + 417.5 \times 10^{-7} (600 + 300) - 5.567 \times 10^{-9} [(600 + 300)^2 - (600 \times 300)]$$

$$O_2 \quad C_{300-600} = 0.2154 + 0.0000095 (600 + 300)$$

$$N_2 \quad C_{300-600} = 0.2343 + 0.0000105 (600 + 300)$$

$$H_2O \quad C_{300-600} = 0.4541 + 16 \times 10^{-7} (600 + 300) + 942 \times 10^{-11} [(600 + 300)^2 - (600 \times 300)]$$

or over the temperature range assumed the specific heats of the various constituents are:

$$CO_2 = 0.2324$$
$$O_2 = 0.2240$$
$$N_2 = 0.2438$$
$$H_2O = 0.4614$$

The specific heat of the gas as a whole is then:

CO_2	$18.239 \times 0.2324 =$	4.239
O_2	$2.108 \times 0.2240 =$	0.472
N_2	$72.902 \times 0.2438 =$	17.774
H_2O	$6.751 \times 0.4614 =$	3.114
		25.599

and the mean specific heat is:

$$25.599 \div 100 = 0.2560$$

In the fuel chosen for the example, oil was purposely assumed as being high in hydrogen content, to indicate that in economizer practice, even where the temperatures are low, the assumption of a specific heat of flue gases of 0.24 may lead to an error of at least 6.5 per cent.

Air may carry an appreciable amount of water vapor, which is frequently as high as 3 per cent of the total weight. This fact is of importance in problems relating to heating, drying and the compressing of air. Table 29 gives the amount of vapor required to saturate air at different temperatures. its weight. expansive force. etc.. and gives sufficient data for the solution of most of the problems of this nature that may arise.

TABLE 29

WEIGHTS OF AIR, VAPOR OF WATER, AND SATURATED MIXTURES OF AIR AND VAPOR AT DIFFERENT TEMPERATURES, UNDER THE ORDINARY ATMOSPHERIC PRESSURE OF 29.921 INCHES OF MERCURY

Temperature Degrees Fahrenheit	Volume of Dry Air at Different Tempera- tures, the Volume at 32 Degrees being 1.000	Weight of Cubic Foot of Dry Air at the Different Tempera- tures Pound	Elastic Force of Vapor in Inches of Mercury (Regnault)	Mixtures of Air Saturated with Vapor						Cubic Feet of Vapor from One Pound of Water at its own Pressure in Column 4
				Elastic Force of the Air in the Mixture of Air and Vapor in Inches of Mercury	Weight of Cubic Foot of the Mixture of Air and Vapor			Weight of Vapor Mixed with One Pound of Air in Pounds	Weight of Dry Air Mixed with One Pound of Vapor in Pounds	
					Weight of the Air in Pound	Weight of the Vapor in Pound	Total Weight of Mixture in Pound			
1	2	3	4	5	6	7	8	9	10	11
0	.935	.0864	.044	29.877	.0863	.000079	.086379	.00092	1092.4	. . .
12	960	.0842	.074	29.849	.0840	000130	.084130	00155	646.1	. . .
22	.980	.0824	.118	29.803	.0821	.000202	.082302	00245	406.4	. . .
32	1.000	.0807	.181	29.740	.0802	.000304	.080504	.00379	263.81	3289
42	1.020	0791	.267	29.654	0784	000440	.078840	00561	178.18	2252
52	1.041	.0776	.388	29.533	.0766	.000627	.077227	.00810	122.17	1595
62	1.061	.0761	.556	29.365	.0747	000881	.075581	.01179	84.79	1135
72	1.082	.0747	.785	29.136	.0727	.001221	.073921	.01680	59.54	819
82	1.102	.0733	1.092	28.829	.0706	.001667	.072267	.02361	42.35	600
92	1.122	.0720	1.501	28.420	.0684	.002250	.070717	.03289	30.40	444
102	1.143	.0707	2.036	27.885	.0659	.002997	.068897	.04547	21.98	334
112	1.163	.0694	2.731	27.190	.0631	.003946	.067046	.06253	15.99	253
122	1.184	.0682	3.621	26.300	.0599	.005142	.065042	.08584	11.65	194
132	1.204	.0671	4.752	25.169	.0564	.006639	.063039	.11771	8.49	151
142	1.224	.0660	6.165	23.756	.0524	008473	.060873	16170	6.18	118
152	1.245	.0649	7.930	21.991	.0477	.010716	.058416	.22465	4.45	93.3
162	1.265	.0638	10.099	19.822	.0423	013415	.055715	.31713	3.15	74.5
172	1.285	.0628	12.758	17.163	.0360	.016682	.052682	.46338	2.16	59.2
182	1.306	.0618	15.960	13.961	.0288	.020536	.049336	.71300	1.402	48.6
192	1.326	.0609	19.828	10.093	.0205	025142	.045642	1.22643	.815	39.8
202	1.347	.0600	24.450	5.471	.0109	.030545	.041445	2.80230	.357	32.7
212	1.367	.0591	29.921	0.000	.0000	.036820	.036820	Infinite	.000	27.1

Column 5 = barometer pressure of 29.921, minus the proportion of this due to vapor pressure from column 4.

COMBUSTION

COMBUSTION may be defined as the phenomenon resulting from any reaction evolving heat. Oxygen is the sole supporter of combustion, and a combustible may therefore be defined as a substance capable of combining with oxygen to produce heat. The speed of combustion, which is primarily dependent upon the affinity of a combustible element for oxygen, covers a range from very slow, as in the case of rust formation, to instantaneous, as in the explosion of confined powder.

As applied to heat production for steam-generating purposes, combustion may be defined as the rapid combination of the combustible elements of a fuel with oxygen, while in this sense the term combustible implies the capacity of an element for rapid combination with oxygen for the evolving of heat.

The elements and compounds encountered in commercial combustion work, together with their molecular symbols and weights, are given in Table 30.

TABLE 30

ELEMENTS AND COMPOUNDS ENCOUNTERED IN COMBUSTION

Substance	Molecular Symbol	Atomic Weight		Molecular Weight		Form
		Accurate	Approximate	Accurate	Approximate	
Carbon	C*	12.005	12	†	.	Solid
Hydrogen	H_2	1.008	1	2.015	2	Gas
Oxygen	O_2	16.00	16	32.00	32	Gas
Sulphur	S_2	32.07	32	64.14	64	Solid
Nitrogen‡	N_2	14.01	14	28.02	28	Gas
Carbon Monoxide	CO	. .	. . .	28.01	28	Gas
Carbon Dioxide	CO_2	. .	. . .	44.01	44	Gas
Methane	CH_4	. .	. . .	16.03	16	Gas
Acetylene	C_2H_2	. . . .	. .	26.03	26	Gas
Ethylene	C_2H_4	. .	. .	28.03	28	Gas
Ethane	C_2H_6	.		30.05	30	Gas
Sulphur Dioxide	SO_2	. .	. . .	64.07	64	Gas
Hydrogen Sulphide	H_2S	. .	. . .	34.08	34	Gas
Water Vapor	H_2O	. . .	. . .	18.02	18	Vapor
Air	. . .	. . .	. . .	28.94	29	Gas

*Atomic symbol.

†The molecular weight of C has not been definitely determined. Carbon exists in a number of forms each of which probably has its own molecular weight. The latest investigations indicate that a molecule of carbon in any form consists of at least 12 atoms.

‡Atmospheric nitrogen as distinguished from chemically pure nitrogen, which has an atomic weight slightly less than 14.01.

The chemical reaction of combustion, as in the case of any chemical reaction, results from a rearrangement of the atoms of the constituent elements into a new combination of molecules. Such reactions occur in accordance with fixed and invariable weight relations, which are characteristic of the elements involved, and definite volumetric changes based upon the number of gaseous molecules reacting and produced. The chemical reactions and resulting combinations of combustion are given in Table 31.

POWER STATION OF THE ADIRONDACK POWER & LIGHT CORPORATION, AMSTERDAM, N. Y. OPERATING 6020 HORSE POWER OF BABCOCK & WILCOX BOILERS EQUIPPED WITH BABCOCK & WILCOX SUPERHEATERS

For the reactions to be as indicated in this table, it is necessary that the amount of oxygen supplied for combustion be the exact amount required and be wholly utilized. Such a condition necessitates a distinction between perfect and complete combustion. Combustion may be complete when an excess of oxygen over that actually required is supplied, such excess appearing as unused oxygen in the products of combustion. Combustion is only perfect when complete without the presence of oxygen in the products resulting from such combustion. An example of combustion that is neither complete nor perfect is that of carbon to carbon monoxide. Since carbon monoxide is capable of a further combination with oxygen to form carbon dioxide, combustion is obviously incomplete. Combustion that is complete but not perfect is discussed hereafter under the heading of excess air.

TABLE 31

CHEMICAL REACTIONS OF COMBUSTION

Combustible Substance	Reaction
Carbon (to CO) .	$2C + O_2 = 2CO$
Carbon (to CO_2) .	$C + 2O_2 = 2CO_2$
Carbon Monoxide	$2CO + O_2 = 2CO_2$
Hydrogen . .	$2H_2 + O_2 = 2H_2O$
Sulphur (to SO_2) .	$S + O_2 = SO_2$
Sulphur (to SO_3) .	$2S + 3O_2 = 2SO_3$
Methane	$CH_4 + 2O_2 = CO_2 + 2H_2O$
Acetylene	$2C_2H_2 + 5O_2 = 4CO_2 + 2H_2O$
Ethylene . .	$C_2H_4 + 3O_2 = 2CO_2 + 2H_2O$
Ethane	$2C_2H_6 + 7O_2 = 4CO_2 + 6H_2O$
Hydrogen Sulphide	$2H_2S + 3O_2 = 2H_2O + 2SO_2$

IGNITION TEMPERATURE—While, as stated, the speed of combustion is primarily dependent upon the affinity of the combustible for combination with oxygen, it is also dependent upon the conditions under which combustion takes place, and the chief of such conditions is that of temperature. Introducing a combustible into the presence of oxygen does not of necessity result in combustion.

Every combustible substance has a temperature called its "ignition temperature," to which it must be brought before it will unite with oxygen in chemical combination, and below which such combination will not take place. This ignition temperature must exist with oxygen present or no combustion will occur. The ignition temperature of various fuels and of the combustible constituents of such fuels is given in Table 32.

TABLE 32

IGNITION TEMPERATURES

Combustible Substance	Molecular Symbol	Ignition Temperature Degrees Fahrenheit
Sulphur	S_2	470
Fixed Carbon—Bituminous Coal . .	. .	766
Fixed Carbon—Semi-bituminous Coal	. . .	870
Fixed Carbon—Anthracite Coal . .	. . .	925
Acetylene	C_2H_2	900
Ethane	C_2H_6	1000
Ethylene	C_2H_4	1022
Hydrogen	H_2	1130
Methane	CH_4	1202
Carbon Monoxide	CO	1210

The temperature of ignition of the gases of a coal varies and is higher than the ignition temperature of the fixed carbon constituent. The ignition temperature of coal is that of its fixed carbon content, the gaseous constituents being ordinarily distilled off, though not ignited, before this temperature is reached.

When combustion has started, the heat evolved in the oxidization of the combustible matter will, under proper conditions, maintain sufficiently high temperatures for a continuity of ignition.

Heat of Combustion—Elements entering into a direct combination to form a compound either evolve or absorb a definite amount of heat, called the "heat of combination," which, from its definition, may be positive or negative. Since the term combustion, as applied to boiler practice, is limited to the rapid chemical combination of the combustible elements of a fuel with oxygen, with a resulting production of heat, the heat of combustion of a fuel is the heat evolved in the complete oxidization of the combustible elements through their union with oxygen. Thus, the heat of combustion of a fuel is the heat of combination of a specific set of elements and compounds, the combination of which with oxygen always results in the production of heat. It follows that the heat of combination of a compound which results from the union of a single combustible element with oxygen to produce heat is the same as the heat of combustion of that element.

In the commercial evolving of heat for steam generation, the "second law" of Berthelot is of importance. This law may be summarized as follows:

The heat energy evolved in any chemical change in the boiler furnace, where no mechanical work is done, *i. e.*, evolved through the union of combustible elements with oxygen, is dependent upon the final products of combustion and in no way upon any intermediate combination or combinations that may have occurred in reaching the final result.

The application of this law may be readily shown by example:

A coal fire from which all of the volatile constituents have been driven and which consists of incandescent coke may, for the present purpose, be considered as consisting entirely of carbon. If air is introduced under the fire, the oxygen immediately breaks its mechanical union with nitrogen and enters into chemical combination with carbon to form carbon dioxide ($C + 2O = CO_2$). Each unit of carbon has combined with the maximum amount of oxygen with which it can exist as a compound. The oxygen, on the other hand, is capable of uniting with additional carbon and as the unit of carbon dioxide passes upward through the fuel bed under the influence of draft it encounters other free carbon with which it unites to form carbon monoxide ($CO_2 + C = 2CO$), thus "satisfying" the affinity of oxygen for carbon. If no additional oxygen is encountered in the further passage through the fuel bed, these particular molecules, as representative of the products of combustion, will issue from the fuel bed as carbon monoxide. If no additional oxygen is encountered in the furnace, the total heat available for later absorption by the boiler is that due to the combustion of carbon to carbon monoxide, regardless of the fact that at one stage of the process the carbon had been completely oxidized and carbon dioxide had been produced. If, on the other hand, additional oxygen is encountered in the furnace, the temperature is above the ignition point of carbon monoxide, and this temperature is maintained a sufficient length of time for further combustion, *i. e.*, if the gases are not cooled below the ignition temperature by the boiler-heating surface before further combustion can be completed, the carbon of the carbon monoxide will unite with additional oxygen to form carbon dioxide ($2CO + 2O = 2CO_2$). The total heat evolved and available for absorption in such cases will be that due to the burning of carbon to carbon dioxide regardless of the two intermediate steps.

The methods of determining the heat of combustion of solid and liquid fuels are discussed in a later chapter. The heat of combustion of elements and combustible compounds entering into gaseous fuels has been determined by so many authorities that definite values may be accepted as correct without determination. Such values are given in Table 33.

TABLE 33*

HEAT OF COMBUSTION

BY CALORIMETRIC DETERMINATION

Combustible	Molecular Symbol	Heat Value—B.t.u. per Pound		Per Cubic Foot‡
		Higher	Lower or Net†	Higher
Hydrogen . .	H_2	62000	52920	348
Carbon (to CO) .	C	4380	. . .	. . .
Carbon (to CO_2)	C	14540	.	. . .
Carbon Monoxide	CO	4380	. . .	342
Carbon in CO§ .	C	10160	. . .	. . .
Methane . .	CH_4	23850	21670	1073
Acetylene . . .	C_2H_2	21460	21020	1590
Ethylene .	C_2H_4	21450	20420	1675
Ethane . . .	C_2H_6	22230	20500	1883
Sulphur (to SO_2) .	S_2	4050	. . .	. . .
Sulphur (to SO_3) .	S_2	5940	. . .	. . .

† There is a considerable discrepancy between lower heat values as given by different authorities, the variation being due to methods of computation and assumptions. (See text.) The values given are those of G. A. Goodenough.

‡ At 32 degrees Fahrenheit and atmospheric pressure.

§ Per pound of carbon in carbon monoxide, *i. e.*, 2.33 pounds of CO

* Heating Value by Calorimetry, see discussion, page 198.

While this table gives two heat values, the higher and net, the former is that which should be used. The difference is discussed hereafter in connection with the determination of the heating values of fuels.*

AIR AND COMBUSTION—In the foregoing more or less abstract consideration of combustion, it has been assumed that there existed a supply of oxygen sufficient for combination with the combustible elements of a fuel and a temperature sufficient to bring about the chemical combinations of combustion. In practice, it is the physical introduction of oxygen to the combustible elements or compounds of a fuel, in such manner as to assure complete oxidization and at the same time to assure the utilization of all or the maximum proportion so supplied, that presents the most difficult problem in the burning of all fuels.

Oxygen, as stated, is the sole supporter of combustion. The source of supply of oxygen is the air. From the proportionate parts by weight of oxygen and nitrogen as given, namely, oxygen 23.15 per cent and nitrogen 76.85 per cent, in order to supply one pound of oxygen for combustion it is necessary to supply $1 \div 0.2315 = 4.320$ pounds of air, and such weight represents, for each pound of oxygen supplied, 3.320 pounds of nitrogen, which serves no useful function in combustion and is a source of direct loss in combustion work.

From the manner of chemical combinations occurring in the union of oxygen with combustible elements and compounds, as indicated in Table 31, and the atomic weights of the elements involved, the proportionate parts by weight of the elements entering into the resulting compounds and the weights of the products of combustion may be readily computed. With the amount of oxygen required for

* See page 195.

combustion thus determined, the amount of air required is directly indicated by the oxygen-nitrogen ratio of air.

The methods of such computations are clearly indicated by example, and since the relation of the products of combustion to the combustible elements of the fuel is the most important factor in the determination of the efficiency of combustion, it appears advisable to illustrate such computations fully.

Consider first carbon. From Table 31 it was seen that one atom of carbon united with two atoms of oxygen to form carbon dioxide, $C + 2O = CO_2$. From the atomic weights, $12 + (2 \times 16) = 44$, or in the burning of one pound of carbon to carbon dioxide, twelve parts by weight of carbon combine with thirty-two parts by weight of oxygen to form forty-four parts by weight of carbon dioxide. Hence, any weight of carbon dioxide must be composed of 27.27 per cent by weight of carbon and 72.73 per cent by weight of oxygen, or 1 pound $CO_2 = 0.2727$ pound $C + 0.7273$ pound O_2.

Since the ratio of carbon to oxygen in carbon dioxide is 1 to 2.667, it is obvious that in burning one pound of carbon to carbon dioxide, 2.667 pounds of oxygen will be required. If one pound of oxygen is contained in 4.32 pounds of air it will be necessary to supply for the complete combustion of one pound of carbon $2.667 \times 4.32 = 11.52$ pounds of air, and since each pound of oxygen is accompanied by 3.32 pounds of nitrogen, there will pass off with the carbon dioxide $2.667 \times 3.32 = 8.85$ pounds of nitrogen.

In the complete combustion of one pound of carbon, then, the resulting products of combustion will be

$$1 \text{ pound } C + 2.667 \text{ pounds } O_2 = 3.667 \text{ pounds } CO_2$$
$$2.667 \times 3.32 \text{ pounds } N_2 = 8.885 \text{ pounds } N_2$$

Again, consider hydrogen. Table 31 indicates that two atoms of hydrogen will combine with one atom of oxygen to form water vapor, $2H + O = H_2O$. From the atomic weights, $(2 \times 1) + 16 = 18$, or in the burning of one pound of hydrogen to water vapor, one part by weight of hydrogen will combine with eight parts by weight of oxygen to form nine parts by weight of water vapor. Hence in one pound of water vapor we have

$$1 \text{ pound } H_2O = 0.111 \text{ pound } H_2 + 0.889 \text{ pound } O_2$$

Since the ratio of hydrogen to oxygen in water vapor is thus 1 to 8, it will require 8 pounds of oxygen for the complete combustion of one pound of hydrogen. which means, as for the combustion of carbon, $8 \times 4.32 = 26.56$ pounds N_2, and the products of combustion of one pound of hydrogen will be

$$1 \text{ pound } H_2 + 8 \text{ pounds } O_2 = 9 \text{ pounds } H_2O$$
$$8 \times 3.32 \text{ pounds } N_2 = 26.56 \text{ pounds } N_2$$

As typical of the combustible compounds, consider ethylene; the members of the CH series are all computed in a similar manner. $C_2H_4 = 2C + 4H$, or from atomic weights $28 = 24 + 4$. Thus one pound of ethylene is composed of 0.857 pound $C + 0.143$ pound H. To burn 0.857 pound of carbon will require $0.857 \times 2.667 = 2.2856$ pounds O_2. To burn 0.143 pound of hydrogen will require $0.143 \times 8 = 1.144$ pounds O_2. The total oxygen required, then, will be $2.286 + 1.144 = 3.430$ pounds, and the total air $3.43 \times 4.32 = 14.82$ pounds. The products of combustion will be

	CO_2 *Pounds*	H_2O *Pounds*	N_2 *Pounds*
0.857 pound C + 2.286 pounds O_2 =	3.143		
0.143 pound H_2 + 1.144 pounds O_2 =		1.287	. . .
14.82 pounds air × 0.7685 (per cent N in air)			11.39

The methods of computation are simple, but, as stated, are considered at length because of their importance, particularly in the case of gaseous fuels. Table 34 gives the results of such computations in terms of weight for all of the combustible

TABLE 34

COMBUSTION DATA

IN TERMS OF POUNDS PER POUND OF FUEL

	Molecular Symbol	Theoretically Required Pounds		Products of Combustion Pounds				
		O_2	Air	CO_2	H_2O	N_2	CO	SO_2
Carbon (to CO_2)	C	2.667	11.52	3.667	. . .	8.85	. . .	.
Carbon (to CO)	C	1.333	5.76	. . .	. . .	4.43	2.333	. .
Carbon Monoxide	CO	0.572	2.46	1.57	. . .	1.89	. . .	. .
Sulphur	S	1.000	4.32	. . .	. . .	3.32	. . .	2.00
Hydrogen	H_2	8.000	34.56	. . .	9.00	26.56	.	
Methane	CH_4	4.000	17.28	2.75	2.25	13.28	. . .	. .
Acetylene	C_2H_2	3.077	13.29	3.39	0.69	10.21	. . .	.
Ethylene	C_2H_4	3.429	14.81	3.14	1.29	11.38	. . .	.
Ethane	C_2H_6	3.733	16.13	2.93	1.80	12.40	. . .	. .
Hydrogen Sulphide	H_2S	1.412	6.10	. . .	0.53	4.69	.	1.88

TABLE 35

COMBUSTION DATA

IN TERMS OF CUBIC FEET PER CUBIC FOOT OF FUEL

	Molecular Symbol	Theoretically Required Cubic Feet		Products of Combustion Cubic Feet				
		O_2	Air	CO_2	H_2O	N_2	CO	SO_2
Carbon Monoxide	CO	0.5	2.391	1	.	1.891	. .	. . .
Hydrogen	H_2	0.5	2.391	. . .	1	1.891	. . .	. . .
Methane	CH_4	2.0	9.564	1	2	7.564	. . .	. . .
Acetylene	C_2H_2	2.5	11.955	2	1	9.455	. .	. . .
Ethylene	C_2H_4	3.0	14.346	2	2	11.346	. .	. . .
Ethane	C_2H_6	3.5	16.737	2	3	13.237		. . .
Hydrogen Sulphide	H_2S	1.5	7.173	. . .	1	5.673	.	1

elements and compounds encountered in commercial fuels. Table 35 gives such values in terms of volume.

The methods of computing the products of perfect combustion may best be illustrated by example. If we assume a coal having an ultimate analysis as follows:

	Per Cent
Carbon	66.9
Hydrogen	4.4
Oxygen	7.9
Nitrogen	1.2
Sulphur	3.3
Moisture	4.5
Ash	11.8

the weight of oxygen and air theoretically required for combustion and the weight of products of combustion may be computed from Table 34 and are as given in Table 36A. In this table it is assumed that the oxygen content of the fuel is utilized in combustion, and that the sulphur content is burned to sulphur dioxide and appears in the analysis of the products of combustion as carbon dioxide. This method of computation of the products of perfect combustion of any fuel is of importance in connection with analyses of such products as are discussed in the next chapter.

TABLE 36A

AIR AND OXYGEN REQUIRED FOR AND PRODUCTS OF COMBUSTION

	Weight per Pound of Coal Pound	Required Pounds		Products of Combustion—Pounds				
		O_2	Air	CO_2	O_2	N_2	H_2O	SO_2
C	.669	1.784	7.707	2.453	. .	5.921	. . .	. .
H_2	.044	.352	1.521	. . .	. . .	1.169	.396	. .
O_2	.079	. . .	. . .	. . .	.079	. . .	. . .	. .
N_2	.012	. . .	. . .	. .	. .	.012	. .	. .
S	.033	.033	.143	. .	. .	.110	. . .	.066
H_2O	.045	. . .	. .	. . .	. . .	. . .	.045	. .
Ash	.118	. .	. .	.	. .	. .	.	
	1.000	2.169	9.371	2.453	.079	7.212	.441	066
O_2 in Coal		.079	.341		.079	.262		
		2.090	9.030	2.453		6.950	.441	.066
SO_2 as CO_2				.066				.066
		2.090	9.030	2.519		6.950	.441	

Excess Air—While supplying just the proper amount of oxygen to assure complete combustion would appear from the chemical standpoint, as indicated by Table 34, to be a simple matter, it is the difficulty of supplying sufficient oxygen for complete combustion, and the maintaining such supply with its accompanying nitrogen at a point as nearly as possible at the theoretical amount required, that is the primary problem of combustion. In other words, it is the relation of the amount of air actually supplied to that theoretically required for combustion that is the measure of the efficiency of such combustion.

Nitrogen, which almost of necessity must be introduced into the boiler furnace with the oxygen required for the combustion of any fuel, is, as stated, an inert gas which performs no function in combustion. It passes through the furnace and boiler without change, except in temperature and volume, dilutes the air, absorbs heat and reduces the temperature of the products of combustion. For this reason and because of the high proportion of nitrogen to oxygen in air, it is the chief source of heat loss in combustion. Any oxygen supplied to the furnace in excess of that required for combustion results in the same losses as in the case of nitrogen, and further, such excess oxygen is accompanied by additional nitrogen which accentuates the combustion loss. On the other hand where sufficient oxygen is not supplied for complete combustion the loss due to the presence of the inert nitrogen is inappreciable as compared to that resulting from incomplete combustion. The figures of Table 33, indicating the difference in the amount of heat evolved in the burning of carbon to carbon monoxide and to carbon dioxide represent the possible combustion losses resulting from incomplete combustion. This factor is discussed hereafter in connection with flue gas analyses and combustion losses.

THE FIFTY-NINTH STREET STATION OF THE INTERBOROUGH RAPID TRANSIT COMPANY, NEW YORK, OPERATING 43 160 HORSE-POWER OF BABCOCK & WILCOX BOILERS EQUIPPED WITH BABCOCK & WILCOX SUPERHEATERS

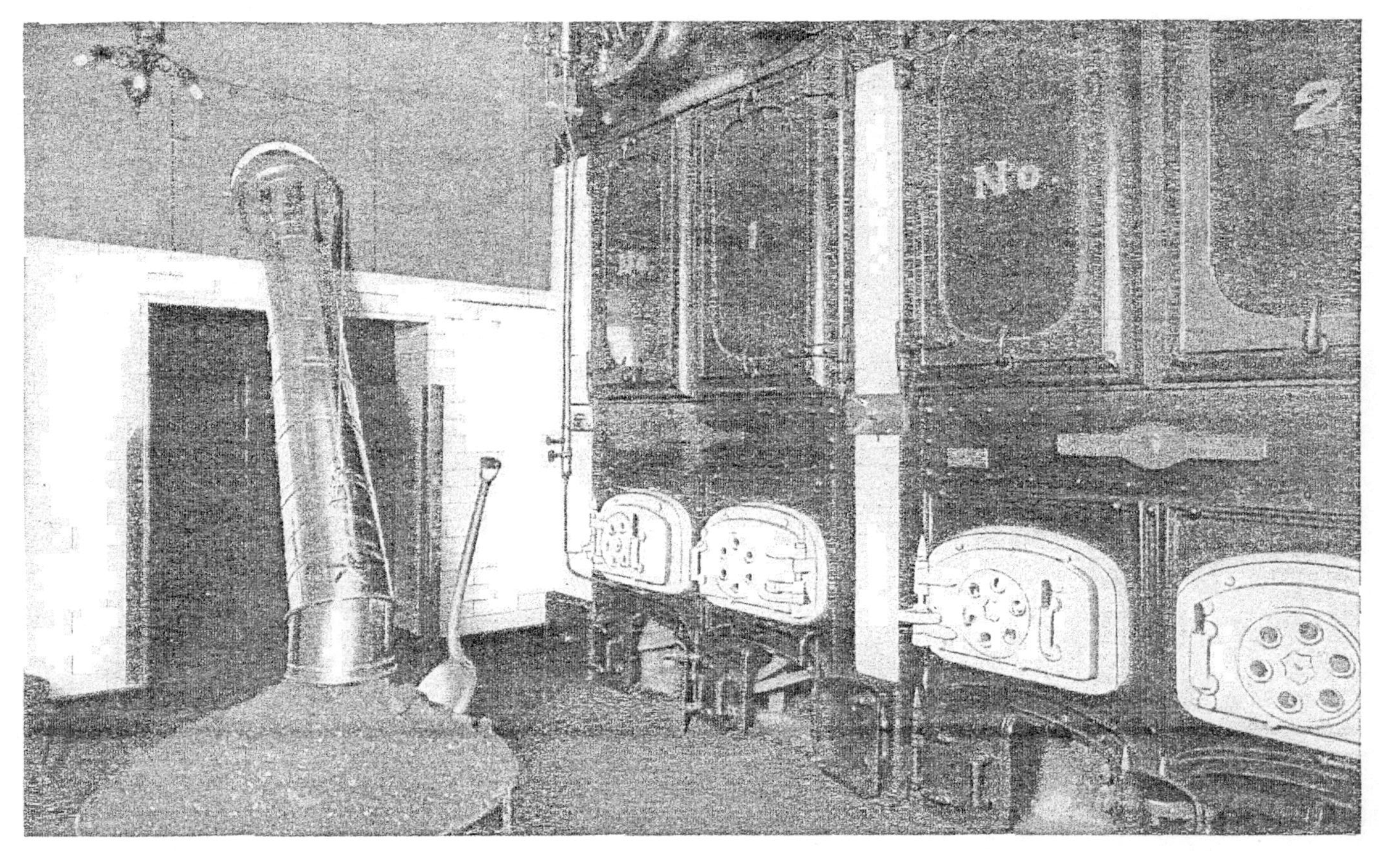

381 Horse Power Installation of Babcock & Wilcox Boilers, B. F. Keith's Theatre, Boston, Mass.

FLUE GASES AND THEIR ANALYSIS

WITH the data given in the chapter on "Combustion," formulæ for the amount of air theoretically required for the combustion of different fuels may be readily developed. The amount of air actually supplied for combustion, however, cannot be suitably or reliably measured or weighed in commercial practice, and such amount can only be determined from the analysis of the products of combustion, ordinarily known as flue gases. As indicated, it is the relation of these products of combustion to the amount of air theoretically required for a given fuel that is the measure of the efficiency of combustion.

APPARATUS FOR FLUE-GAS ANALYSIS—In the ordinary commercial gas analysis, the proportionate parts by volume of carbon dioxide, carbon monoxide, and oxygen are determined, the difference between the sum of these constituents and 100 per cent being assumed as nitrogen. Where accuracy, carried, perhaps, beyond the known limitations of the ability in boiler practice to obtain true average gas samples, is required, the gas sampling apparatus also determines the percentage by volume of hydrogen and methane, the latter representing the most probable hydrocarbon content of the flue gases.

For the ordinary analysis of flue gases the Orsat apparatus, illustrated in Fig. 11, is most commonly used. The burette *A* is graduated in cubic centimeters up to 100, and is surrounded by a water jacket to prevent any change in temperature from affecting the density of the gas being analyzed.

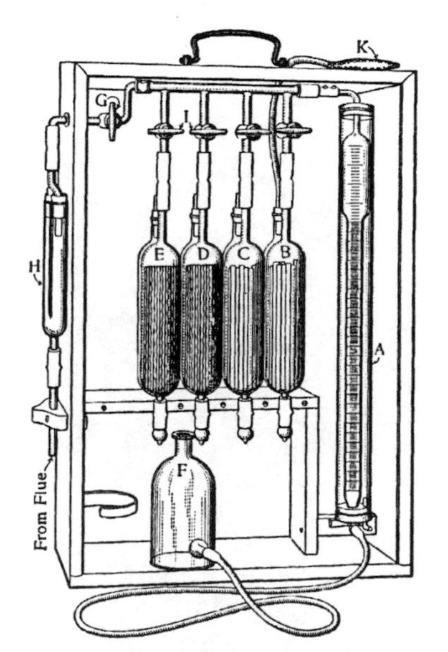

FIG. 11. ORSAT APPARATUS

For accurate work, it is advisable to use four pipettes, *B, C, D, E,* the first containing a solution of caustic potash for the absorption of carbon dioxide, the second an alkaline solution of pyrogallol for the absorption of oxygen, and the remaining two an acid solution of cuprous chloride for absorbing the carbon monoxide. Each pipette contains a number of glass tubes, to which some of the solution clings, thus facilitating the absorption of the gas. In the pipettes *D* and *E*, copper wire is placed in these tubes to re-energize the solution as it becomes weakened. The rear half of each pipette is fitted with a rubber bag, one of which is shown at *K,* to protect the solution from the action of the air. The solution in each pipette should be drawn up to the mark on the capillary tube.

The gas is drawn into the burette through the U-tube *H,* which is filled with spun glass or similar material to clean the gas. To discharge any air or gas in the apparatus, the cock *G* is opened to the air and the bottle *F* is raised until the water in the burette reaches the 100 cubic centimeters mark. The cock *G* is then turned so as to close the air opening and allow gas to be drawn through *H*, the bottle *F* being lowered for this purpose. The gas is drawn into the burette to a point below the

The Magnolia Building, Dallas, Texas. Operating 1579 Horse-power of Babcock & Wilcox Cross-drum Boilers with Cast-iron Headers and Babcock & Wilcox Oil Burners

zero mark, the cock G then being opened to the air and the excess gas expelled until the level of the water in F and in A is at the zero mark. This operation is necessary in order to obtain the zero reading at atmospheric pressure.

The apparatus should be carefully tested for leakage, as well as all connections leading to it. Simple tests can be made; for example, if after the cock G is closed, the bottle F is placed on top of the frame for a short time and again brought to the zero mark, the level of the water in A is above the zero mark, then a leak is indicated.

Before taking a final sample for analysis, the burette A should be filled with gas and emptied once or twice, to make sure that all the apparatus is filled with the new gas. The cock G is then closed and the cock I in the pipette B is opened, and the gas driven over into B by raising the bottle F. The gas is drawn back into A by lowering F, and when the solution in B has reached the mark in the capillary tube, the cock I is closed and a reading is taken on the burette, the level of the water in the bottle F being brought to the same level as the water in A. The operation is repeated until a constant reading is obtained, the number of cubic centimeters being the percentage of CO_2 in the flue gases.

The gas is then driven over into the pipette C and a similar operation is carried out. The difference between the resulting reading and the first reading gives the percentage of oxygen in the flue gases.

The next operation is to drive the gas into the pipette D, the gas being given a final wash in E, and then passed into the pipette C to neutralize any hydrochloric acid fumes which may have been given off by the cuprous chloride solution, which, especially if it be old, may give off such fumes, thus increasing the volume of the gases and making the reading on the burette less than the true amount.

The process must be carried out in the order named, as the pyrogallol solution will also absorb carbon dioxide and the cuprous chloride solution will also absorb oxygen.

The Orsat apparatus is ordinarily used for taking what may be called "snap" samples of gas at fixed intervals during the period of a boiler's operation, though it may be used in analyzing a continuous gas sample. Where an analysis is extended beyond the limitations of the Orsat apparatus as to constituent gas content, it is usually made of a continuous sample and a Hemphill apparatus is used.

While the errors resulting from the proper use of flue-gas analyses in the computation of combustion data are probably well within the error of boiler testing as a whole, there are, however, sources of possible error in the actual making of analyses that must be guarded against. These may be summarized as follows:

1st. Care should be taken that the sample of gas for analysis is an average sample. This is the feature which should be most carefully watched and is perhaps the most difficult of achievement. No hard and fast rules can be laid down for the methods of obtaining such average sample and it is largely a question of common sense. The sample should be drawn from the main body of the gases at a location where the possibility of dilution through air infiltration is a minimum.

2nd. Absorption reagents should be reasonably fresh. Each reagent is capable of absorbing a definite amount of one of the constituent gases, this amount ordinarily being expressed in terms of volume of the absorbing medium, and a check should be kept on the total absorption. Where solutions are weak and absorption

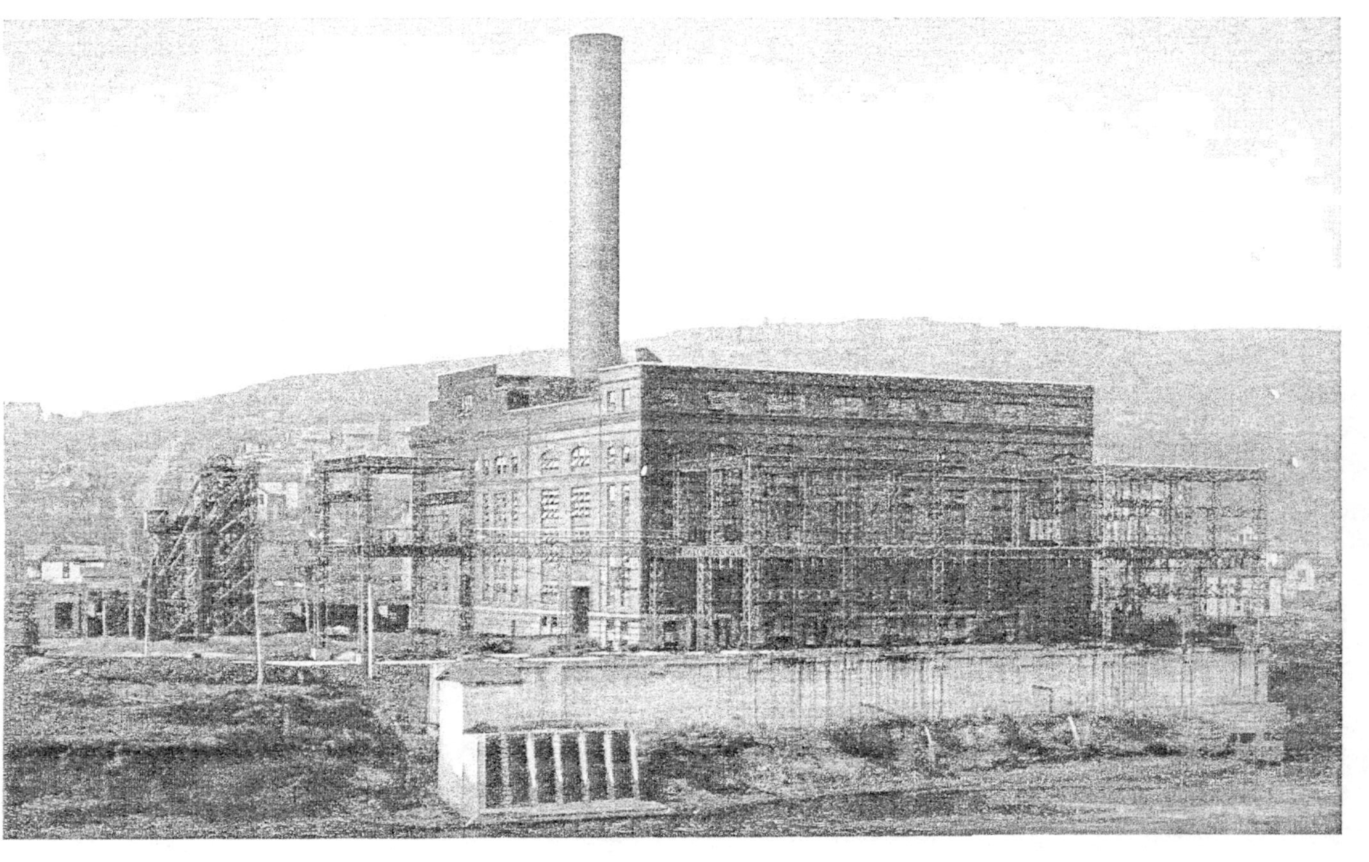

Rivesville Power Station, Monongahela West Penn Public Service Company, Fairmont, W. Va. Where [illegible]208 Horse Power of Babcock & Wilcox Boilers and Superheaters Have Been Installed

is not accomplished within a relatively short time, there is a tendency to accept the absorption as complete, which results in an inaccurate analysis.

3rd. There is a tendency, particularly in the case of inexperienced operators, to attempt to force the absorption. With reasonably fresh solutions, the gas should be brought into contact with the absorption tubes at least twice, and oftener as the solutions become weaker. In the case of oxygen, where, through attempting to force the rapidity of the analysis, absorption is not complete, erroneous results as to both oxygen and carbon monoxide content will be obtained, since the absorption reagent for the latter will also absorb oxygen.

4th. Analyses should be completed. Too frequently it is assumed that the carbon dioxide content alone, or the carbon dioxide and oxygen content, is sufficient, but often the efficiency seemingly indicated by a high carbon dioxide content alone would be more than offset by the fact that appreciable amounts of carbon monoxide were present and not analyzed.

Various tables have been published as indicative of the losses due to decreasing percentages of carbon dioxide in the flue gases. Such tables, if based on the combustion of carbon alone, would be an approximately correct measure of such loss, but with the range of analyses of different fuels no table of carbon dioxide, as representing combustion losses, may be accepted as indicative of the extent of such loss.

TABLE 36B

VARIATION IN CO_2 WITH DIFFERENT PERCENTAGES OF EXCESS AIR AND DIFFERENT FUELS

Constituent	Class of Fuel					
	Coal*	Wood*	Oil*	Natural Gas†	By-product Gas†	Blast Furnace Gas†
C	79.86	50.31	84.00	. . .	. . .	. .
H_2	5.02	6.20	12.70	1.82	53.00	3.50
O_2	4.27	43.08	1.20	0.35	. . .	. .
N_2	1.86	0.04	1.70	3.40	12.10	58.60
S	1.18	. .	0.40	. . .	. .	. . .
Ash	7.81	0.37	. . .	. .	. . .	. . .
CO	. . .	. . .	. . .	0.45	6.00	25.40
CO_2	. . .	. . .	. . .	0.22	0.75	12.50
CH_4	. . .	. .	. . .	93.33	28.15	. . .
C_2H_4	. . .	. . .	. . .	0.25	. . .	. . .
H_2S	. . .	. . .	. . .	0.18	. .	.
Excess Air Per Cent	Percentage of CO_2 Corresponding to Different Amounts of Excess Air					
0	18.43	20.10	15.40	11.65	9.36	25.08
20	15.29	16.72	12.69	9.54	7.67	22.97
40	13.06	14.31	10.79	8.07	6.49	21.20
60	11.40	12.51	9.38	6.99	5.63	19.67
80	10.11	11.12	8.30	6.17	4.97	18.38
100	9.09	10.00	7.45	5.52	4.48	17.20

*Analysis by weight. †Analysis by volume.

The variation in percentage of carbon dioxide with different amounts of excess air with different fuels is shown in Table 36B; and if a minimum of excess air, provided combustion is complete, is the measure of efficient combustion, the range of such combustion conditions, with different fuels, as shown by the carbon dioxide content of the products of combustion, is wide.

COMBUSTION FORMULÆ—When combustion is complete, regardless of the amount of air supplied for combustion, the only products that can result from the combustion of any fuel are carbon dioxide, oxygen, sulphur dioxide or sulphur trioxide, water vapor and nitrogen. Since commercial gas analyses indicate only the first two constituents, they are in reality only a measure of the completeness of combustion of the carbon content of the fuel.

AIR REQUIRED FOR COMBUSTION—With carbon, hydrogen and sulphur, either free or in combination, the only combustible elements of the fuels used for steam generation, the amount of air theoretically required for complete combustion may be readily computed from the values of Table 34. This amount may be expressed as:

Pounds air required per pound fuel =

$$11.52C + 34.56\left(H - \frac{O_2}{8}\right) + 4.32S \qquad (16)$$

or reduced to simpler form:

Pounds air required per pound fuel =

$$34.56\left(\frac{C}{3} + (H_2 - \frac{O_2}{8}) + \frac{S}{8}\right) \qquad (17)$$

where C, H_2, O_2, and S represent the percentages by weight of carbon, hydrogen, oxygen and sulphur, as determined by ultimate analysis. These formulæ, as in the case of all combustion formulæ, assume that the oxygen constituent of the fuel is free to unite with the hydrogen to form water vapor, such assumption leading to an error in the computation of the amount of air required for combustion that is negligible. The constants of the formulæ are taken directly from Table 34. With gaseous fuels, in order to make use of formula *(17)*, it would be necessary to break the hydrocarbons into their constituent elements, and it is simpler to develop a formula based on the values of Table 34, which would give:

Pounds air required per pound fuel =

$$2.46CO + 34.56H_2 + 17.28CH_4 + 13.29C_2H_2 + 14.81C_2H_4 + 16.13C_2H_6 + 6.10H_2S - 4.32O_2 \qquad (18)$$

If it is desired to compute combustion data for gaseous fuels on a volumetric basis, a similar formula may be readily developed from the data of Table 35.

PRODUCTS OF COMBUSTION—The weight of the products of combustion resulting from the theoretically perfect combustion of any fuel has been indicated by example, and may be directly computed from the values of Table 34. To emphasize such values, it may be stated again that for perfect combustion the weight is:

Products of combustion, pounds, per pound of fuel:

$$CO_2 = 3.667C$$

$$H_2O = 9\left(H_2 - \frac{O_2}{8}\right) + H_2O$$

$$SO_2 = 2S$$

$$N_2 = 8.85C + 26.56H_2 + 3.32S + N_2$$

The weights of the theoretical products of combustion as thus determined may be expressed as percentage weights and from such percentages may be readily transformed into percentages by volume. The products of combustion of gaseous fuels may be computed from the values of Table 34 for weight, or from Table 35 in terms of volume.

COMBUSTION DATA FROM FLUE-GAS ANALYSIS—While a flue-gas analysis may be used in connection with the analysis resulting from the perfect combustion of any fuel for the determination of combustion losses its generally accepted use is its application in a formula which is assumed to give directly the weight of dry gas per pound of carbon burned, this weight being used in combustion loss computations. This formula is:

$$\frac{11CO_2 + 8O_2 + 7\,(CO + N_2)}{3(CO_2 + CO)} \qquad (19)^*$$

where the symbols represent the percentages by volume of the constituents of the gas analyzed.

The principal assumption of this formula is that the analysis as determined is of dry gas. As a matter of fact, all flue gases as analyzed contain a greater or less amount of water vapor, depending upon the moisture content of the air supplied for combustion and of the fuel, and resulting from the burning of the hydrogen constituent of the fuel. From this latter alone, with certain fuels, the moisture content may be as high as 15 per cent of the total weight of flue gases. The presence of the moisture in the flue gases, together with the ordinary use of water as the displacement element in the commercial gas analysis apparatus, tends toward saturation of the gas analyzed. If such saturation exists, and from actual determinations such an assumption seems warranted, proportionate parts of the water vapor in the gases will be absorbed with the different other constituents, and the resulting analysis may be assumed to be that of dry gas.

A further source of error arising from the use of formula *(19)* results from the presence of sulphur in a number of fuels. Such portion of the sulphur content as is burned appears ordinarily in the gases to be analyzed as sulphur dioxide, though it is absorbed with carbon dioxide and appears as such in the flue-gas analysis. Where the sulphur constituent of a fuel is low the error arising from this source is so small as to be safely neglected although the most accurate results are desired. Even where the sulphur content of the fuel is high, it is at best questionable whether in commercial boiler practice an attempt to make a sulphur correction is warranted. Any such correction would, of necessity, assume that all of the sulphur is burned to sulphur dioxide, whereas in practice some of the sulphur probably appears in compounds in the furnace refuse. Further, since the gases are in contact with the water used as a displacement agent in which sulphur dioxide is highly soluble, it is probable that some of the sulphur dioxide is absorbed before the gases come into contact with the absorption agents with which the analysis is accomplished.

Computations show that in burning a coal containing 70 per cent carbon, 5 per cent hydrogen, 7 per cent oxygen, 1 per cent nitrogen and 5 per cent sulphur, burned with as little as 10 per cent excess air, the sulphur dioxide appearing in the products of combustion, assuming all of the sulphur burned, is only 0.43 per cent by volume. For any smaller percentage of sulphur in the fuel, and 5 per cent

* For derivation of this formula, see "Principles of Combustion," The Babcock & Wilcox Company, 1920.

THE HARVARD STATION OF THE BOSTON ELEVATED RAILWAY COMPANY OPERATING FOUR 600 HORSE POWER BABCOCK & WILCOX BOILERS

represents the approximate maximum even in the case of the poorest fuels, the percentage by volume of sulphur dioxide in the flue gases would be appreciably less under the same conditions of combustion, and if the amount of excess air supplied is greater than 10 per cent, and such excess is probably far below the average minimum, the percentage would be correspondingly less. In view of the small percentage of sulphur dioxide appearing in a volumetric flue-gas analysis in what may be considered the extreme case of commercial combustion, any attempt to make a sulphur correction in the ordinary gas analysis would appear to be a refinement not warranted from the standpoint of limits of error of such analysis.*

Formula *(19)*, as stated, gives the weight of dry products of combustion per pound of carbon burned. By multiplying this quantity by the actually *burned* carbon in one pound of fuel, the weight of dry products per pound of fuel may be determined, this expression being represented by the formula:

Weight of Dry Products per Pound of Fuel =

$$\frac{11CO_2 + 8O_2 + 7(CO + N_2)}{3(CO_2 + CO)} C \qquad (20)$$

It is necessary to emphasize here that the weight of carbon as given by an analysis of the fuel and the weight of carbon burned, *i. e.*, the weight by which the result of the application of formula *(19)* is to be multiplied, is not of necessity the same. Obviously, if a coal containing 70 per cent carbon (or 0.70 pound of carbon per pound of coal) is burned in such manner that the resulting ash contains 20 per cent of unconsumed carbon, the value of C in formula *(20)* cannot be taken as 0.7 per pound of fuel. The variation in the weight of carbon as given by a fuel analysis and in the weight of carbon actually burned is best shown by example:

Assume a coal containing 70 per cent carbon and 10 per cent *chemical* ash. If such coal is burned to show 20 per cent unconsumed carbon in the ash, the ash as determined by test is:

$$10.0 \div (1.0 - 0.2) = 12.5 \text{ per cent}$$

which, expressed in terms of one pound of fuel, consists of 0.10 pound of chemical ash and 0.025 pound of unconsumed carbon included in the total ash. The carbon per pound of coal *burned* would thus be 0.700 — 0.025 = 0.675 pound, and it is this value that should be used in multiplying the weight of dry products per pound of carbon in the fuel, as given by formula *(19)*, for the determination of the weight of dry products of combustion per pound of fuel.

It is to be remembered that the unconsumed carbon content of the ash varies with every boiler test, even of a single boiler and stoker unit, and if but a single fuel analysis is made for a series of tests, a different unconsumed carbon correction must be made for each test.

Aside from the moisture content of the air supplied for combustion, which in terms of the total moisture in the ordinary flue gas is small, the total weight

* A formula developed by Mr. C. H. Berry of the Detroit Edison Company, for sulphur correction that gives results probably as accurate as any offered, where most accurate results are required, and where fuel and gas analyses are known to be accurate *and average*, is obtained by adding to the weight of carbon burned per pound of fuel 3/8 of the weight of sulphur burned, it being necessary to assume that all of the sulphur *is* burned. This formula may be expressed for the weight of dry products of combustion per pound of fuel (*W*).

W=Weight of combustion products per pound of C[by(*19*)] × C† + 3/8S†.

† Weight of C and S burned per pound of fuel.

of moisture appearing in the products of combustion, resulting from the moisture in the fuel and from that due to the combustion of the hydrogen content, is a constant for any fuel, regardless of the amount of excess air supplied for combustion. The loss due to the moisture in the air supplied for combustion. except where every effort is made to distribute combustion losses, is ordinarily included in the unaccounted loss. Neglecting, then, the moisture in the air supplied for combustion, the total weight of water vapor per pound of fuel burned appearing in the products of combustion may be expressed:

$$\text{Weight of Water Vapor per Pound of Fuel} = 9H_2 + H_2O \qquad (21)$$

where H_2 and H_2O represent the weights of hydrogen and moisture per pound of fuel burned.

The sum of the dry products per pound of fuel burned from formula *(20)* and of the water vapor products from formula *(21)* evidently represents the weight of the total products of combustion per pound of fuel.

The total air supplied for combustion per pound of fuel burned must be represented by the total weight of products per pound, less the weight of fuel appearing in the flue gases. In the case of solid fuels, this weight is the unit weight of fuel less the total ash, *i. e.,* the ash by test and not the chemical ash. In the case of liquid or gaseous fuels, where the assumption is that all of the constituents pass up the stack in the products of combustion, the weight of air supplied is the total weight of the flue gases per pound of fuel less one.

With an ultimate analysis of the fuel burned available, the weight of air actually supplied as determined by this method may be compared with the weight of air theoretically required as given by the methods of Table 34, and the amount, expressed in percentage, of excess air, may be computed.

$$\text{Percentage Excess Air} = 100 \; \frac{\text{Air Supplied} - \text{Air Required}}{\text{Air Required}}$$

$$\text{Dry air supplied per pound of fuel} = \frac{3.032N_2}{CO_2 + CO}C \qquad (22)$$

where the symbols represent the percentages by volume of the constituents of the gas. This formula is based on a constant nitrogen content of one per cent in the fuel burned, and for solid and liquid fuels where the nitrogen percentage is reasonably close to the assumed value, results accurate within the limits of accuracy of commercial gas analyses may be determined by the use of this formula. With gaseous fuels, on the other hand, where the amount of nitrogen in the fuel itself is, under ordinarily good combustion conditions, almost as great as in the air supplied for combustion, formula *(22)* is useless. The error in the case of a fuel like blast-furnace gas may be as great as 70 per cent. For gaseous fuels, then, the only method of determining accurately the amount of excess air supplied is from the actual weight as determined from the flue gas analysis and the theoretical weight required as computed by Table 34.

It is frequently desirable to compute the percentage of excess air corresponding to a given amount of carbon dioxide with any fuel the analysis of which is given.

A method of such determination has been developed that is, perhaps, best illustrated by example:

Let us assume a coal having an ultimate analysis as given in Table 36. The products of perfect combustion, per pound of fuel, as determined by this table, are:

	Pounds
CO_2	2.453
SO_2	0.066
O_2	0.000
CO	0.000
N_2	6.950
H_2O	0.441

which, assuming that the moisture content of the flue gases is proportionately absorbed, gives a dry gas analysis corresponding to a total weight of flue gas of 9.469 pounds.

The volumes of these different constituents under standard conditions are, from Table 25:

$$\begin{aligned} CO_2 \quad 2.453 \times 8.103 &= 19.877 \text{ cubic feet } CO_2 \\ SO_2 \quad 0.066 \times 5.473 &= 0.361 \text{ cubic foot } SO_2 \\ & \overline{20.238} \text{ cubic feet } CO_2 + SO_2 \ (a) \\ N_2 \quad 6.950 \times 12.809 &= 89.023 \text{ cubic feet } N_2 \\ & \overline{109.261} \text{ cubic feet total } (b) \end{aligned}$$

The total volume of products of perfect combustion under standard conditions is, then, 109.261 cubic feet.

If x equals the weight of excess oxygen supplied per pound of fuel, $3.32x$ will equal the weight of nitrogen accompanying such excess oxygen, or expressed volumetrically:

$$\text{Excess } O_2 = 11.209x \text{ cubic feet,}$$
$$\text{Excess } N_2 = 12.809 \times 3.32x \text{ cubic feet,}$$

and the total volume of excess air supplied is:

$$11.209x + (12.809 \times 3.32x) = 53.73x \text{ cubic feet.}$$

The percentage of carbon dioxide by volume is then:

$$\frac{a}{b + 53.73x}$$

or if, for the coal assumed, combustion is such as to show 13.5 per cent of carbon dioxide, we have

$$\frac{20.238}{109.261 + 53.73x} = \frac{13.5}{100}$$

or

$$x = \frac{20.238 - (0.135 \times 109.261)}{0.135 \times 53.73}$$
$$= 0.7565$$

The weight of excess oxygen, x, corresponding to a gas analysis showing 13.5 per cent of carbon dioxide is thus 0.7565 pound, and the weight of excess nitrogen, $3.32x$, is 2.5116 pounds, or a total weight of excess air supplied for combustion of

3.2681 pounds. The total weight of dry products of combustion per pound of fuel is then

$$9.030 + 3.268 = 12.298 \text{ pounds}$$

and the percentage of excess air supplied

$$3.268 \div 9.030 = 36.2 \text{ per cent.}$$

As has been stated, the amount of excess air supplied for combustion affects directly the loss in the flue gases leaving the boiler. This loss is given by the formula:

$$\text{Loss in B.t.u. per pound of fuel} = W \times 0.24(T - t) \qquad (23)$$

Where W = weight of dry products of combustion per pound of fuel,
T = temperature of flue gases,
t = temperature of atmosphere,
0.24 = the assumed specific heat of the flue gases (see below).

Obviously, for the combustion conditions just considered, the loss as computed from formula *(23)* would be 36.2 per cent greater with a percentage of carbon dioxide in the flue gases of 13.5 than if combustion had been perfect.

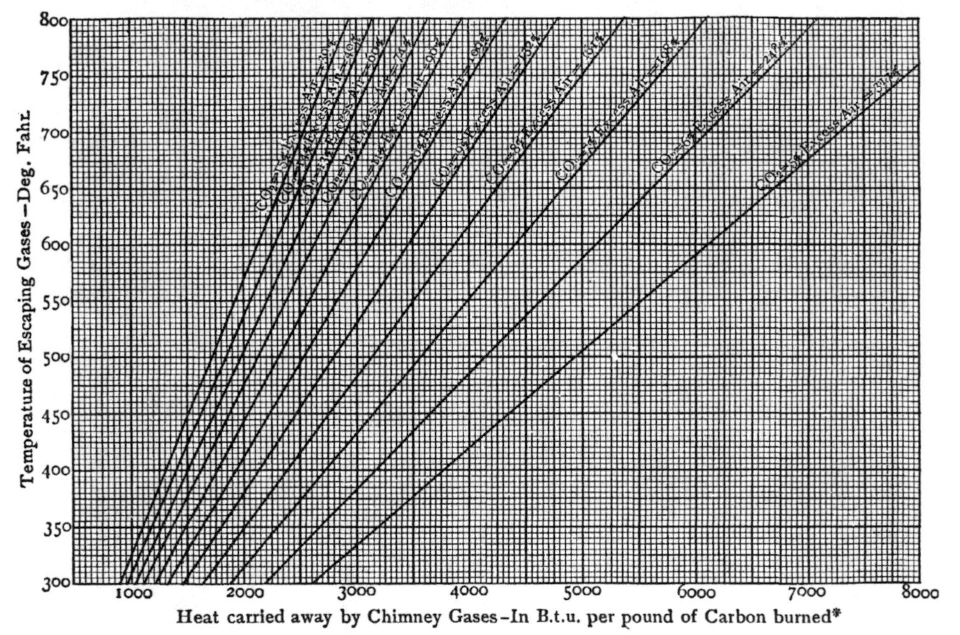

Fig. 12A. Loss Due to Heat Carried Away by Chimney Gases for Varying Percentages of Carbon Dioxide. Based on Boiler Room Temperature = 80 Degrees Fahrenheit. Nitrogen in Flue Gas = 80.5 Per Cent. Carbon Monoxide in Flue Gas = 0. Per Cent

Fig. 12A represents graphically the loss due to heat carried away in the dry flue gases for various percentages of carbon dioxide and different flue gas temperatures. The percentage loss depends upon the carbon content and heat value of the fuel. The nitrogen content of the flue gases has been assumed constant, regardless of variation in carbon dioxide. As a matter of fact, the percentage of

*For loss per pound of coal, multiply by percentage of carbon in coal by ultimate analysis.

nitrogen varies with different percentages of carbon dioxide, but such variation is small and, since Fig. 12A can be used only for approximate results, may be neglected.

The value 0.24 given in formula *(23)* as the specific heat of flue gases is rather an arbitrary value that custom rather than fact has brought into common use. The specific heat of the flue gases varies with the analysis of the gases and with the temperature, and where gas analyses and temperatures are known to be correct, for the most accurate results the specific heat should be computed for every set of conditions encountered in testing.

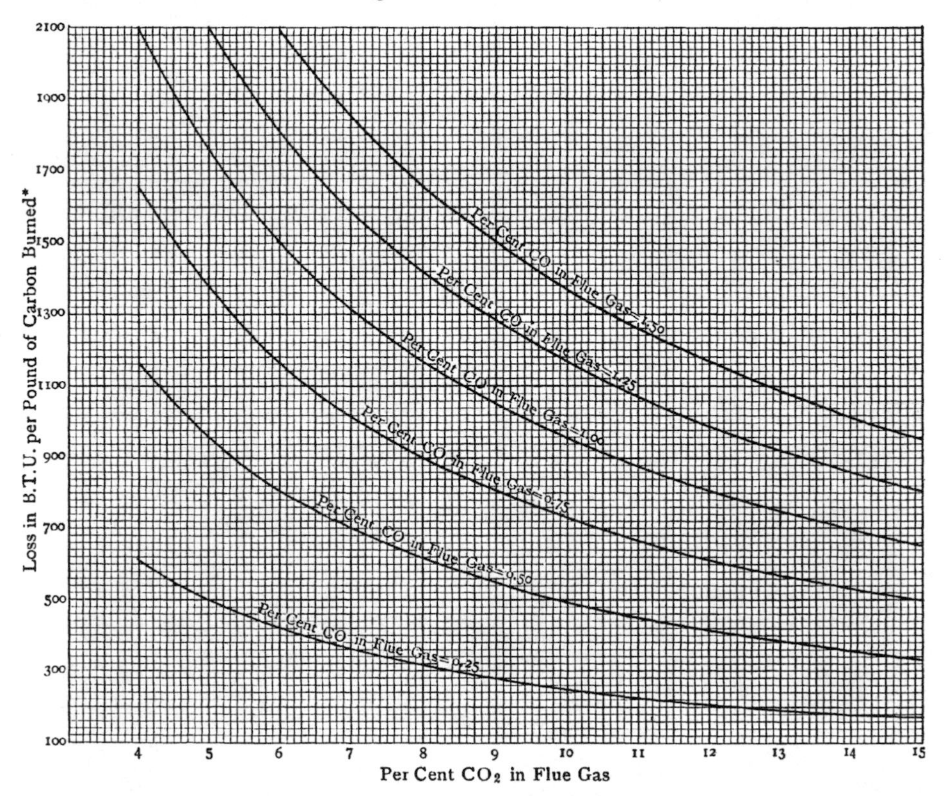

FIG. 12B. LOSS DUE TO UNCONSUMED CARBON CONTAINED IN THE CO IN THE FLUE GASES

If combustion is not complete, and some part of the carbon constituent of the fuel is burned to carbon monoxide instead of carbon dioxide, the loss due to such incomplete combustion is very appreciable. Such loss is expressed by the formula:

Loss in B.t.u. per pound of fuel burned =

$$10{,}160 \left(\frac{CO}{CO + CO_2}\right) C \qquad (24)$$

where, as before, CO and CO_2 are the percentages by volume of these constituents as shown by the flue-gas analysis, and C is the weight of carbon per pound of fuel

*For loss per pound of coal, multiply by percentage of carbon in coal by ultimate analysis.

THE MILLERS FORD STATION OF THE DAYTON POWER & LIGHT COMPANY, WHERE 5,420 HORSE POWER OF BABCOCK & WILCOX BOILERS ARE OPERATED

that is burned and appears in the flue gases. The value 10,160 represents the difference in the amount of heat evolved in the combustion of one pound of carbon to carbon monoxide and to carbon dioxide, as indicated by the heats of combustion as given in Table 33. Fig. 12B represents graphically the loss due to such carbon constituent in the fuel as is not completely burned and appears in the flue gases as carbon monoxide.

UNRELIABILITY OF CO_2 READINGS ALONE—The possibility of error in the ordinary gas analysis has been indicated. It is perhaps well to add here a caution against taking a carbon dioxide reading, alone, as indicative of the efficiency of combustion. Ordinarily, a high carbon dioxide content in the flue gases is assumed as showing proper combustion conditions. Such an assumption is correct in so far as it indicates an amount of excess air approaching that theoretically necessary for combustion, but as the percentage of carbon dioxide increases, there is a tendency toward the presence of carbon monoxide in the flue gases, which an automatic carbon dioxide recorder does not show and which is frequently difficult to detect with an Orsat apparatus. For this reason, any automatic recording device showing carbon dioxide alone should be periodically checked by another method of gas analysis. and the solution for the absorption of carbon monoxide in the Orsat or other gas analysis apparatus should be kept fresh and known to be capable of absorption.

Any comparison of losses as indicated by flue-gas analyses showing comparatively low carbon dioxide, without carbon monoxide, as against a high carbon dioxide with carbon monoxide present does, of course, depend upon the fuel burned. Roughly, it may be stated that with a gas temperature leaving an economizer of 300 degrees, the efficiency of combustion, as represented by 13 per cent carbon dioxide and no carbon monoxide, is as high as that indicated by a carbon dioxide content of 15 per cent with a carbon monoxide content of 0.25 per cent. Such a relation may be readily computed from the values of comparative losses as indicated in Figs. 12A and 12B.

CONVERSION OF ANALYSES—In combustion work, gas analyses are frequently given in terms of volume when analyses in terms of weight are required. To convert a volumetric analysis to one by weight, the percentage by volume of each constituent gas should be multiplied by its relative density, each product being divided by the sum of the products. To convert an analysis by weight to one in terms of volume, the percentage by weight of each constituent should be divided by its relative density. and each quotient so obtained be divided by the sum of the quotients. Since the molecular weights of the various gases bear the same relations to each other as the relative densities, these molecular weights may be used in transforming analyses instead of the relative densities.

Another method of converting volumetric analyses to analyses in terms of percentage by weight is through the use of the weights per cubic foot of the various constituent gases. The percentages by volume are multiplied by the weights per cubic foot, and each product is divided by the sum of the products. This method has an advantage in that it gives directly, in the sum of the products. the weights of the gas as a whole cubic foot.

As an example of these methods of conversion, assume a flue gas having a volumetric analysis as follows:

	Per Cent
Carbon Dioxide	13.5
Carbon Monoxide	0.1
Oxygen	5.9
Nitrogen	80.5

(a) Conversion by relative densities:

Gas	Relative Density		Product				Weight Per Cent
13.5	11	=	148.5	÷	759.9	=	19.54
0.1	7	=	0.7	÷	759.9	=	0.09
5.9	8	=	47.2	÷	759.9	=	6.21
80.5	7	=	563.5	÷	759.9	=	74.16
			759.9				100.00

(b) Conversion by weights:

Gas	Volume per Cubic Foot		Weight per Cubic Foot		Weight Pound				Weight Per Cent
CO_2	0.135	×	0.12341	=	0.016660	÷	0.084847	=	19.64
CO	0.001	×	0.07806	=	0.000078	÷	0.084847	=	0.09
O_2	0.059	×	0.08921	=	0.005263	÷	0.084847	=	6.21
N_2	0.805	×	0.07807	=	0.062846	÷	0.084847	=	74.06
					0.084847				100.00

The weight per cubic foot of the flue gas as determined by this second method is thus 0.08485 pound.

The slight discrepancy in the two sets of weight percentages is due to the fact that in the first computations relative densities based on approximate instead of accurate molecular weights are used.

Station of the New Bedford Gas & Edison Light Company

CLASSIFICATION OF FUELS

(WITH PARTICULAR REFERENCE TO COAL*)

FUELS for steam boilers may be classified as solid, liquid and gaseous. Of the solid fuels, anthracite and bituminous coals are the most common, but in this class must also be included lignite, peat, wood, bagasse and the refuse from certain industrial processes, such as sawdust, tan bark and the like. Straw, corn, coffee and rice husks are utilized in isolated cases.

The class of liquid fuel is represented chiefly by petroleum, though coal tar and water-gas tar are used to a limited extent.

Gaseous fuels are limited to natural gas, blast furnace gas and coke oven gas, the first being a natural product and the others by-products from industrial processes. Distinction must be made between these fuels and waste gases. While the latter are used for steam generation, they cannot be classified as fuels, since they are incombustible.

As coal is by far the fuel in most general use, this chapter will be devoted entirely to the formation, composition and distribution of its different grades. The other fuels will be discussed in succeeding chapters and their composition dealt with in connection with their combustion.

FORMATION OF COAL—All coals are of vegetable origin and are the remains of prehistoric forests. They are the result of the transformation of woody fiber and other vegetable matter by the elimination of oxygen and hydrogen in proportionately larger quantities than the removal of carbon, or a devolatilization of the original matter. The causes and methods of the first stage of transformation are not exactly defined but were probably due to the action of bacterial ferment in transforming cellulose into peaty matter. Subsequent changes were the results of great pressures and temperatures which, by destructive distillation, resolved the organic matter into its invariable ultimate constituents, carbon, hydrogen, oxygen and other substances in varying proportions. The factors of time, depth of beds, disturbances of beds and introduction of foreign matter resulting from such disturbances, have produced the variation in the degree of evolution from vegetable fiber to hard coal. While age is an important factor in the degree of evolution, it is not, as ordinarily considered, the primary factor in transformation. This primary factor is the presence or absence of disturbance in the rocky crust of the earth's surface during the period of transformation, either in the field where the coal was being formed, or close to such field. Lignites represent the coals of most recent formation and are considered, to an extent correctly, low in the rank of coals because of this fact. On the other hand, there are coal beds in the Western States, where disturbances of the earth's crust were great, which are of the same age as the lignitic beds of other sections of the country and still are representative of the highest rank coals of that portion of the country.

The amount of foreign and incombustible matter found in coal depends upon two factors in its formation: the degree of disturbance of the bed in the formative stage as affected by the introduction of rock, sand, etc., and the conditions which permitted mineral-bearing waters to circulate through the beds in process of formation. Geologically, these conditions are difficult, if not impossible, of

* The data in this chapter are taken largely from the various publications of the U. S. Geological Survey.

determination, and the grade of coal cannot be foretold simply from the area in which it is to be mined.

COMPOSITION OF COAL—The analyses of coal and the methods of analyses are discussed in detail in the following chapter. For an understanding of the classification of coals, however, analyses as ordinarily given must be briefly considered here. The uncombined carbon of coal is known as the fixed carbon. A certain percentage of the total carbon constituent is combined with hydrogen and, in combination, burns as a hydrocarbon. This, together with the other gaseous substances driven off by the application of heat, forms that portion of the coal known as volatile matter. The fixed carbon and the volatile matter constitute the combustible. The oxygen and nitrogen contained in the volatile matter are not combustible, but custom has applied this term to that portion of the coal free from moisture and ash, thus including oxygen and nitrogen. The other important substances entering into the composition of coal are moisture and the refractory earths which form the ash. The ash varies in different coals from 3 to 30 per cent and the moisture from 0.75 to 45 per cent of the total weight of the coal, depending upon the grade and the locality in which it is mined.

A large percentage of ash is undesirable, as it not only reduces the heat value of the coal but chokes the air passages in the furnace and in the fuel bed, thus preventing the rapid combustion necessary for high efficiency. If the sulphur content of the coal is high, it unites with the ash to form a fusible slag or clinker which will cause furnace difficulties.

Moisture in coal may be more detrimental than ash. It absorbs heat in evaporation and in being brought to the furnace temperature, and thus reduces such temperature. With some classes of coal, however, a certain amount of water produces a mechanical action that assists in combustion and leads to better results.

CLASSIFICATION OF COAL—The U. S. Geological Survey classifies coal by "ranks," a term used to designate differences in coal in various states of transformation from lignite to anthracite, such changes being marked by a decrease in the moisture, oxygen and volatile matter. and usually an increase in fixed carbon, sulphur, and possibly ash. The Survey uses the term "grade" in distinguishing between two coals as to ash and sulphur content. A high-grade coal would be one relatively pure and a high-rank coal one high in the scale of coal or one in which devolatilization had occurred to a high degree.

The second Pennsylvania Geological Survey used a method of coal classification based on what it termed "fuel ratios," *i. e.,* the quotient of the fixed carbon divided by the volatile matter as given by a proximate analysis of the ash-free coal. The U. S. Geological Survey, believing it to be impossible to classify all ranks of coal by chemical composition, makes use of a modification of the fuel ratio basis for high-rank coals, and supplements chemical composition of the lower ranks by physical characteristics.

The ranks of coal recognized by the U. S. Geological Survey are Lignite. Sub-bituminous, Bituminous. Semi-bituminous. Semi-anthracite and Anthracite While the chemical composition of any rank may vary considerably, the figures of Table 37 indicate the approximate variation among the different ranks. Since the ash content is a matter of grade rather than rank the analyses are given on an ash-free basis.

Lignite—Lignite is coal in process of transformation from peat to sub-bituminous. It is brown in color and either woody or clay-like in appearance. The moisture content is high, varying from 30 to 45 per cent, and the available heat on combustion is correspondingly low. Lignite readily yields its moisture when exposed and disintegrates more readily than any other rank of coal. For this reason, unless care is taken in its storage, spontaneous combustion is very likely to occur. Further, the tendency toward ready slacking limits the markets for it largely to the localities where it is mined.

Lignite has been successfully used in gas producers and some experiments have been made burning it in pulverized form. It may be this latter field in which lignite may be best used.

Sub-bituminous—This rank is given to those coals ordinarily known as "black lignites." Sub-bituminous coals differ from lignites in color and in the absence of the distinctly woody texture found in the latter. While this coal approaches in value the poorer grades of bituminous coal, it differs from bituminous in its slacking characteristics, which tendency is great. As in the case of lignites, care must be taken in its storage to minimize the danger of spontaneous combustion. It is used ordinarily in the localities where mined and makes a clean domestic fuel.

TABLE 37

CHEMICAL COMPOSITION AND HEAT VALUE OF VARIOUS RANKS BASED ON ASH-FREE COAL

Rank	Fixed Carbon Per Cent	Volatile Matter Per Cent	Moisture Per Cent	Heat Value B.t.u.
Lignite	37.8	18.8	43.4	7400
Sub-bituminous	42.4	34.2	23.4	9720
Low-rank Bituminous	47.0	41.4	11.6	12880
Medium-rank Bituminous	54.2	40.8	5.0	13880
High-rank Bituminous	64.6	32.2	3.2	15160
Low-rank Semi-bituminous	75.0	22.0	3.0	15480
High-rank Semi-bituminous	83.4	11.6	5.0	15360
Semi-anthracite	83.8	10.2	6.0	14880
Anthracite	95.6	1.2	3.2	14440

Bituminous Coal—This is the lowest rank to which the Geographical Survey gives a fuel ratio, this being as a maximum 3, and in general the volatile and fixed carbon being approximately the same. This description does not, however, definitely classify bituminous coal, since the same fuel ratio is found in some of the coals of lower rank. It differs from lignite and sub-bituminous coal primarily in the effect of weathering, being only slightly affected by exposure except over extended periods.

Bituminous coal represents the largest division of classified coals and is widest in range of chemical composition. Numerous attempts have been made to subdivide this rank, but none of the schemes advanced has ever been generally accepted. The possibility or lack of possibility of coking is a classification frequently attempted, but since the chemical composition of coal has apparently no definite relation to its coking property, such a classification fails. In this rank are included the so-called gas coals, cannel coal and scattered local groups known to the trade as "block" or "splint" coals. Inasmuch as the bituminous rank includes all grades of coal, the logical subdivision from the commercial standpoint would appear to be a classification by heat value, say above or below 12,500 B.t.u. per pound.

The Seward Station of the Penn Public Service Corporation, Johnstown, Pa. Where Four 1000 Horse Power Babcock & Wilcox Boilers Are Operated

SEMI-BITUMINOUS — Mr. Marius B. Campbell, of the U. S. Geological Survey, refers to the term used for this rank of coal as "exceedingly unfortunate, as literally it implies that this coal is half the rank of bituminous, whereas it is . . . really super-bituminous." The Survey defines this coal definitely as having a fuel ratio of from 3 to 7, the narrowest limits of any rank. The high ratio of fixed carbon allows nearly smokeless combustion to be accomplished, and the better grades have a higher heat value than coal of any other rank. Semi-bituminous coal is friable and in transportation the size of individual pieces becomes relatively small. Burned commercially for steam generation, provided the size is relatively uniform, this fact is of but little importance. While some coal of this rank is found in scattered districts of the West, it is primarily an Eastern field coal.

SEMI-ANTHRACITE is defined as a "hard" coal having a fuel ratio of from 6 to 10, the lower limit being questionable and probably overlapping semi-bituminous. There is but little semi-anthracite in this country and it is of only small importance in the trade.

ANTHRACITE may be defined as a hard coal having a fuel ratio of not to exceed 50 or 60 and never less than 10. It is found principally in eastern Pennsylvania, and may be attributed to the peculiar crushing stresses of transformation affecting this particular locality. It is found in occasional deposits in the West, ordinarily in beds with coal of other ranks. Until quite recently anthracite was used almost entirely for domestic purposes or where smokeless combustion was necessary. For such purposes, particularly the former, fuel sizes had to be relatively large, and even for hand-fired anthracite furnaces the sizes required were at least No. 3 Buckwheat.* The selection of sizes for early commercial and domestic purposes resulted in a percentage of coal mined but not marketed that was appreciable. Improvements in methods of combustion, however, have opened a field for the commercial use of small sizes of anthracite that earlier were considered unavailable for fuel purposes.

CLASSIFICATION OF COAL AREAS — Numerous attempts to classify coal distribution geographically have been made, but owing to the almost unlimited number of possible subdivisions or combinations of areas, they have proven ordinarily unserviceable. The U. S. Geological Survey now recognizes four classes of areas which, beginning with the smallest, it calls district, field, region and province.

COAL DISTRICT is a term restricted to areas in which mines are developed continuously in given beds, or to developed areas around a fairly definite center. Coal from a district is ordinarily sold under a trade name, usually taken from that of the town forming the center of development. No definite limits can be set to the areas that can be classed as districts.

COAL FIELDS are the areas next in size to districts, though the term is rather loosely used, and while larger, a field still applies to a well-defined compact area. In some cases the term is applied to an area because of the class of coal produced and is not limited to a center of production.

COAL REGION — Fields are grouped into larger divisions called regions, the coals of which have some characteristic or characteristics in common. The anthracite region of Pennsylvania illustrates an area of this classification.

COAL PROVINCE — Coal regions are grouped into provinces largely because of geologic considerations, quality of coal as a class, and questions of transportation.

* See page 209.

The Marion Station of the Columbus Delaware & Marion Electric Company Operating Four 800 Horse-power Babcock & Wilcox Boilers Equipped with Babcock & Wilcox Superheaters

There is a certain degree of similarity in the physical features of the fields of a province or in the quality of coal. Some provinces contain all ranks of coal, the fields being grouped because of geographical position, structural features or age of beds.

The provinces classified by the Geographical Survey are Eastern, Interior, Gulf, Northern Great Plain, Rocky Mountain and Pacific Coast.

THE EASTERN PROVINCE contains approximately 90 per cent of the high-rank coal of the United States and includes the anthracite regions of Pennsylvania and Rhode Island, the Atlantic Coast region of Virginia and North Carolina and the Appalachian region. The anthracite region of Rhode Island is of but little importance.

The anthracite region of Pennsylvania is divided by the trade into four fields, these and the principal districts being:

Northern or Wyoming Field
- Carbondale District
- Scranton District
- Pittston District
- Wilkes-Barre District
- Plymouth District

Eastern Middle or Lehigh Field
- Green Mountain District
- Black Creek District
- Hazelton District
- Beaver Meadow District
- Panther Creek District

Western Middle or Mahanoy Field
- East Mahanoy District
- West Mahanoy District

Southern or Schuylkill Field
- East Schuylkill District
- West Schuylkill District
- Lorberry District
- Lykens Valley District
- Shamokin District

The Atlantic Coast region is of but little importance, the amount of coal mined being negligible.

The Appalachian region contains the greatest amount of high-rank coal in the United States and probably in the world. Beginning near the northern boundary of Pennsylvania, in the western portion of the State, it extends southwestward through West Virginia, touching Maryland and Virginia on their western borders, passing through southeastern Ohio, eastern Kentucky and central Tennessee, and ending in western Alabama, 900 miles from its northern extremity.

Along the eastern margin of the Appalachian region are a number of isolated natural fields known as "outliers." The best known of these are Broadtop, in south central Pennsylvania, Georges Creek and Upper Potomac in Maryland and West Virginia, Lookout Mountain in Georgia and Alabama, and Cahaba in Alabama.

The coal of the Appalachian region is generally of high rank but of appreciable variation in chemical composition, the fixed carbon increasing and the volatile matter decreasing in a west to east direction. The semi-bituminous coals are found on the eastern margin of this region.

THE INTERIOR PROVINCE includes the fields and regions bordering the Great Lakes, in the Mississippi Valley, and in Texas. The four general regions are the Northern (Michigan), the Eastern (Illinois, Indiana and western Kentucky), the Western (Iowa, Missouri, Kansas, Oklahoma and Arkansas), and the Southern (Texas).

Part of a Total Installation of 8552 Horse Power of Babcock & Wilcox Boilers in the Power Station of the Dallas Power & Light C[illegible]

Practically all of the coal of the Interior Province is of low rank, the pressures bringing about devolatilization in formation being due only to the weight of overlying rocks and not to violent changes in the contour of the fields, such as occurred in the Appalachian region. Where the borders of the individual fields approach mountainous regions, as in Arkansas, the coal is of higher rank, even approaching semi-anthracite.

The coals of the southern portion of the Interior Region are high in moisture and low in heat value.

THE GULF PROVINCE is of but little importance. The coal is practically all lignite, and while it is mined for local use it is not extensively used, due to better grades being available from adjacent fields.

THE NORTHERN GREAT PLAINS PROVINCE includes all of the fields in the great plains lying east of the Front Range of the Rocky Mountains. As in the Interior Province, due to the absence of earth upheavals during formation, the coals are of low rank, being mostly lignite and sub-bituminous except in the regions bordering the mountains.

The largest region of this province is the Fort Union region in North and South Dakota, Montana and Wyoming. In the northern portion of this region the coal is lignite of a very poor quality, the moisture content varying from 40 to 45 per cent. The quality improves toward the southern portion of the region, reaching its best development at Sheridan, Wyoming, where the moisture content is approximately 25 per cent. This region is largely one of farming and cattle raising, and but little coal is mined except at Sheridan, Wyoming.

Other fields of this province are the Bull Mountain, Great Falls and Lewiston in Montana, Denver region in Colorado, Cañon City Field and the Raton Mesa region, in southern Colorado and northern New Mexico. That portion of the Raton Mesa region lying in Colorado is known as the Trinidad field and the section in New Mexico as the Raton field, the coals of these fields being the highest rank coal of the province.

THE ROCKY MOUNTAIN PROVINCE contains a greater variety of coal than any other province, including all ranks from lignite to anthracite, though it is for the most part lignite and sub-bituminous. A partial list of the fields of the province and their location follows:

Red Lodge Field	Montana
Wind River Basin	Western Wyoming
Green River Region	Southwestern Wyoming
Hams Fork Region	Southwestern Wyoming
Hanna Field	Wyoming
Uinta Region	Western Colorado
Uinta Region	Eastern Utah
San Juan River Region	Northwestern New Mexico

There are numerous other fields in this province that have been investigated but which, however, have not been developed. The greatest centers of development in the province are Red Lodge, Mont., Rock Springs and Kemmerer, Wyo., Crested Butte and Durango, Colo., Castlegate and Sunnyside, Utah, and Gallup, N. M.

PACIFIC COAST PROVINCE—While there are small coal fields of low grade and low rank in California and Oregon, they are of no importance, and the coals of this

Partial View of [illegible],500 Horse Power Installation of Babcock & Wilcox Boilers at the Philadelphia, Pa., Plant of the Baldwin Locomotive Works. This Company Operates in its Various Plants a Total of 17,700 Horse Power of Babcock & Wilcox Boilers

province may be considered largely limited to Washington. The disturbances during formation in this state were so great that it is not possible to separate fields definitely, and the Geological Survey follows local practice in classifying the coal in accordance with the counties in which it is mined.

In King county both sub-bituminous and bituminous coals are found, though the former, because of location relative to tidewater, is more extensively mined. In Pierce county, though the coal areas are not as large as in King, is found the best coal in the State. In this county the Wilkeson-Carbon Hill districts produce high-grade bituminous and semi-bituminous coals, while the Melmont, Fairfax and Montezuma districts produce coals having fuel ratios of from 2.5 to 6. The Roslyn field, in Kittitas county, produces a hard blocky coal of good quality well adapted to shipping. The fields of Thurston and Lewis counties are not extensively developed and are sub-bituminous or lignitic in character. Whatcom and Skagit counties contain fields of smaller area than the other counties discussed, the rank varying from sub-bituminous at Bellingham to anthracite at Glacier. But little mining is done in either of these counties.

ALASKA COALS—Coal has been found in Alaska and undoubtedly is of great value. though the extent and character of the fields have probably been exaggerated. Great quantities of lignite are known to exist, and in quality the coal ranges in character from lignite to anthracite. There are at present, however, only two fields of high-grade coals known, these being the Bering River Field, near Controllers Bay, and the Matanuska Field, at the head of Cook Inlet. Both of these fields are known to contain both anthracite and high-grade bituminous coals, though as yet they cannot be said to have been opened up.

WEATHERING OF COAL—The storage of coal has become within the last few years a necessity to a certain extent, due to market conditions, danger of labor difficulties at the mines and in the railroads, and the crowding of transportation facilities. The first cause is probably the most important, and this is particularly true of anthracite coals where a sliding scale of prices is used according to the season of the year. While market conditions serve as one of the principal reasons for coal storage, most power plants and manufacturing plants feel compelled to protect their coal supply from the danger of strikes, car shortages and the like, and it is customary for large power plants, railroads and coal companies to store bituminous coal. Naval coaling stations are also an example of what is done along these lines.

Anthracite is the nearest approach to the ideal coal for storing. It is not subject to spontaneous ignition, and for this reason is unlimited in the amount that may be stored in one pile. With bituminous coals, however. the case is different. Most bituminous coals will ignite if placed in large enough piles and all suffer more or less from disintegration. Coal producers only store such coals as are least liable to ignite, and which will stand rehandling for shipment.

The changes which take place in stored coal are of two kinds: 1st, the oxidization of the inorganic matter such as pyrites, and, 2nd, the direct oxidization of the organic matter of the actual coal.

The first change will result in an increased volume of the coal, and sometimes in an increased weight, and a marked disintegration. The changes due to direct oxidization of the coal substances usually cannot be detected by the eye, but as they involve the oxidization of the carbon and available hydrogen and the absorption of

Installing One of the Eleven Babcock & Wilcox Boilers 4105 Horse Power Operated at the Botany Worsted Mills Passaic, N. J.

the oxygen by unsaturated hydrocarbons, they are the chief cause of the weathering losses in heat value. Numerous experiments have led to the conclusion that this is also the cause of spontaneous combustion.

Experiments to show loss in calorific heat values due to weathering indicate that such loss may be as high as 10 per cent when the coal is stored in the air, and 8.75 per cent when stored under water. It would appear that the higher the volatile content of the coal, the greater will be the loss in calorific value and the more subject to spontaneous ignition.

Some experiments made by Messrs. S. W. Parr and W. F. Wheeler, published in 1909 by the Experiment Station of the University of Illinois, indicate that coals of the nature found in Illinois and neighboring states are not affected seriously during storage from the standpoint of weight and heating value. the latter loss averaging about 3½ per cent for the first year of storage. They found that the losses due to disintegration and to spontaneous ignition were of greater importance. Their conclusions agree with those deduced from the other experiments, viz., that the storing of a larger size coal than that which is to be used will overcome, to a certain extent, the objection to disintegration, and that the larger sizes, besides being advantageous in respect to disintegration, are less liable to spontaneous ignition. Storage under water will, of course, entirely prevent any fire loss and, to a great extent, will stop disintegration and reduce the calorific losses to a minimum.

To minimize the danger of spontaneous ignition in storing coal, the piles should be thoroughly ventilated.

Pressed or Briquetted Fuels—The term briquette is now generally accepted as applying to any manufactured solid fuel, whether with or without a binder. The use of this class of fuel is extremely old and in European countries, because of lack of fuels of better quality, extensive. In this country, because of the availability of natural fuels, briquettes have never come into common use. With increasing fuel costs, however, and improved methods of briquette manufacture, there will undoubtedly be an increase in the field for such fuel, particularly for the briquettes manufactured from the low rank of fuels.

In this country the factors militating against briquetted fuels have been the cost and difficulty of manufacture, which were such as to enable no real competition with the natural fuels available, even in the selected sizes. The art, however, has progressed in the past few years and today briquettes are available that in certain localities, and particularly for domestic purposes, can successfully compete with natural fuels.

The primary advantage of briquetted fuel claimed by the manufacturers is the increased heat value per unit of volume. This comes, not from a change in the composition of the coal briquetted, but from the addition of the binder, which is tar, coal-tar pitch, oil or asphaltum.

The principal factors in obtaining proper combustion are apparently in the size and form of the briquettes. For commercial reasons, the shape of the briquettes must be such as to allow handling in spouts, buckets, shovels, etc., in the same manner as coal, with as few corners or rough edges as possible to eliminate breakage or abrasion. The size is of importance as affecting the air space through the fuel bed. For domestic purposes, sizes representing a weight of from 1⅝ to 2½ ounces have been shown to give the best results, and while sizes corresponding to as high as 5 ounces have been manufactured for commercial steam generation, smaller sizes apparently give better results. The larger the furnace, the larger the size of briquetted fuel that can be satisfactorily handled.

The briquettes of present-day manufacture are offered by their makers, of the proper size and proper hardness and possessing the ability to withstand weathering, which may or may not lead to the more common adoption of this class of fuel, at least for domestic purposes.

COKE—Mr. F. W. Sperr, Jr., of the Mellon Institute of Industrial Research, defines coke as "the coherent cellular residue of the destructive distillation of coal at high temperatures." The same definition, however, may be applied to the residues from the distillation of coal-tar pitch, petroleum, asphalt, and other bituminous bases. The latter classes of coke, while important in the electrochemical field, are of no importance in that of steam generation.

The smaller sizes of coke, pea and nut, have a ready market for domestic use. As a fuel for the commercial generation of steam, however, coke may be considered as limited to breeze. Coke breeze is that portion of coke which passes through a ½-inch screen with square openings, average sieve tests of a by-product plant showing

	Per Cent
Through ½ and on ¼ inch . .	23.8
Through ¼ and on ⅛ inch	26.0
Through ⅛ inch	50.2

Coke breeze is difficult to burn and it is only within the past few years that methods of burning have been developed that have made it a commercial fuel. Today it is being satisfactorily handled on blast chain-grate stokers, and inasmuch as it has a high heat value, approximately 12,500 B.t.u., it is a valuable class of fuel.

CENTRAL JARONU CUBA, WITH TEN BABCOCK & WILCOX BOILERS, 11,120 HORSE-POWER, EQUIPPED WITH BABCOCK & WILCOX SUPERHEATERS, OIL BURNERS AND BAGASSE BURNERS

ANALYSIS AND HEATING VALUE OF FUELS

THE heating value of a fuel may be determined either by the direct measurement of the heat evolved during combustion in a calorimeter, or by computation based on a chemical analysis of the fuel. The former method is decidedly preferable where accuracy is desired and should be used wherever possible.

A fuel analysis may be reported in two different forms. A proximate analysis, which is the more common, determines only the percentages of moisture, fixed carbon, volatile matter and ash, without determining the ultimate composition of the volatile matter. Such an analysis is possible only with solid fuels.

An ultimate analysis reduces the fuel to its elementary constituents of carbon, hydrogen, oxygen, nitrogen, sulphur, ash and moisture. Such an analysis requires the services of a competent chemist and laboratory equipment that should include a calorimeter for the direct determination of the heat value of the fuel. On the other hand, there are apparently many instances in which an ultimate fuel analysis is available without a heat value for the particular fuel used, and for this reason the methods of computation of heat value are given here. The accuracy of a computed as compared with a calorimetric heat value is given hereafter, and is appreciably greater than the heat value computed from a proximate analysis. The methods employed in an ultimate analysis may be found in books on engineering chemistry, or in Technological Paper No. 8 of the Bureau of Mines.

An ultimate analysis, while resolving the fuel into its elementary constituents, does not reveal how these may have been combined in the fuel as analyzed. The manner of their combination undoubtedly has a direct effect upon the calorific value of the fuel, since fuels having almost identical ultimate analysis show a difference in heating values as determined by calorimetric test.

The relative values of the constituents of solid fuels, as reported in a proximate or ultimate analysis, are indicated by Table 38, taken from the analyses made in connection with the coal testing plant of the United States Geological Survey at the Louisiana Purchase Exposition, St. Louis.

Proximate and ultimate fuel analyses are reported on both a wet and a dry fuel basis. Inasmuch as the latter is the basis ordinarily accepted for comparative results, it would appear to be the proper basis on which results should be given. When an analysis is given on the wet basis, it may be readily converted to the dry basis by dividing the percentage of the different constituents by one minus the percentage of moisture, reporting the moisture content separately. This method, which is the same for either a proximate or an ultimate analysis, is represented by an example of the latter:

	Moist Fuel	*Dry Fuel*
C	83.95	84.45
H	4.23	4.25
O	3.02	3.04
N	1.27	1.28
S	0.91	0.91
Ash	6.03	6.07
		100.00
Moisture	0.59	0.59
	100.00	

Portion of 1800 Horse-power Installation of Babcock & Wilcox Boilers at the Billings Sugar Co., Billings, Mont. 604 Horse-power of These Boilers Are Equipped with Babcock & Wilcox Chain Grate Stokers

TABLE 38

SHOWING RELATION BETWEEN PROXIMATE AND ULTIMATE ANALYSES OF COAL

State	Field or Bed	Mine	Proximate Analysis		Ultimate Analysis					Common to Proximate and Ultimate Analyses	
			Volatile Matter	Fixed Carbon	Carbon	Hydrogen	Oxygen	Nitrogen	Sulphur	Ash	Moisture
Ala.	Horse Creek	Ivy Coal & Iron Co., No. 8	31.81	53.90	72.02	4.78	6.45	1.66	.80	14.29	2.56
Ark.	Huntington	Central C. & C. Co., No. 3	18.99	67.71	76.37	3.90	3.71	1.49	1.23	13.30	1.99
Ill.	Pana or No. 5	Clover Leaf, No. 1	37.22	45.64	63.04	4.49	10.04	1.28	4.01	17.14	13.19
Ind.	No. 5, Warrick Co.	Electric	41.85	44.45	68.08	4.78	7.56	1.35	4.53	13.70	9.11
Ky.	No. 11, Hopkins Co.	St. Bernard, No. 11	41.10	49.60	72.22	5.06	8.44	1.33	3.65	9.30	7.76
Pa.	"B" or Lower Kittanning	Eureka, No. 31	16.71	77.22	84.45	4.25	3.04	1.28	.91	6.07	.56
Pa.	Indiana Co.	. . .	29.55	62.64	79.86	5.02	4.27	1.86	1.18	7.81	2.90
W. Va.	Fire Creek	Rush Run	22.87	71.56	83.71	4.64	3.67	1.70	.71	5.57	2.14

CALCULATIONS FROM AN ULTIMATE ANALYSIS—The first formula for the calculation of heating values from the composition of a fuel as determined from an ultimate analysis is due to Dulong, and this formula, slightly modified, is the most commonly used today. Other formulæ have been proposed, some of which are more accurate for certain specific classes of fuel, but all have their basis in Dulong's formula, the accepted modified form of which is:

Heat units in B.t.u. per pound of dry fuel =

$$14{,}600C + 62{,}000\left(H - \frac{O}{8}\right) + 4000S \quad (25)$$

where C, H, O, and S are the proportionate parts by weight of carbon, hydrogen, oxygen and sulphur.

Assume a coal of the composition given. Substituting in this formula (25):

Heating value per pound of dry coal:

$$= 14{,}600 \times 0.8445 + 62{,}000 \left(0.0425 - \frac{0.0304}{8}\right) + 4000 \times 0.0091 = 14{,}765 \text{ B.t.u.}$$

This coal, when tested in a calorimeter, showed a heat value of 14,843 B.t.u. per pound, and from a comparison of these values the accuracy of computation as against calorimeter determination is indicated. The investigations of Lord and Haas in this country, Mahler in France, and Bunte in Germany, all indicate that the use of Dulong's formula gives results which approximate calorimeter tests closely, except in the case of cannel coal, lignite, peat and wood. The difference with these fuels unquestionably results from the manner of combination of the hydrocarbons in the fuel, previously discussed.

A number of methods of computing heat values of solid fuels based on proximate analyses have been offered. William Kent's method is based upon the relation between heat value and fixed carbon in the combustible matter, while Goutal gives the fixed carbon a fixed value and the heat value of

Station of the Buffalo General Electric Company, with Eleven [illegible] Horse-power Babcock & Wilcox Boilers with Babcock & Wilcox Superheaters

the volatile matter is considered as a function of its percentage referred to combustible. These two methods give results that approximate each other, but the accuracy, as compared with calorimetric determinations, is less than that of the use of Dulong's formula, particularly with fuels containing less than 60 per cent of fixed carbon in the combustible matter. Where no method of determining the heat value of the fuel other than the proximate analysis is possible, perhaps the use of the isothermal-isocalorific proposed by the Bureau of Mines Technical Paper No. 93, "Graphic Studies of Ultimate Analyses of Coal." will be found helpful.

HIGH AND LOW HEAT VALUES—The heat value of a fuel, as defined, is known as the "higher" heat value and is ordinarily accepted as the standard in this country. In the case of fuel containing hydrogen, and this includes practically all fuels in commercial use, there is another value known as the "lower," "net" or "available" heat value, in the determination of which an attempt is made to allow for the latent heat recovered in the condensation of the water vapor formed in the combustion of hydrogen. For example, in the calorimetric determination of the heat value of a fuel containing hydrogen, the products of combustion are cooled to approximately the temperature of the original mixture, say 62 degrees Fahrenheit. In cooling the products to this temperature, the water vapor formed by the combustion of hydrogen is condensed, and the result, expressed in B.t.u., after being corrected for sulphur and like factors, *i. e.*, the higher heat value, includes the latent heat of water vapor given up in such condensation.

If the lower value be represented by H_p' and the weight of water produced per pound of fuel by w, the lower heat value may be determined from

$$H_p' = H_p - wr \qquad (20)$$

where H_p equals the higher heat value and r is a factor which varies with the percentage of hydrogen in the fuel, the amount of air or oxygen used in combustion, the moisture in the air and the temperature to which the products of combustion are cooled in the calorimeter. Too frequently r is simply taken as the latent heat of steam, either at 32 degrees or 212 degrees, though in calorimetric work neither of these temperatures is likely to occur.

With the lower heat value so defined, the difference between the higher and the net value will obviously be the total heat of the steam or water vapor as it escapes less the sensible heat of an equivalent weight of water at the temperature of the fuel and of the oxygen before combustion takes place.

The lower heat value is in common use in Great Britain and in most foreign countries. In this country, the higher value is almost universally accepted, and this is the standard recommended by The American Society of Mechanical Engineers.

Any attempt to make use of the lower heat value introduces a source of possible error in the proper temperature for use in computation, and advocates of the use of this value are not in entire agreement as to the proper methods of such computations.

To sum up, a theoretically perfect absorption of heat after combustion would condense all of the vapor formed in the burning of hydrogen, and since the efficiency of any apparatus is based upon the performance of a theoretically perfect machine, it appears only logical to charge against the apparatus what would be secured from the theoretically perfect. Further, in the report of the performance of any apparatus, a heat balance offers a method of determining and expressing any loss due to the burning of hydrogen, and no such test or performance report can be accepted as reliable unless accompanied by a heat balance or by data from which a heat balance may be computed.

HEAT VALUE OF GASEOUS FUELS—The heat value of gaseous fuels, as in the case of solid fuels, should for accuracy be determined by actual combustion in a calorimeter, of the continuous or constant flow type. The Junker calorimeter is ordinarily accepted as standard for the determination of the heat value of gaseous fuels. Such value may be computed by Dulong's formula, but, because of the necessity of breaking up the gases as ordinarily reported in the analysis into their constituent elements, there is some danger of arithmetical error. The heating values of the combustible elements and compounds entering into gaseous fuels have been definitely determined by calorimetric tests, and it is simpler to use such values in the determination of the value for the fuel as a whole than to make use of Dulong's formula. The heat values of the gases ordinarily encountered in combustion work are given in Table 33.

The method of computing such a heat value is as follows:

Assume a natural gas having the volumetric analysis as given below, with corresponding weights per cubic foot as determined by the methods of converting from volumetric to weight analysis which have been indicated.

	Volume per Cubic Foot	*Percentage by Weight*	*Weight per Pound*		*B.t.u. per Pound*		*B.t.u.*
CO	0.0045	0.762	0.00762	×	4380	=	33
H_2	0.0182	0.221	0.00221	×	62000	=	137
CH_2	0.9333	91.188	0.91188	×	23850	=	21748
C_2H_4	0.0025	0.423	0.00423	×	21450	=	91
H_2S	0.0018	0.376	0.00376	×	7458	=	28
O_2	0.0035	0.677	. . . . .		. . . .		. . . .
CO_2	0.0022	0.591	. . . .		. . . .		. . . .
N_2	0.0340	5.762	. . . .		. . . .		. . . .
		Total B.t.u. per pound			. . .	=	22037

The gas in question, under standard conditions, weighs 0.046058 pound per cubic foot and the heat value per cubic foot is thus:

$$22037 \times 0.46058 = 1015 \text{ B.t.u.}$$

The heat value per cubic foot could, of course, be computed directly from the values of Table 33.

PROXIMATE ANALYSIS—The proximate analysis of a fuel, as stated, reports the percentage by weight of moisture, fixed carbon, volatile matter and ash. The laboratory sample of fuel as obtained during the boiler trial,* reduced to about one quart, should be carefully sealed and shipped for analysis. The method of such an analysis is as follows:

Moisture. The moisture content is determined in one of two ways. In the first method, a differentiation is made between surface moisture as represented by that driven off by air drying (1) and inherent moisture, while the second (2) determines only total moisture.

* For methods of sampling see chapter on Efficiency and Capacity of Boilers.

(1) The sample is dried in an oven at a temperature of from 30 to 35 degrees Centigrade (86 to 98 degrees Fahrenheit), through which a current of warm air is drawn, until a constant weight of the sample is obtained. Such loss represents the air-drying loss, and is assumed to correct for difference in conditions between the time the sample is taken and that of this moisture determination.

The sample, after this drying, is quickly pulverized, quartered, and a portion reduced to 60 mesh. One gram of this portion is used for the determination of the residual or inherent moisture, by drying the sample in a shallow dish placed in a glycerine-jacketed drying oven through which dry air at a temperature of 105 degrees Centigrade (221 degrees Fahrenheit) is drawn. By this method, the loss so determined represents the loss at 105 degrees Centigrade, while the sum of the air drying loss and this loss represents the total moisture loss.

This method is that recommended by the Bureau of Mines, but if the results of a boiler test are to be reported, as has been recommended, on a dry fuel basis, it has no advantage over the second method.

(2) The entire sample is dried at a temperature of 105 degrees Centigrade (221 degrees Fahrenheit) to constant weight, the loss in weight representing moisture. The dry sample is then crushed, quartered through a riffle, and a portion of it ground to 60 mesh in an Abbé ball mill using flint pebbles. This sample is used for the analysis.

Volatile Matter. The determination of the volatile constituent of the coal is best accomplished by the use of an electric furnace of special design, the temperature of which can be maintained constant and which will give a crucible temperature of 950 degrees Centigrade (1742 degrees Fahrenheit). The temperature should be controlled by means of a thermocouple, the tip of which almost touches the bottom of the crucible. The sample is placed in a 10-cubic centimeter capsule-covered platinum crucible and placed in the furnace described, the heating period being seven minutes. The crucible is then removed from the furnace, cooled and weighed. If the sample used in this determination was of the fuel air dried, as described under the first method of moisture determination, the loss in weight represents the volatile matter plus the moisture at 105 degrees Centigrade, while if the sample was completely dried, as described in the second method of moisture determination, the loss in weight represents directly the volatile content.

Ash. If the first method of moisture determination has been used, the ash constituent is determined from the same sample used in such determination and not from that used for volatile matter. If the second moisture method has been used it is best to take a fresh sample for ash. The sample, in a crucible, is placed in a cool muffle furnace and the temperature gradually raised to 750 degrees Centigrade (1382 degrees Fahrenheit) and ignited. The ash should be occasionally stirred, and the temperature maintained until a constant weight is reached.

Fixed Carbon. The fixed carbon is the difference between 100 and the sum of the ash and volatile matter on a dry basis, determined as described.

Table 39 gives the proximate analysis and calorific value of a number of representative coals found in the United States*

---

* Bulletins Nos. 22, 123 and 193 of the Bureau of Mines give many thousands of analyses of mine and car samples collected between 1904 and 1919, together with descriptions of the beds where mined.

CALORIMETRIC DETERMINATION OF HEAT VALUE (SOLID AND LIQUID FUELS)—While an ultimate or a proximate analysis is useful in the determination of the general characteristics of a fuel and may be used for an approximate determination of its heating value, wherever the accurate determination of the efficiency of a steam generating unit is involved, such heat value should be determined by means of a fuel calorimeter. In such an apparatus the fuel is completely burned under oxygen pressure, the heat evolved in combustion absorbed by water, and the amount of such heat calculated from the rise in water temperature.

There are a number of fuel calorimeters available for heat value determinations, which vary somewhat in design and construction, but which in principle of operation and manipulation are essentially the same. The following description of the parts and operation of a calorimeter applies in general to any of the standard types.

Calorimeter Bomb. The combustion bomb is made of steel, lined with material which is not attacked by nitric and sulphuric acid or other products of combustion. The bomb is equipped with a small pan attached internally, in which the sample of the fuel to be tested is placed, and with two connections to which are attached leads, passing through the jacket cover to a source of current by which ignition of the sample tested is accomplished. Actual ignition is brought about by the use of a fuse wire passing through the fuel sample in the pan, such wire completing the electric circuit.

Calorimeter Bucket. The bucket is a vessel in which the bomb is placed. It is filled with a definite weight of water, the rise in temperature of which, when the sample is burned, is the measure of the heat evolved.

Calorimeter Jacket. The calorimeter is equipped with an outer water jacket which has a cover to prevent air currents from reaching the calorimeter. The jacket is either a plain jacket filled with water at approximately room temperature, or a so-called adiabatic water jacket. Where the former is used, a radiation correction must be made in the computation of results from the observed data.

Where an adiabatic water jacket is used, radiation is eliminated by raising the temperature of the jacket water at the same rate as the temperature of the water in the calorimeter increases after combustion of the fuel sample. With the temperature of the water in the jacket kept the same as that of the water in the calorimeter, as the temperature of the latter increases following ignition of the sample, there is no exchange of heat between the calorimeter bucket and its surrounding jacket and the necessity for a cooling or radiation correction is obviated. This is of importance in scientific calorimetric work as tending toward increased accuracy, while in commercial work it is of advantage in that it eliminates the necessity of thermometer readings prior to ignition and following the maximum temperature of calorimeter water after combustion, which are necessary in the computation of the cooling correction.

The temperature of the cooling jacket is changed with the temperature of the calorimeter, either by means of an electric current or by an adjustable hot and cold water supply. A form of adiabatic water jacket developed by Dr. Farrington Daniels, of the Worcester Polytechnic Institute, utilizes the water itself as the heating medium, and by passing an electric current through such water its temperature is raised simultaneously with that of the calorimeter. With this form of jacket there is no lag due to the use of heating coils or necessity for stirring jacket water, since it is heated uniformly throughout the jacket.

State	County	Field, Bed or Vein	Mine	Size	Moisture	Proximate Analysis (Dry Coal)			B.t.u. Per Pound Dry Coal	Authority
						Volatile Matter	Fixed Carbon	Ash		
		ANTHRACITES								
Pa.	Carbon	Lehigh	Beaver Meadow	. . .	1.50	2.41	90.30	7.29	. . .	Gale
Pa.	Dauphin	Schuylkill	. . .	Buckwheat	2.15	12.88	78.23	8.89	13137	Whitham
Pa.	Lackawanna	Wyoming	Belleview	No. 2 Buck	8.29	7.81	77.19	15.00	12341	Sadtler
Pa.	Lackawanna	Wyoming	Johnson	Culm	13.90	11.16	65.96	22.88	10591	B. & W. Co.
Pa.	Luzerne	Wyoming	Pittston	No. 2 Buck.	3.66	4.40	78.96	16.64	12865	B. & W. Co.
Pa.	Luzerne	Wyoming	Mammoth	Large	4.00	3.44	90.59	5.97	13720	Carpenter
Pa.	Luzerne	Wyoming	Exeter	Rice	0.25	8.18	79.61	12.21	12400	B. & W. Co.
Pa.	Northumberland	Schuylkill	Treverton	. . .	0.84	6.73	86.39	6.88	. . .	Isherwood
Pa.	Schuylkill	Schuylkill	Buck Mountain	. . .	. . .	3.17	92.41	4.42	14220	Carpenter
Pa.	Schuylkill	. . .	York Farm	Buckwheat	0.81	5.51	75.90	18.59	11430	. . .
Pa.	. . .	. . .	Victoria	Buckwheat	4.30	0.55	86.73	12.72	12642	B & W Co.
Pa.	Carbon	Lehigh	Lehigh & Wilkes.C.Co.	Buck. and Pea	1.57	6.27	66.53	27.20	12848	B. & W. Co.
Pa.	Carbon	Lehigh	. . .	Buckwheat	. . .	5.00	81.00	14.00	11800	Carpenter
Pa.	Lackawanna	. . .	Del & Hudson Co	No. 1 Buck.	6.20	. . .	. . .	11.60	12100	Denton
		SEMI-ANTHRACITES								
Pa.	Lycoming	Loyalsock	. . .	. . .	1.30	8.72	84.44	6.84	. . .	. . .
Pa.	Sullivan	. . .	Lopez	. . .	5.48	7.53	81.00	11.47	13547	B. & W. Co.
Pa.	Sullivan	Bernice	. . .	. . .	1.29	8.21	84.43	7.36	. . .	. . .
		SEMI-BITUMINOUS								
Md.	Alleghany	Big Vein, George's Crk.	. . .	. . .	3.50	21.33	72.47	6.20	14682	B. & W. Co.
Md.	Alleghany	George's Creek	. . .	. . .	3.63	16.27	76.93	6.80	14695	B. & W. Co.
Md.	Alleghany	George's Creek	. . .	. . .	2.28	19.43	77.44	6.13	14793	B. & W. Co.
Md.	Alleghany	George's Creek	Ocean No. 7	Mine run	1.13	. . .	. . .	. . .	14451	B. & W. Co.
Md.	Alleghany	Cumberland	. . .	. . .	1.50	17.26	76.65	6.09	14700	. . .
Md.	Garrett	. . .	Washington No. 3	Mine run	2.33	14.38	74.93	10.49	14033	U. S Geo. S.*
Pa.	Bradford	. . .	Long Valley	. . .	1.55	20.33	68.38	11.29	12965	. . .
Pa.	Tioga	. . .	Antrim	. . .	2.19	18.43	71.87	9.70	13500	. . .
Pa.	Cambria	"B" or Miller	Sonman Shaft C. Co.	. . .	3.40	20.70	71.84	7.46	14484	N. Y. Ed. Co.
Pa.	Cambria	"B" or Miller	Henrietta	. . .	1.23	18.37	75.28	6.45	14770	So. Eng. Co.
Pa.	Cambria	"B" or Miller	Penker	. . .	3.64	21.34	70.48	8.18	14401	B. & W. Co.
Pa.	Cambria	"B" or Miller	Lancashire	. . .	4.38	21.20	70.27	8.53	14453	B. & W. Co.
Pa.	Cambria	Lower Kittanning	Penn. C. & C. Co. No. 3	Mine run	3.51	17.43	75.69	6.88	14279	U. S. Geo. S.
Pa.	Cambria	Upper Kittanning	Valley	Mine run	3.40	14.89	75.03	10.08	14152	B. & W. Co.
Pa.	Clearfield	Lower Kittanning	Eureka	Mine run	5.90	16.71	77.22	6.07	14843	U. S. Geo. S.
Pa.	Clearfield	. . .	Ghem	Mine run	3.43	17.53	69.67	12.80	13744	B. & W. Co.
Pa.	Clearfield	. . .	Osceola	. . .	1.24	25.43	68.56	6.01	13589	B. & W. Co.
Pa.	Clearfield	Reynoldsville	. . .	. . .	2.91	21.55	69.03	9.42	14685	B. & W. Co.
Pa.	Clearfield	Atlantic-Clearfield	. . .	Mine run	1.55	23.36	71.15	5.94	13963	Whitham
Pa.	Huntington	Barnet & Fulton	Carbon	Mine run	4.50	18.34	73.06	8.60	13770	B. & W. Co.
Pa.	Huntington	. . .	Rock Hill	Mine run	5.91	17.58	73.44	8.99	14105	B. & W. Co.
Pa.	Somerset	Lower Kittanning	Kimmelton	Mine run	3.09	17.84	70.47	11.69	13424	U. S. Geo. S.
Pa.	Somerset	"C" Prime Vein	Jenner	Mine run	9.37	16.47	75.76	7.77	14507	P. R. R.

* U. S. Geological Survey.

APPROXIMATE COMPOSITION AND CALORIFIC VALUE OF CERTAIN TYPICAL AMERICAN COALS—CONTINUED

State	County	Field, Bed or Vein	Mine	Size	Moisture	Proximate Analysis (Dry Coal)			B.t.u. Per Pound Dry Coal	Authority
						Volatile Matter	Fixed Carbon	Ash		
W. Va.	Fayette	New River	Rush Run	Mine run	2.14	22.87	71.56	5.57	14959	U. S. Geo. S.
W. Va.	Fayette	New River	Loup Creek	. . . .	0.55	19.36	78.48	2.16	14975	Hill
W. Va.	Fayette	New River	. . . .	Slack	6.66	20.94	73.16	5.90	14412	B. & W. Co.
W. Va.	Fayette	New River	. . . .	Mine run	2.16	17.82	75.66	6.52	14786	B. & W. Co.
W. Va.	Fayette	New River	Rush Run	Mine run	0.94	22.16	75.85	1.99	15007	B. & W. Co.
W. Va.	McDowell	Pocahontas No. 3	Zenith	Mine run	4.85	17.14	76.54	6.32	14480	U. S. Geo. S.
W. Va.	McDowell	Tug River	Big Sandy	Mine run	1.58	18.55	76.44	4.91	15170	U. S. Geo. S.
W. Va.	Mercer	Pocahontas	Mora	Lump	1.74	18.55	75.15	6.30	15015	U. S. Geo. S.
W. Va.	Mineral	Elk Garden	. . . .	. . . .	2.10	15.70	75.40	8.90	14195	B. & W. Co.
W. Va.	McDowell	Pocahontas	Flat Top	Mine run	0.52	24.02	74.59	1.39	14490	B. & W. Co.
W. Va.	McDowell	Pocahontas	Flat Top	Slack	3.24	15.33	77.60	7.07	14653	B. & W. Co.
W. Va.	McDowell	Pocahontas	Flat Top	Lump	3.63	16.03	78.04	5.93	14956	B. & W. Co.
		BITUMINOUS								
Ala.	Bibb	Cahaba	Hill Creek	Mine run	6.19	28.58	55.60	15.82	12576	B. & W. Co.
Ala.	Jefferson	Pratt	Pratt No. 13	. . . .	4.29	25.78	67.68	6.54	14482	B. & W. Co.
Ala.	Jefferson	Pratt	Warner	Mine run	2.51	27.80	61.50	10.70	13628	U. S. Geo. S.
Ala.	Jefferson	. . . .	Coalburg	Mine run	0.94	31.34	65.65	3.01	14513	B. & W. Co.
Ala.	Walker	Horse Creek	Ivy C. & I. Co. No. 8	Nut	2.56	31.82	53.89	14.29	12937	U. S. Geo. S.
Ala.	Walker	Jagger	Galloway C. Co. No. 5	Mine run	4.83	34.65	51.12	14.03	12976	U. S. Geo. S.
Ark.	Franklin	Denning	Western No. 4	Nut	2.22	12.83	75.35	11.82	. . .	U. S. Geo. S.
Ark.	Sebastian	Jenny Lind	Mine No. 12	Lump	1.07	17.04	74.45	8.51	14252	U. S. Geo. S.
Ark.	Sebastian	Huntington	Cherokee	Mine run	0.97	19.87	70.30	9.83	14159	U. S. Geo. S.
Col.	Boulder	South Platte	Lafayette	Mine run	19.48	38.80	49.00	12.20	11939	B. & W. Co.
Col.	Boulder	Laramie	Simson	Mine run	19.78	44.69	48.62	6.69	12577	U. S. Geo. S.
Col.	Fremont	Cañon City	Chandler	Nut and Slack	9.37	38.10	51.75	10.15	11850	B. & W. Co.
Col.	Las Animas	Trinidad	Hastings	Nut	2.15	31.07	53.40	15.53	12547	B. & W. Co.
Col.	Las Animas	Trinidad	Moreley	Slack	1.88	28.47	55.58	15.95	12703	B. & W. Co.
Col.	Routt	Yampa	Oak Creek	. . . .	6.67	42.91	55.64	1.45	. . .	Hill
Ill.	Christian	Pana	Penwell Col.	Lump	8.05	43.67	49.97	6.36	10900	Jones
Ill.	Franklin	No. 6	Benton	Egg	8.31	34.52	54.05	11.43	11727	U. S. Geo. S.
Ill.	Franklin	Big Muddy	Zeigler	¾ inch	13.28	31.97	57.37	10.66	12857	U. S. Geo. S.
Ill.	Jackson	Big Muddy	. . . .	.	4.85	31.55	62.19	6.26	11466	Breckenridge
Ill.	La Salle	Streator	. . . .	. . . .	8.40	41.76	51.42	6.82	11727	Breckenridge
Ill.	La Salle	Streator	Marseilles	Mine run	12.98	43.73	49.13	7.14	10899	B. & W. Co.
Ill.	Macoupin	Nilwood	Mine No. 2	Screenings	13.34	34.75	44.55	20.70	10781	B. & W. Co.
Ill.	Macoupin	Mt. Olive	Mine No. 2	Mine run	13.54	41.28	46.30	12.42	10807	U. S. Geo. S.
Ill.	Madison	Belleville	Donk Bros.	Lump	13.47	38.69	48.07	13.24	12427	U. S. Geo. S.
Ill.	Madison	Glen Carbon	. . . .	Mine run	9.78	38.18	51.52	10.30	11672	Bryan
Ill.	Marion	. . . .	Odin	Lump	6.20	42.91	49.06	8.03	11880	Breckenridge
Ill.	Mercer	Gilchrist	. . . .	Screenings	8.50	36.17	41.64	22.19	10497	Breckenridge
Ill.	Montgomery	Pana or No. 5	Coffeen	Mine run	11.93	34.05	49.85	16.10	10303	U. S. Geo. S.
Ill.	Peoria	No. 5	Empire	. . . .	17.64	31.91	46.17	21.92	10705	B. & W. Co.
Ill.	Perry	Du Quoin	Number 1	Screenings	9.81	33.67	48.36	17.97	11229	B. & W. Co.

State	County	Field, Bed or Vein	Mine	Size	Moisture	Proximate Analysis (Dry Coal)			B.t.u. Per Pound Dry Coal	Authority
						Volatile Matter	Fixed Carbon	Ash		
Ill.	Perry	Du Quoin	Willis	Mine run	7.22	33.06	53.97	12.97	11352	U. S. Geo. S.
Ill.	Sangamon	. . . .	Pawnee	Slack	4.81	41.53	39.62	18.85	10220	Jones
Ill.	St. Clair	Standard	Nigger Hollow	Mine run	14.39	32.90	44.84	22.26	11059	B. & W. Co.
Ill.	St. Clair	Standard	Maryville	Mine run	15.71	38.10	41.10	20.80	10999	B. & W. Co.
Ill.	Williamson	Big Muddy	Daws	Mine run	8.17	34.33	52.50	13.17	12643	U. S. Geo. S.
Ill.	Williamson	Carterville or No. 7	Carterville	. . . .	4.66	35.65	56.86	7.49	12286	Univ. of Ill.
Ill.	Williamson	Carterville or No. 7	Burr	Nut, Pea and Sl.	11.91	33.70	55.90	10.40	12932	B. & W. Co.
Ind	Brazil	Brazil	Gartside	Block	2.83	40.03	51.97	8.00	13375	Stillman
Ind.	Clay	. . . .	Louise	Block	0.83	39.70	52.28	8.02	13248	Jones
Ind.	Green	Island City	. . . .	Mine run	6.17	35.42	53.55	11.03	11916	Dearborn
Ind.	Knox	Vein No. 5	Tecumseh	Mine run	10.73	35.75	54.46	9.79	12911	B. & W. Co.
Ind.	Parke	Vein No. 6	Parke Coal Co.	Lump	10.72	44.02	46.33	9.65	11767	U. S. Geo. S.
Ind.	Sullivan	Sullivan No. 6	Mildred	Washed	16.59	42.17	48.44	9.59	13377	U. S. Geo. S.
Ind	Vigo	Number 6	Fontanet	Mine run	2.28	34.95	50.50	14.55	11920	Dearborn
Ind	Vigo	Number 7	Red Bird	Mine run	11.62	41.17	46.76	12.07	12740	U. S. Geo. S.
Iowa	Appanoose	Mystic	Mine No. 3	Lump	13.48	39.40	43.09	17.51	11678	U. S. Geo. S.
Iowa	Lucas	Lucas	Inland No. 1	Mine run	16.01	37.82	46.24	15.94	11963	U. S. Geo. S.
Iowa	Marion	Big Vein	Liberty No. 5	Mine run	14.88	41.53	39.63	18.84	11443	U. S. Geo. S.
Iowa	Polk	Third Seam	Altoona No. 4	Lump	12.44	41.27	40.86	17.87	11671	U. S. Geo. S.
Iowa	Wapello	Wapello	. . . .	Lump	8.69	36.23	43.68	20.09	11443	U. S. Geo. S.
Kan.	Cherokee	Weir Pittsburgh	Southwestern Dev. Co.	Lump	4.31	33.88	53.67	12.45	13144	U. S. Geo S.
Kan.	Cherokee	Cherokee	. . . .	Screenings	6.16	35.56	46.90	17.54	10175	Jones
Kan.	Cherokee	Cherokee	. . . .	Lump	1.81	34.77	52.77	12.46	12557	Jones
Kan.	Linn	Boicourt	. . . .	Lump	4.74	36.59	47.07	16.34	10392	Jones
Ky.	Bell	Straight Creek	Str. Ck. C. & C. Co.	Mine run	2.89	36.67	57.24	6.09	14362	U. S. Geo. S.
Ky.	Hopkins	Bed No. 9	Earlington	Lump	6.89	40.30	55.16	4.54	13381	Ky. State Col.
Ky.	Hopkins	Bed No. 9	Barnsley	Mine run	7.92	40.53	48.70	10.77	13036	U. S. Geo. S.
Ky.	Hopkins	Vein No. 14	Nebo	Pea and Slack	8.02	31.91	54.02	14.07	12448	B. & W. Co.
Ky.	Johnson	Vein No. 1	Miller's Creek	Mine run	5.12	38.46	58.63	2.91	13743	U. S. Geo S.
Ky.	Mulenburg	Bed No. 9	Pierce	Pea and Slack	9.22	33.94	52.18	13.88	12229	B. & W. Co.
Ky.	Pulaski	. . . .	Greensburg	. . . .	2.80	26.54	63.58	9.88	14095	N. Y. Ed. C.
Ky.	Webster	Bed No. 9	. . . .	Pea and Slack	7.30	31.08	60.72	8.20	13600	B. & W. Co.
Ky.	Whitley	. . . .	Jellico	Nut and Slack	3.82	31.82	58.78	9.40	13175	B. & W. Co.
Mo.	Adair	. . . .	Danforth	Mine run	9.00	30.55	46.26	23.19	9889	B. & W. Co.
Mo.	Bates	Rich Hill	New Home	Mine run	7.28	37.62	43.83	18.55	12109	U. S. Geo. S.
Mo.	Clay	Lexington	Mo. City Coal Co.	. . . .	12.45	39.39	48.47	12.14	12875	Univ. of Mo.
Mo.	Lafayette	Waverly	Buckthorn	. . . .	8.58	41.78	45.99	12.23	12735	Univ. of Mo.
Mo.	Lafayette	Waverly	Higbee	. . . .	10.84	31.72	55.29	12.99	12500	Univ. of Mo.
Mo.	Linn	Bevier	Marceline	. . . .	9.45	36.72	52.20	11.08	13180	Univ. of Mo.
Mo.	Macon	Bevier	Northwest Coal Co.	. . . .	13.09	37.83	42.95	19.22	11500	U. S. Geo. S.
Mo.	Morgan	Morgan Co.	Morgan Co. Coal Co	Mine run	12.24	45.69	47.98	6.33	14197	U. S. Geo. S.
Mo.	Putnam	Mendotta	Mendotta No. 8	. . . .	20.78	39.36	50.00	10.64	12602	U. S. Geo. S.
N. Mex	McKinley	Gallup	Gibson	Pea and Slack	12.17	36.31	51.17	12.52	12126	B. & W. Co.

APPROXIMATE COMPOSITION AND CALORIFIC VALUE OF CERTAIN TYPICAL AMERICAN COALS — CONTINUED

State	County	Field, Bed or Vein	Mine	Size	Moisture	Proximate Analysis (Dry Coal)			B.t.u. Per Pound Dry Coal	Authority
						Volatile Matter	Fixed Carbon	Ash		
Ohio	Athens	Hocking Valley	Sunday Creek	Slack	12.16	34.64	53.10	12.26	12214	
Ohio	Belmont	Pittsburg No. 8	Neff Coal Co.	Mine run	5.31	38.78	52.22	9.00	12843	U. S. Geo. S.
Ohio	Columbiana	Middle Kittanning	Palestine		2.15	37.57	51.80	10.63	13370	Lord & Haas
Ohio	Coshocton	Middle Kittanning	Morgan Run	Mine run		41.76	45.24	13.00	13239	B. & W. Co.
Ohio	Guernsey	Vein No. 7	Little Kate		6.19	33.02	59.96	7.02	13634	B. & W. Co.
Ohio	Hocking	Hocking Valley		Lump	6.45	39.12	50.08	10.80	12700	Lord & Haas
Ohio	Hocking	Hocking Valley			2.60	40.80	47.60	11.60	12175	Jones
Ohio	Jackson	Brookville	Superior Coal Co.	Mine run	7.59	38.45	43.99	17.56	11704	U. S. Geo. S.
Ohio	Jackson	Lower Kittanning	Superior Coal Co.	Mine run	8.99	41.43	50.06	8.51	13113	U. S. Geo. S.
Ohio	Jackson	Quakertown	Wellston		3.38	35.26	54.18	7.56	12506	Hill
Ohio	Jefferson	Pittsburg or No. 8	Crow Hollow	¾ inch	4.04	40.08	52.27	9.65	13374	U. S. Geo. S.
Ohio	Jefferson	Pittsburg or No. 8	Rush Run No. 1	¾ inch	4.74	36.08	54.81	9.11	13532	U. S. Geo. S.
Ohio	Perry	Hocking	Congo		6.41	38.33	46.71	14.96	12284	B. & W. Co.
Ohio	Stark	Massillon		Slack	6.67	40.02	46.46	13.52	11860	B. & W. Co.
Ohio	Vinton	Brookville or No. 4	Clarion	Nut and Slack	2.47	42.38	50.39	6.23	13421	U. S. Geo. S.
Okla.	Choctaw	McAlester	Edwards No. 1	Mine run	4.79	39.18	49.97	10.85	13005	U. S. Geo. S.
Okla.	Choctaw	McAlester	Adamson	Slack	4.72	28.54	58.17	13.29	12105	B. & W. Co.
Okla.	Creek		Henrietta	Lump and Slack	7.65	36.77	50.14	13.09	12834	U. S. Geo. S.
Pa.	Allegheny	Pittsburgh 3rd Pool		Slack	1.77	32.06	57.11	10.83	13205	Carpenter
Pa.	Allegheny	Monongahela	Turtle Creek		1.75	36.85	53.94	9.21	13480	Lord & Haas
Pa.	Allegheny	Pittsburgh	Bertha	¾ inch	2.61	35.86	57.81	6.33	13997	U. S. Geo. S.
Pa.	Cambria		Beach Creek	Slack	3.01	32.87	55.86	11.27	13755	B. & W. Co.
Pa.	Cambria	Miller	Lincoln	Mine run	5.39	30.83	61.05	8.12	13600	B. & W. Co.
Pa.	Clarion	Lower Freeport			0.54	35.93	57.66	6.41	13547	
Pa.	Fayette	Connellsville		Slack	1.85	28.73	63.22	7.95	13775	Whitham
Pa.	Greene	Youghiogheny		Lump	1.25	32.60	54.70	12.70	13100	B. & W. Co.
Pa.	Greene	Westmoreland		Screenings	11.12	31.67	55.61	12.72	13100	P. R. R.
Pa.	Indiana		Iselin	Mine run	2.70	29.33	63.56	7.11	14220	B. & W. Co.
Pa.	Jefferson		Punxsutawney	Mine run	3.38	29.33	64.93	5.73	14781	B. & W. Co.
Pa.	Lawrence	Middle Kittanning			0.70	37.06	56.24	6.70	13840	Lord & Haas
Pa.	Mercer	Taylor			4.18	32.19	55.55	12.26	12820	B. & W. Co.
Pa.	Washington	Pittsburgh	Ellsworth		2.46	35.35	58.46	6.19	14013	U. S. Geo. S.
Pa.	Washington	Youghiogheny	Anderson	¾ inch	1.00	39.29	54.80	5.91	13729	Jones
Pa.	Westmoreland	Pittsburgh	Scott Haven	Lump	4.06	32.91	59.78	7.31	13934	B. & W. Co.
Tenn.	Campbell	Jellico			1.80	37.76	62.12	1.12	13846	U. S. Navy
Tenn	Claiborne	Mingo			4.40	34.31	59.22	6.47		U. S. Geo. S.
Tenn	Marion		Etna		3.16	32.98	56.59	10.43		
Tenn.	Morgan	Brushy Mt.			1.77	33.46	54.73	11.87	13824	B. & W. Co.
Tenn	Scott	Glen Mary No. 4	Glen Mary		1.53	40.80	56.78	2.42	14625	Ky. State Col.
Tex.	Maverick		Eagle Pass		5.41	33.73	44.89	21.38	10945	B. & W. Co.
Tex.	Paolo Pinto		Thurber	Mine run	1.90	36.01	49.09	14.90	12760	B. & W. Co.
Tex.	Paolo Pinto		Strawn	Mine run	4.19	35.40	52.98	11.62	13202	B. & W. Co.
Va.	Henrico		Gayton		0.82	17.14	74.92	7.94	14363	B. & W. Co.

State	County	Field, Bed or Vein	Mine	Size	Moisture	Proximate Analysis (Dry Coal)			B.t.u. Per Pound Dry Coal	Authority
						Volatile Matter	Fixed Carbon	Ash		
Va.	Lee	Darby	Darby	1½ inch	4.35	38.46	56.91	4.63	13939	U. S. Geo. S.
Va.	Lee	McConnel	Wilson	Mine run	3.35	36.35	57.88	5.77	13931	U. S. Geo. S.
Va.	Wise	Upper Banner	Coburn	3½ inch	3.05	32.65	62.73	4.62	14470	U. S. Geo. S.
Va.	Rockingham		Clover Hill			31.77	57.98	10.25	13103	
Va.	Russel	Clinchfield			2.00	35.72	56.12	8.16	14200	
Va.		Monongahela	Bernmont			32.00	59.90	8.10	13424	Carpenter
W. Va.	Harrison	Pittsburg	Ocean	Mine run	2.47	39.35	52.78	7.87	14202	U. S. Geo. S.
W. Va.	Harrison		Girard	Nut, Pea and Sl.		36.66	57.49	5.85	14548	B. & W. Co.
W. Va.	Kanawha	Winifrede	Winifrede		1.05	32.74	64.38	2.88	14111	Hill
W. Va.	Kanawha	Keystone	Keystone	Mine run	2.21	33.29	58.61	8.10	14202	U. S. Geo. S.
W. Va.	Logan	Island Creek		Nut and Slack	1.12	38.61	55.91	5.48	14273	Hill
W. Va.	Marion	Fairmont	Kingmont		1.90	35.31	57.34	7.35	14198	U. S. Geo. S.
W. Va.	Mingo	Thacker	Maritime		0.68	31.89	63.48	4.63	14126	Hill
W. Va.	Mingo	Glen Alum	Glen Alum	Mine run	3.02	33.81	59.45	6.74	14414	U. S. Geo. S.
W. Va.	Preston	Bakerstown			4.14	29.09	63.50	7.41	14546	U. S. Geo. S.
W. Va.	Putnam	Pittsburg	Black Betsy	Bug Dust	7.41	32.84	53.96	13.20	12568	B. & W. Co.
W. Va.	Randolph	Upper Freeport	Coalton	Lump and Nut	2.11	29.57	59.93	10.50	13854	U. S. Geo. S.
LIGNITES AND LIGNITIC COALS										
Col.	Boulder		Rex		16.05	42.12	47.97	9.91	10678	B. & W. Co.
Col.	El Paso		Curtis		23.25	42.11	49.38	8.51	11090	B. & W. Co.
Col.	El Paso		Pike View		23.77	48.70	41.47	9.83	10629	B. & W. Co.
Col.	Gunnison	South Platte	Mt. Carbon		20.38	46.38	47.50	6.12		
Col.	Las Animas		Acme		16.74	47.90	44.60	7.50		Col. Sc. of M.
Col.		Lehigh			18.30	45.29	44.67	10.04		
N. Dak.	McLean		Eckland	Mine run	29.65	45.56	47.05	7.39	10553	Lord
N. Dak.	McLean		Wilton	Lump	35.96	49.84	38.05	12.11	11036	U. S. Geo. S.
N. Dak.	McLean		Casino		29.65	46.56	38.70	14.74		Lord
N. Dak.	Stark	Lehigh	Lehigh	Mine run	35.84	43.84	39.59	16.57	10121	U. S. Geo. S.
N. Dak.	William	Williston		Mine run	41.76	39.37	48.09	12.54	10121	B. & W. Co.
N. Dak.	William	Williston		Mine run	42.74	40.83	47.79	11.38	10271	B. & W. Co.
Tex.	Bastrop	Bastrop	Glenham		32.77	42.76	36.88	20.36	8958	B. & W. Co.
Tex.	Houston	Crockett			23.27	40.95	38.37	20.68	10886	U. S. Geo. S.
Tex.	Houston		Houston C. & C. Co.		31.48	46.93	34.40	18.87	10176	B. & W. Co.
Tex.	Milam	Rockdale	Worley		32.48	43.04	41.14	15.82	10021	B. & W. Co.
Tex.	Robertson	Calvert	Coaling No. 1		32.01	43.70	43.08	13.22	10753	B. & W. Co.
Tex.	Wood	Hoyt	Consumer's Lig. Co.		33.98	46.97	41.40	11.63	10600	U. S. Geo. S.
Tex.	Wood	Hoyt			30.25	43.27	41.46	15.27	10597	
Wash.	King		Black Diamond		3.71	48.72	46.56	4.72		Gale
Wyo.	Carbon	Hanna		Mine run	6.44	51.32	43.00	5.68	11607	B. & W. Co
Wyo.	Crook	Black Hills	Stilwell Coal Co.		19.08	45.21	46.42	8.37	12641	U. S. Geo. S.
Wyo.	Sheridan	Sheridan	Monarch		21.18	51.87	40.43	7.70	12316	U. S. Geo. S.
Wyo	Sweetwater	Rock Spring		Screenings	7.70	38.57	56.99	4.44	12534	B. & W. Co.
Wyo.	Uinta	Adaville	Lazeart		19.15	45.50	48.11	6.39	9868	U. S. Geo. S.

Stirring Device. In order that temperature changes in the calorimeter may be represented by consistent thermometer readings, the water in the calorimeter must be stirred continuously, and for this purpose a motor-driven screw or turbine device should be used. The speed of stirring should be such that the temperature of the calorimeter water is not increased more than 0.01 degree Centigrade during a period of 10 minutes before firing, and such speed should be kept constant and not irregular.

Thermometers. The accuracy of a calorimeter depends largely upon the accuracy of the thermometers with which changes in temperature are measured. For such work, either special calorimetric thermometers or Beckman thermometers graduated to 0.01 or 0.02 degree Centigrade should be used. All thermometers should have a Bureau of Standards calibration certificate, and readings should be corrected in accordance with the corrections given by such certificate.

Oxygen. The oxygen used to bring about rapid and complete combustion should be free from all combustible matter. The amount should be not less than 5 grams per gram of coal, which amount corresponds to a pressure of approximately 20 atmospheres for the larger bombs or 30 atmospheres for small bombs. Completeness of combustion is indicated by the absence of sooty matter on the interior of the bomb after the test.

Firing Wire. The wire used for ignition may be either of iron or platinum. If the former, it should be of about No. 34 B & S gauge, and not more than 10 centimeters (preferably 5 centimeters) should be used at a time. A correction of 1600 calories per gram weight of iron burned must be deducted from the observed number of calories.

Calorimeter Standardization. During every calorimetric test, the immersed parts of the calorimeter are carried through the same range of temperature as the water in the calorimeter bucket. The weight of water, the temperature of which would be increased one degree for an increase of one degree in the temperature of the combined parts of the immersed calorimeter, is known as the "water equivalent" of the calorimeter. Expressed in another manner, the water equivalent is the amount of heat absorbed by the immersed calorimeter for one degree rise in the water contained therein. While the water equivalent can be computed from the weight of the various parts of the apparatus and the specific heat of the different materials entering into the construction of such different parts, it is best determined by the combustion of standard combustion samples supplied by the Bureau of Standards. The required water equivalent, as determined by such method, is equal to the weight of the sample multiplied by its heat of combustion per gram and divided by the corrected rise in calorimeter water temperature. The methods to be followed in the burning of these standard combustion samples as to amount of water, oxygen, firing wire, radiation correction, etc., are the same as for the combustion of solid or liquid fuel samples.

*Manipulation of Calorimeter.** (1) The sample, ground as described in the methods of making proximate analyses, should be thoroughly mixed in the containing bottle and approximately one gram taken out and weighed in the pan or crucible in which it is to be burned.

(2) Bomb Preparation. The ignition wire, if of iron, should be weighed and, whether of iron or platinum, connected to the platinum terminals. This wire must

* Some of the methods of manipulation given apply only to calorimeters not equipped with adiabatic water jacket. Where a difference in the two types occurs they are discussed hereafter.

be so placed that it is in contact with the fuel sample, but should not touch the fuel pan. The wire should be clean. Approximately 0.5 cubic centimeter of water is placed in the bottom of the bomb to saturate with moisture the oxygen used for combustion and for titration after the test.

(3) Oxygen. The oxygen from the supply tank should be admitted to the bomb slowly to prevent blowing the sample from the fuel pan or crucible. The pressure, indicated by a gauge forming part of the calorimeter equipment, should be from 20 to 30 atmospheres, depending upon the size of the bomb.

(4) Calorimeter Water. The calorimeter bucket should be filled with the amount of water corresponding to the water equivalent of the particular apparatus in use. Such amount naturally must be that determined in the standardization of the calorimeter.

(5) Temperature. The initial temperature in the calorimeter should be such that the maximum temperature attained after combustion will not be more than one degree (and preferably not more than 0.5 degree) above that of the jacket. Under such temperature conditions, the radiation correction, while necessary, is small and tends toward accuracy in the final computation of heat value.

(6) Firing Current. The current used for firing, obtained from dry or storage cells, must be of a voltage lower than that which would result in arcing between firing connections. The circuit should be closed by a switch, which is kept in contact not over two seconds. An ammeter in the circuit or some other means should be used for the indication of the burning out of the firing wire.

(7) Observation Methods. The combustion bench in which the sample for testing has been placed, as described, is placed in the calorimeter bucket, the cover of the water jacket placed, and the stirring apparatus and thermometers so placed as not to be in contact with the bomb or bucket. The stirrer is then started, and after thermometer readings have become steady, temperatures are read for a period of five minutes at one-minute intervals. The charge is then fired by means of the switch, the exact time of firing being noted. Temperature observations after firing are observed at intervals that vary with the method used in the determination of the radiation loss. When the temperature of the calorimeter has reached its maximum, a second series of readings is taken over a five-minute period, at one-minute intervals, to determine the final cooling or radiation loss.

(8) Titration. After combustion, the bomb is opened. If incomplete combustion is indicated by traces of unconsumed combustible or sooty deposit, the test should be discarded. If combustion has been completed, the bomb should be rinsed with the water placed within it before the test and other water placed in the bomb, and such water titrated with a standard alkali solution (one cubic centimeter = 0.02173 gram HNO_3 = 5 calories), using methylorange to determine the amount of sulphuric acid formed. A correction of 230 calories per gram of nitric acid should be deducted from the total heat observed, and an additional correction of 1300 calories per gram of sulphur in the coal should be made for the excess of heat of formation in the combustion of sulphur dioxide, as occurring in the combustion of sulphur in practice, over that of aqueous nitric acid occurring in a calorimetric test.

Test Observations and Methods of Computation—The following methods of observation of thermometer readings during a calorimetric test, and those of computing heat values from such observations, are those recommended by a joint committee of the American Society for Testing Materials and the American Chemical Society, 1917. The observations are:

(1) The rate of rise in temperature in degrees per minute (r_1) for five minutes before firing.

(2) The time a, at which the last temperature reading is made prior to firing.

(3) The time b, when the rise in temperature has reached six-tenths of the total rise after firing. Such temperature may generally be determined by adding to the temperature observed before firing 60 per cent of the expected temperature rise and noting the time when this temperature is reached.*

(4) The time c of temperature measurements when the temperature change of the calorimeter water has become uniform, some five minutes after firing; and

(5) The final rate of cooling, r_2, as represented by temperature change per minute for the five-minute period after firing.

The rate r_1 is to be multiplied by the time $(b-a)$ expressed in minutes and tenths of a minute, and this product added (or, if the temperature at the time a was falling, subtracted) to the thermometer reading taken at the time a. The rate r_2 is to be multiplied by the time $(c-b)$ and this product added (subtracted if the temperature was rising at the time c and later) to the thermometer reading taken at the time c. The difference in the two temperatures so determined, with proper calibration corrections made, gives the total rise in temperature resulting from combustion. Such rise multiplied by the water equivalent of the calorimeter gives the total amount of heat, expressed in calories, liberated. This latter result corrected for the heat of the formation of nitric acid and sulphuric acid, as determined by titration, and for the heat of combustion of the firing wire, when this is included, is divided by the weight of the sample to determine the heat of combustion in calories per gram. Calories per gram multiplied by 1.8 give the heat in B.t.u. per pound.

The results determined from the above computations should be given in calories per gram or B.t.u. per pound of *dry fuel*, the methods of conversion from fuel as fired to a dry basis being the same as described for the conversion of ultimate or proximate fuel analyses.

The method of computation of the heat value of a fuel from observed data is illustrated in the following example: Assume:

Water equivalent of calorimeter, 2550 grams.

Weight of full sample tested, 1.0535 grams.

Approximate rise of calorimeter temperature expected 3.2 degrees Centigrade.

Six-tenths of approximate rise, 1.9 degrees Centigrade.

Thermometer readings as given at the top of the next page.

$$r = 15.272 - 15.244 = 0.028 \div 5 = 0.0056 \text{ degree per minute,}$$

$$(b - a) = 10{:}272 - 10{:}26 = 1.2 \text{ minutes.}$$

The corrected initial temperature is then:

$$15.276 + 0.0067 = 15.283 \text{ degrees,}$$

$$r_2 = 18.500 - 15.493 = 0.007 \div 5 = 0.0014 \text{ degree per minute,}$$

$$(c - b) = 10{:}310 - 10{:}272 = 3.8 \text{ minutes.}$$

The correct final temperature is:

$$18.497 + (0.0014 \times 3.8) = 18.502 \text{ degrees.}$$

* In practice the difference in temperature before firing and that when six-tenths of the temperature rise has occurred $(b-a)$ is approximately a constant, and the value of b need be determined only occasionally

Time of Observation	*Thermometer Readings Degrees Centigrade*	*Corrected Temperature from Bureau of Standards Certificate*
10:21	15.244	......
10:22	15.250	......
10:23	15.255	......
10:24	15.261	. .
10:25	15.266	......
10:26 (*a*)	15.272	15.276
10:27.2 (*b*)	17.200*	..
10:31 (*c*)	18.500	18.497
10:32	18.498	... ..
10:33	18.497	......
10:34	18.498	......
10:35	18.494	... ..
10:36	18.493	......

* The initial temperature *a* is 15.272 degrees; 60 per cent of the expected rise is 1.9 degrees. The reading for observation is thus 15.27 + 1.9 = 17.2 degrees.

The total rise in temperature, as represented by the corrected initial and final readings, is 18.502 — 15.283 = 3.219 degrees, and from the water equivalent of the calorimeter, the heat value less a deduction of heat evolved in the combustion of the firing wire and that from titration of, say, 7 calories, is, for the sample tested,

$$(2550 \times 3.219) - 7 = 8202 \text{ calories.}$$

The heat value per gram is this value divided by the weight in grams of the sample, or 8202 ÷ 1.0535 = 7785 calories or, expressed in B.t.u. per pound, 7785 × 1.8 = 14,103 B.t.u.

The methods of operation of the calorimeter equipped with the ordinary water jacket, described above, in general hold for a calorimeter equipped with an adiabatic jacket. With this type of jacket the water in it must be originally at a temperature lower than that of the calorimeter, and then be brought to the latter temperature by use of the electric or other means of temperature adjustment. The temperature of the jacket water should be held within 0.1 degree of that of the calorimeter and after ignition this is accomplished by adjustment of such regulating device as is furnished with the calorimeter. When the temperature of the calorimeter, after ignition of the fuel sample, approaches the maximum, such maximum may be accurately determined by readings at sufficiently short intervals, and, because of the absence of radiation loss, no further readings are necessary, the firing and maximum temperature indicated being the only temperatures necessary for the computation of the heat evolved in combustion of the sample, though, of course, the corrections due to firing wire, products of combustion other than those found in practice, etc., must be made.

Where the water available for use in an adiabatic jacket of the Daniels type is "soft," and for that reason a poor electrical conductor, a small amount of soluble salts should be added to the water, preferably by passing the water for the jacket through a container of such salts before it enters the jacket.*

* For more detailed methods of analysis and determination of heat value, it is suggested that the complete report of the joint committee of the American Society of Testing Materials and the American Chemical Society on the "Methods for the Analysis of Coal," the Bureau of Mines Technical Paper No. 76, "Notes on the Sampling and Analyzing of Coal," and the same Bureau's Technical Paper No. 8, "Methods of Analyzing Coal and Coke," be studied.

Two Units of 8128 Horse-power of Babcock & Wilcox Boilers and Superheaters at the Fisk Street Station of the Commonwealth Edison Co., Chicago, Ill. [illegible] Horse-power Being Installed in [illegible] Station. The Commonwealth Edison Co. Operates in Its Various Stations a Total of 136,000 Horse-power of Babcock & Wilcox Boilers Fitted with Babcock & Wilcox Superheaters

COMBUSTION OF COAL

THE composition of coal varies over such a wide range, and the methods of firing have to be altered so greatly to suit the various coals and the innumerable types of furnaces in which they are burned, that any instructions given for the handling of different fuels must of necessity be of the most general character. For each kind of coal there is some method of firing which will give the best results for each individual set of conditions. General rules can be suggested, but the best results can be obtained only by following such methods as experience and practice show to be the best suited to the specific conditions.

The question of draft is an all important factor. If this be insufficient, proper combustion is impossible, as the suction in the furnace will not be great enough to draw the necessary amount of air through the fuel bed and the gases may pass off only partially consumed. On the other hand, an excessive draft may cause losses due to the excess quantities of air drawn through holes in the fire. Where coal is burned, however, there are rarely complaints from excessive draft, as this can be and should be regulated by the boiler damper to give only the draft necessary for the particular rate of combustion desired. The draft required for various kinds of fuel is treated in detail in the chapter on "Chimneys and Draft." In this chapter it will be assumed that the draft is at all times ample and that it is regulated to give the best results for each kind of coal.

TABLE 40

ANTHRACITE COAL SIZES

Trade Name	Size of Opening	
	Through Inches	Over Inches
Broken	4½	3¼
Egg	3¼	2 5/16
Stove	2 5/16	1 5/8
Chestnut	1 5/8	7/8
Pea	7/8	9/16
No. 1 Buckwheat	9/16	5/16
No. 2 Buckwheat or Rice	5/16	3/16
No. 3 Buckwheat or Barley	3/16	3/32
Culm	3/32	. .

ANTHRACITE—Anthracite coal is ordinarily marketed under the names and sizes given in Table 40.

The larger sizes of anthracite are rarely used for commercial steam generating purposes, as the demand for domestic use now limits the supply. In commercial plants the sizes generally found are Nos. 1, 2 and 3 buckwheat. In some plants where the finer sizes are used, a small percentage of bituminous coal, say, 10 per cent, is sometimes mixed with the anthracite and beneficial results secured both in economy and capacity.

Anthracite coal should be fired evenly, in small quantities and at frequent intervals. If this method is not followed, dead spots will appear in the fire, and if the fire gets too irregular through burning in patches, nothing can be done to remedy it until the fire is cleaned as a whole. After this grade of fuel has been fired it should be left alone, and the fire tools used as little as possible. Owing to the difficulty of igniting this fuel, care must be taken in cleaning fires. The intervals of cleaning will, of course, depend upon the nature of the coal and the rate of combustion. With the small sizes and moderately high combustion rates, fires will have to be cleaned twice on each eight-hour shift. As the fires become dirty, the thickness of the fuel bed will increase until this depth may be 12 or 14 inches just before a cleaning period. In cleaning, the following practice is usually followed: The good coal on the forward

A Section of the Installation of Babcock & Wilcox Boilers at the Open-Hearth Plant of the Cambria Steel Co., Johnstown, Pa.
This Company Operates a Total of [illegible] Horse-power of Babcock & Wilcox Boilers

half of the grate is pushed to the rear half, and the refuse on the front portion either pulled out or dumped. The good coal is then pulled forward onto the front part of the grate and the refuse on the rear section dumped. The remaining good coal is then spread evenly over the whole grate surface and the fire built up with fresh coal.

A ratio of grate surface to heating surface of 1 to from 35 to 40 will, under ordinary conditions, develop the rated capacity of a boiler when burning anthracite buckwheat. Where the finer sizes are used or where overloads are desirable, however, this ratio should preferably be 1 to 25 and a forced blast should be used. Grates 10 feet deep with a slope of 1½ inches to the foot can be handled comfortably with this class of fuel, and grates 12 feet deep with the same slope can be successfully handled. Where grates over 8 feet in depth are necessary, shaking grates or overlapping dumping grates should be used. Dumping grates may be applied either for the whole grate surface or to the rear section. Air openings in the grate bars should be made from $\frac{3}{16}$ inch in width for No. 3 buckwheat to $\frac{5}{16}$ inch for No. 1 buckwheat. It is important that these air openings be uniformly distributed over the whole surface to avoid blowing holes in the fire, and it is for this reason that overlapping grates are recommended.

No air should be admitted over the fire. Steam is sometimes introduced into the ashpit to soften any clinker that may form, but the quantity of steam should be limited to that required for this purpose. The steam that may be used in a steam jet blower for securing blast will in certain instances assist in softening the clinker, but a much greater quantity may be used by such an apparatus than is required for this purpose. Combustion arches sprung above the grates have proved of advantage in maintaining a high furnace temperature and in assisting the ignition of fresh coal.

Stacks used with forced blast should be of such size as to insure a slight suction in the furnace under any conditions of operation. A blast up to 3 inches of water should be available for the finer sizes, supplied by engine-driven fans automatically controlled by the boiler pressure. The blast required will increase as the depth of the fuel bed increases, and the slight suction should be maintained in the furnace by damper regulation.

The use of blast with the finer sizes causes rapid fouling of the heating surfaces of the boiler, the dust often amounting to over 10 per cent of the total fuel fired. Economical disposal of dust and ashes is of the utmost importance in burning fuel of this nature. Provision should be made in the baffling of the boiler to accommodate and dispose of this dust. Whenever conditions permit, the ashes can be economically disposed of by flushing them out with water.

Bituminous Coals—There is no classification of bituminous coal as to size that holds good in all localities. The American Society of Mechanical Engineers suggests the following grading:

Eastern Bituminous Coals—

(A) Run of mine coal; the unscreened coal taken from the mine.

(B) Lump coal; that which passes over a bar-screen with openings 1¼ inches wide.

(C) Nut coal; that which passes through a bar-screen with 1¼-inch openings and over one with ¾-inch openings.

(D) Slack coal; that which passes through a bar-screen with ¾-inch openings.

Northwest Station of the Commonwealth Edison Co., Chicago, Ill. This installation consists of [illegible] Horse-power of Babcock & Wilcox Boilers and Superheaters Equipped with Babcock & Wilcox Chain Grate Stokers

Western Bituminous Coals—

(E) Run of mine coal; the unscreened coal taken from the mine.

(F) Lump coal; divided into 6-inch, 3-inch and 1¼-inch lump, according to the diameter of the circular openings over which the respective grades pass; also 6 × 3-inch lump and 3 × 1¼-inch lump, according as the coal passes through a circular opening having the diameter of the larger figure and over that of the smaller diameter.

(G) Nut coal; divided into 3-inch steam nut, which passes through an opening 3 inches diameter and over 1¼ inches; 1¼-inch nut, which passes through a 1¼-inch diameter opening and over a ¾-inch diameter opening; ¾-inch nut, which passes through a ¾-inch diameter opening and over a ⅝-inch diameter opening.

(H) Screenings; that which passes through a 1¼-inch diameter opening.

(I) Washed sizes; those passing through or over circular openings of the following diameters, in inches:

Number	*Through*	*Over*
1	3	1¾
2	1¾	1⅛
3	1⅛	¾
4	¾	¼
5	¼	..

As the variation in character of bituminous coals is much greater than that of the anthracites, any rules set down for their handling must be the more general. The difficulties in burning bituminous coals with economy and with little or no smoke increase as the content of fixed carbon in the coal decreases. It is their volatile content which causes the difficulties and it is essential that the furnaces be designed to handle properly this portion of the coal. The fixed carbon will take care of itself, provided the volatile matter is properly burned.

Mr. Kent, in his "Steam Boiler Economy," described the action of bituminous coal after it is fired as follows: "The first thing that the fine fresh coal does is to choke the air spaces existing through the bed of coke, thus shutting off the air supply which is needed to burn the gases produced from the fresh coal. The next thing is a very rapid evaporation of moisture from the coal, a chilling process, which robs the furnace of heat. Next is the formation of water-gas by the chemical reaction, $C + H_2O = CO + 2H$, the steam being decomposed, its oxygen burning the carbon of the coal to carbonic oxide, and the hydrogen being liberated. This reaction takes place when steam is brought in contact with highly heated carbon. This also is a chilling process, absorbing heat from the furnaces. The two valuable fuel gases thus generated would give back all the heat absorbed in their formation if they could be burned, but there is not enough air in the furnace to burn them. Admitting extra air through the fire door at this time will be of no service, for the gases being comparatively cool cannot be burned unless the air is highly heated. After all the moisture has been driven off the coal, the distillation of hydrocarbons begins, and a considerable portion of them escapes unburned, owing to the deficiency of hot air and to their being chilled by the relatively cool-heating surfaces of the boiler. During all this time great volumes of smoke are escaping from the chimney, together with

16,000 Horse Power Installation of Babcock & Wilcox Boilers and Superheaters at the Brunot's Island Plant of the Duquesne Light Co., Pittsburgh, Pa.

unburned hydrogen, hydrocarbons and carbonic oxide, all fuel gases, while at the same time soot is being deposited on the heating surface, diminishing its efficiency in transmitting heat to the water."

To burn these gases distilled from the coal, it is necessary that they be brought into contact with air sufficiently heated to cause them to ignite, that sufficient space be allowed for their mixture with the air, and that sufficient time be allowed for their complete combustion before they strike the boiler heating surfaces, since these surfaces are comparatively cool and will lower the temperature of the gases below their ignition point. The air drawn through the fire by the draft suction is heated in its passage and heat is added by radiation from the hot brick surfaces of the furnace, the air and volatile gases mixing as this increase in temperature is taking place. Thus in most instances is the first requirement fulfilled. The elements of space for the proper mixture of the gases with the air and of time in which combustion is to take place, should be taken care of by sufficiently large combustion chambers.

Certain bituminous coals, owing to their high volatile content, require that the air be heated to a higher temperature than it is possible for it to attain simply in its passage through the fire and by absorption from the side walls of the furnace. Such coals can be burned with the best results under fire-brick arches. Such arches increase the temperature of the furnace and in this way maintain the heat that must be present for ignition and complete combustion of the fuels in question. These fuels, too, sometimes require additional combustion space, and an extension furnace will give this in addition to the required arches.

As stated, the difficulty of burning bituminous coals successfully will increase with the increase in volatile matter. This percentage of volatile matter will affect directly the depth of coal bed to be carried and the intervals of firing for the most satisfactory results. The variation in the fuel over such wide ranges makes it impossible to state definitely the thickness of fires for all classes, and experiment with the class of fuel in use is the best method of determining how that particular fuel should be handled. The following suggestions, which are not to be considered in any sense hard and fast rules, may be of service for general operating conditions for hand firing:

Semi-bituminous coals, such as Pocahontas, New River, Clearfield, etc., require fires from 10 to 14 inches thick; fresh coal should be fired at intervals of 10 to 20 minutes and sufficient coal charged at each firing to maintain a uniform thickness. Bituminous coals from the Pittsburgh Region require fires from 4 to 6 inches thick, and should be fired often in comparatively small charges. Kentucky, Tennessee, Ohio and Illinois coals require a thickness from 4 to 6 inches. Free burning coals from Rock Springs, Wyoming, require from 6 to 8 inches, while the poorer grades of Montana, Utah and Washington bituminous coals require a depth of about 4 inches.

In general, as thin fires are found necessary, the intervals of firing should be made more frequent and the quantity of coal fired at each interval smaller. As thin fires become necessary, due to the character of the coal, the tendency to clinker will increase if the thickness be increased over that found to give the best results.

In hand firing the "spreading" method is ordinarily used. But little fuel is fired at one time and is spread evenly over the fuel bed from front to rear. Where there is more than one firing door, the doors should be fired alternately. The advantage of alternate firing is that the whole surface of the fire is not blanketed with green coal and steam is generated more uniformly than if all doors were fired at one time. Again, a better combustion results due to the burning of more of the volatile matter directly after firing than where all doors are fired at one time.

The tendency in hand-fired practice is to work bituminous coal less than formerly. A certain amount of raking and slicing may be necessary, but in general the less the fire is worked the better the results.

LIGNITES—As the content of volatile matter and moisture in lignites is higher than in bituminous coal, the difficulties encountered in burning them are greater. A large combustion space is required and the best results are obtained where a furnace of the reverberatory type is used, giving the gases a long travel before meeting the tube surfaces. A fuel bed from 4 to 6 inches in depth can be maintained, and the coal should be fired in small quantities by the alternate method. Above certain rates of combustion, clinker forms rapidly, and a steam jet in the ashpit for softening this clinker is often desirable. A considerable draft should be available, but it should be carefully regulated by the boiler damper to suit the condition of the fire. Smokelessness with hand firing with this class of fuel is a practical impossibility. It has a strong tendency to foul the heating surfaces rapidly and these surfaces should be cleaned frequently. Shaking grates, intelligently handled, aid in cleaning the fires, but their manipulation must be carefully watched to prevent good coal being lost in the ashpit.

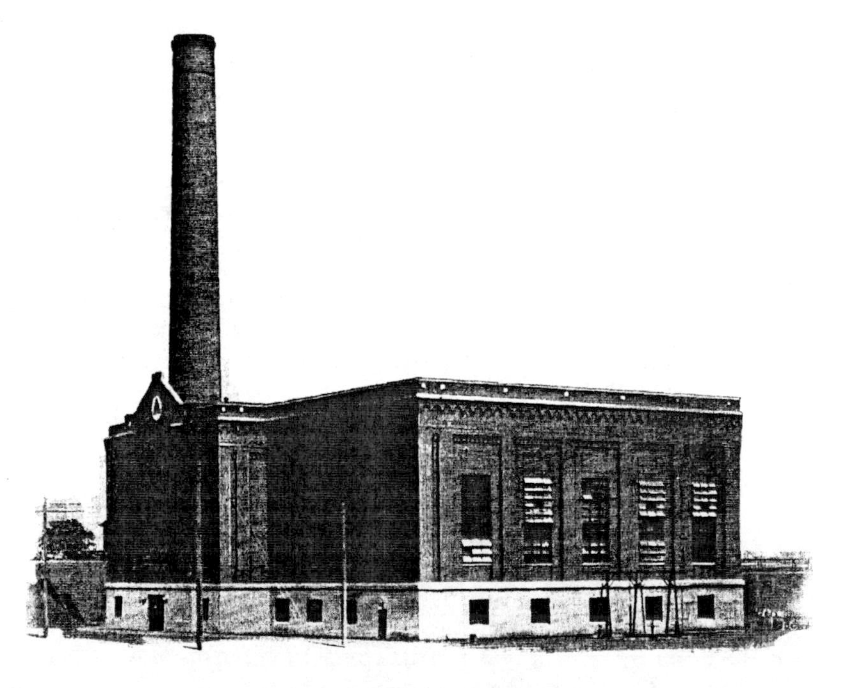

RIVERSIDE STATION OF THE MINNEAPOLIS GENERAL ELECTRIC COMPANY WITH 14,000 HORSE-POWER OF BABCOCK & WILCOX BOILERS

MECHANICAL STOKERS

THE present-day demands for increased rates of steam output have come about through increased fuel costs and the higher cost of equipment resulting from higher pressures and economic conditions. The inherent ability of the properly designed boiler to develop the desired capacities has always been present. Such being the case, the actual production of the high rates of output from a given amount of boiler surface has been dependent upon the development of the methods of burning sufficient fuel to give such rates.

The ratio of grate surface to heating surface is limited in hand-fired boilers, particularly in the larger units, and this, together with the limit of combustion rates with fuel when hand fired, almost regardless of draft available, makes this class of firing impracticable for high rates of output. While other considerations were probably of greater weight in the development of the early methods of automatic firing, these limiting factors have been of the greatest importance in the very marked development of the past few years of stokers and fuel-firing appliances.

The term "automatic stoker" oftentimes gives the erroneous impression that such an apparatus takes care of itself. It is, however, automatic only to the extent that it replaces hand by mechanical labor. Obviously, if coal must be fired to the stoker hopper by hand and ash removed from the ashpit in the same manner, the automatic feature is almost entirely lost. Further, even where coal and ash-handling machinery is installed, the most highly developed stoker is not completely automatic, for the regulation of combustion conditions is largely controlled by hand. There are, with certain stokers, automatic features of control, but these are ordinarily for detached features of operation. In view of these facts, it must be thoroughly understood that the degree of intelligence required for proper stoker operation is as high, if not higher, than in hand-fired practice, and it may be safely stated that stoker regulation may be best secured where intelligent operators are equipped with proper instruments for the indication of combustion conditions.

The advantages of stoker-fired over hand-fired furnaces are many and may be outlined as follows:

By far the most important advantage is in the saving in fuel costs. A stoker-fired furnace, properly operated, shows an efficiency appreciably higher than can be obtained with hand firing. This is not only true with a given fuel but, provided a stoker suited to the fuel is installed, the efficiency of the stoker-fired furnace will be as high or higher, even with a poorer grade and cheaper fuel, than that of the hand-fired furnace with the better grades. Again, assuming the proper type of stoker, it is possible to burn satisfactorily in stoker-fired furnaces certain grades of fuel that could be burned in hand-fired furnaces only with the greatest difficulty, and in many instances not at all. A stoker-fired furnace will show a high efficiency over a wide range of boiler output, while with hand firing, even where a reasonably high efficiency is shown at some fixed rate of output, such efficiency will fall off rapidly with an increase in the rate. All of these conditions clearly indicate the possibility of a saving of fuel costs with stoker-fired as against hand-fired furnaces.

The reasons for the better combustion conditions and the corresponding better efficiencies with stokers may be summarized as follows:

The primary factor in bringing about these better conditions is the control of the air supplied for combustion. Such supply in the case of the stoker-fired furnace is,

Part of [illegible]000 Horse Power Installation of Babcock & Wilcox Boilers Equipped with Babcock & Wilcox Natural Draft Chain Grate Stokers at the Station of the Louisville Gas & Electric Company which also Operates 4,[illegible] Horse Power of Babcock & Wilcox Boilers [illegible] with Babcock & Wilcox Chain Grate Stokers

for a given set of combustion conditions, a constant, while in hand-fired furnaces, even at a constant load, it varies widely due to the necessity for intermittent firing. Further, the air supply in stoker-fired practice can be controlled and regulated under fluctuating and varied load conditions to an extent not possible with hand firing. These conditions make it possible to accomplish combustion in the stoker-fired furnace with a minimum amount of excess air almost regardless of boiler load. Even under the best conditions of hand firing, the amount of excess air used in combustion is higher than in stoker practice, while the variation in the amount of excess air with a fluctuation of load will be much greater. With combustion complete, low excess air tends toward high furnace temperatures, which in turn are indicative of efficient combustion.

The firing with stokers is even and continuous, while with hand firing it is intermittent. There is no necessity in the stoker-fired furnace, as there is with hand firing, for the periodic opening of doors at firing intervals, with a resulting inrush of cold air and a consequent lowering of the furnace temperature. The method of cleaning stoker fires differs with the design of stoker, but even where such cleaning is not continuous and automatic, the duration of cleaning periods is less than in the case of hand-fired furnaces, which results in a higher average efficiency over long periods.

Second in importance among the advantages resulting from the use of stokers is the saving in labor costs. Such saving is, of course, dependent upon the extent to which the automatic features already referred to are carried, and in the small plant not equipped for automatic handling of coal and ash, the labor-saving cost is practically negligible. In the larger station, however, equipped with the modern coal storage and handling and ash-disposal apparatus, this saving is a very appreciable item.

A third advantage in the use of stokers is the ability to develop rates of boiler output greatly in excess of those possible with hand firing; and these high rates, made necessary by increasing cost of fuel and equipment, are obtained with high efficiencies. The ability to develop the higher rates is due in part to the fact that a much greater ratio of grate surface to heating surface can be installed with stokers than it is practicable or possible to handle hand-fired, but mainly because combustion rates per unit of grate surface can be obtained with stokers that cannot be approached with hand firing even under the most favorable conditions of fuel and draft.

A fourth advantage of the stoker-fired over the hand-fired furnace is the ability of the former to meet rapid variations in the plant load. Stoker-fired boilers will satisfactorily carry widely varying and rapidly fluctuating loads that in the case of hand-fired units, if handled at all, could only be with a wide fluctuation in steam pressure. This advantage is of particular importance in central stations where heavy peak loads may be suddenly thrown on the plant, and in manufacturing and other plants where the load, due to the necessities of operation, is subject to wide and rapid fluctuation.

A fifth advantage of the stoker-fired furnace, of importance in thickly populated communities, lies in the fact that a properly designed and properly operated stoker furnace will give more nearly smokeless combustion than can be secured with even the most careful hand firing. While this is in general true, it is not to be understood

3000 Horse-power Installation of Cross-drum Babcock & Wilcox Boilers and Superheaters Equipped with Babcock & Wilcox Natural Draft Chain Grate Stokers at the Washington Terminal Co. Washington D. C.

from this statement that all stoker-fired furnaces are smokeless. The comparative nature of smoke and smokelessness will be discussed at the end of this chapter.

Under the load and operating conditions of the larger plants the question of the advisability of stoker installation is hardly an open one, for the great majority of such plants could not be satisfactorily operated hand fired.

It is in the case of the smaller plant, then, that the question of the *advisability* of stoker installation arises most frequently. Until a few years ago their advisability was seriously questioned, largely because of the lack of automatic features. It was felt that in the small plant the cost of coal and ash-handling apparatus was not warranted, that there would be no labor saving, and that the cost of stokers would be relatively as high or higher than in the large plant. While it was appreciated that the stoker-fired furnace could give better efficiency and more nearly smokeless combustion than would be obtained with hand firing, it was doubtful to what extent improvement in these operating features would be accomplished, due to the class of boiler-room operators ordinarily found in the small plant. Further, in the average small plant of some years ago, the load on the boilers was rarely such as could not readily be handled with hand firing. In general, stokers were not recommended for installation in small plants unless the fuel saving could be shown to be very great or the question of smoke were all important.

With the growing demands for fuel conservation, the difficulty of obtaining grades of coal with which satisfactory results could be obtained when hand fired, the more stringent smoke regulations, and the tendency toward the employment of a better grade of fire-room operator, even in the small plant the number of stoker installations has increased rapidly within the past few years and undoubtedly will increase more rapidly in the years to come.

In determining the advisability of a stoker installation, there are certain features that should be given consideration.

The first cost of stokers is, from the nature of the apparatus, greatly in excess of that of hand-fired furnace material. This cost must be considered from the standpoint of fixed charges.

Maintenance costs will vary with the type of stoker, but will unquestionably be higher than with hand-fired grates.

Furnace brickwork upkeep costs will also be appreciably higher in stoker-fired than in hand-fired furnaces. This and the higher stoker maintenance cost are the result of the more severe service to which stoker-fired units are submitted.

It is to be remembered that stokers require the utilization of power for their operation, and where they are of the blast type the fans furnishing such blast require additional power. The gross efficiency of the stoker-fired unit must be corrected for such power required in arriving at net results.

While these four factors apply to the small rather than to the larger plant in so far as it affects the advisability of a stoker installation, they apply to both the small and the large plant in the consideration of the type of stoker that should be installed.

In respect to efficiency of combustion, other things being equal, there will be no appreciable difference with the various types of stoker, provided the proper stoker is selected for the grade of fuel to be burned and the conditions of operation to be fulfilled. No stoker will satisfactorily handle all classes of fuel, and in making a

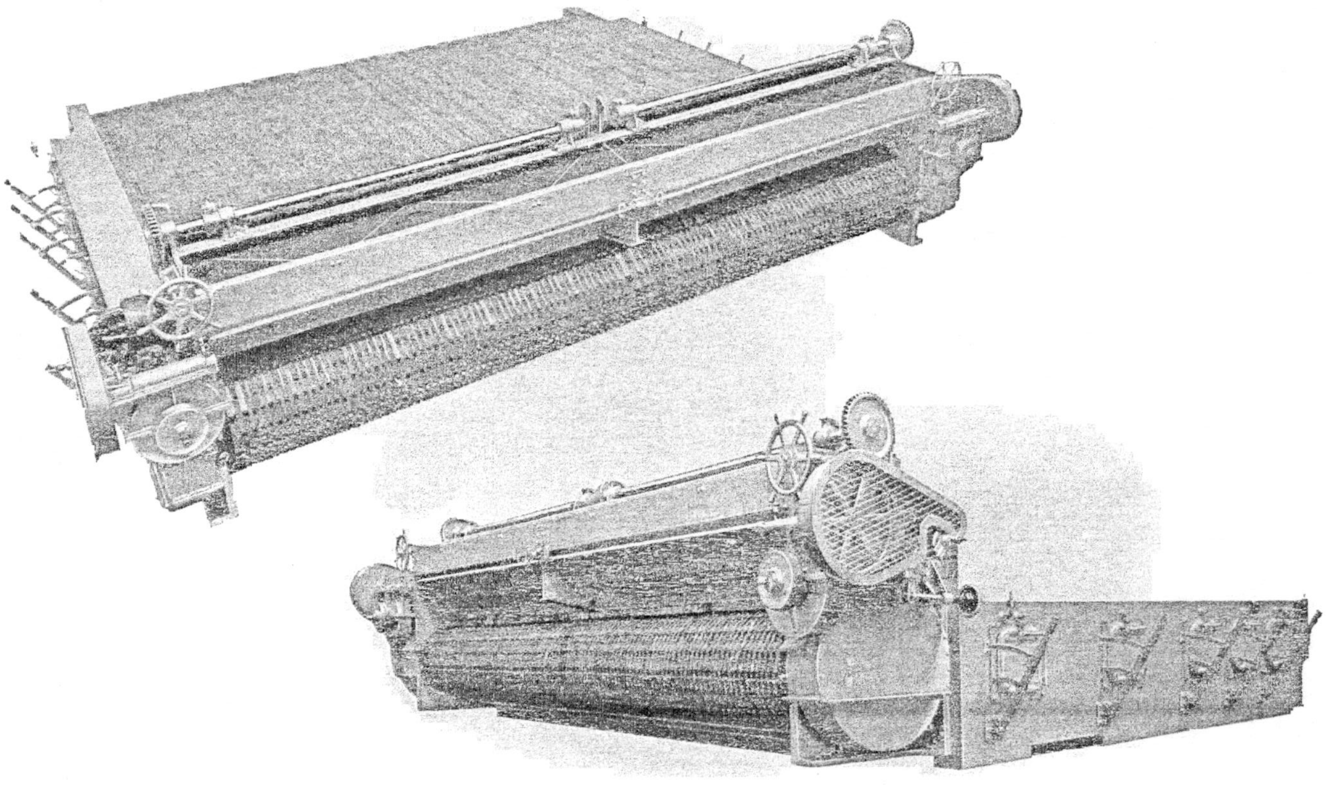

The Babcock & Wilcox Blast Chain Grate Stoker

selection care should be taken that the type should be suited to the fuel and the operating conditions of the particular plant in which the installation is to be made.

Further, the design of the boiler furnace should be proper, not only for the fuel used, but for the type of stoker selected, for even with a stoker suited to the fuel satisfactory results cannot be obtained with a furnace of improper design.

In considering first and maintenance costs, it is to be remembered that the stoker of rugged construction, with low maintenance costs, with a life comparable to that of the boiler under which it is installed, and of dependability as to continuity of service, will prove a better investment than one having a high upkeep cost, a shorter life and a construction necessitating frequent shut-downs for repair, even though the latter, when in operation, shows a somewhat higher efficiency and is of cheaper first cost.

MECHANICAL STOKERS may be classified under three general types, traveling or chain grate, overfeed and underfeed. The first class may be subdivided into natural draft and forced blast chain grates, the second into inclined front-feed and inclined side-feed stokers, and the third into inclined and horizontal underfeed stokers. Another classification of underfeed stokers is into single and multiple retort underfeeds, but inasmuch as more than one single retort units are installed in one furnace such classification does not appear proper.

CHAIN GRATE STOKERS, which probably represent the oldest type of automatic coal-feeding mechanism, were developed to burn the high volatile, high ash, free burning, clinkering coals found in many localities abroad and throughout our own central states. The best results are obtained with free-burning coals if the fuel bed is undisturbed, and this same condition of fuel bed is necessary with coals that clinker. Coals high in ash require the continuous removal of such ash from the furnace.

As the name implies, a chain grate stoker consists of an endless chain running over sprockets at the front and rear of the furnace, these sprockets being keyed to shafts carried by cast-iron side frames. The upper run of the chain between the sprockets is carried by rollers or skids, while the lower or return run is carried on rollers or a pan. The chain itself is made up of small grate bars or links, these being of two kinds, common and driving. The stoker is driven ordinarily through the front sprocket shaft by different methods, depending upon the type of chain grate.

Coal is fed by gravity from a hopper at the forward end of the grate, is ignited as it passes beneath an ignition arch, and is carried toward the rear of the furnace as combustion progresses. When properly operated, combustion is completed as the fire reaches the end of the grate, and the ash and refuse are carried over the rear end as the grate makes the turn over the rear sprocket.

The grate moving through the furnace presents the problem of air leakage at the sides and rear, and it was the development of efficient air seals at these points that had much to do with the success of the chain grate stoker. Air leakage at the sides of the furnace is prevented by means of ledge plates or seals, which make a rubbing seal with the stoker chain. A bridge wall water box installed at the rear of the stoker, and ordinarily connected to the boiler circulation, tends to minimize leakage at this point, though tight ashpits are primarily depended upon to prevent leakage here. The water box, in addition to forming the seal at the rear of the furnace, also protects from the furnace temperature, through cooling, that portion of the grate from which the refuse has been discharged.

2400 Horse-power Installation of Cross-drum Babcock & Wilcox Boilers and Superheaters at the Westinghouse Electric and Manufacturing Co. East Pittsburgh, Pa.

In some of the earlier designs of chain grates, there was an excessive amount of siftings which found their way between links of the upper run of the grate. This difficulty has been overcome by the use of properly designed skids carrying this run of chain.

Arches are used with chain grate stokers, these varying in length and in height above the grate with different classes of coal. Ordinarily the length will be such as will cover approximately from 60 to 65 per cent of the grate.

The manner in which the coal is fed to the stoker, the undisturbed condition of the fuel bed during its passage through the furnace, and the automatic continuous discharge of ash, clearly indicate the manner in which the chain grate stoker meets the requirements of proper combustion of the class of coal for which it was developed.

While natural draft chain grate stokers have been in successful operation in many plants for many years, and without question will continue to be used in many classes of boiler work, they have two disadvantages as compared with certain other types of stoker. These are their inability to carry the high boiler capacities now demanded, 200 per cent of boiler rating representing approximately the maximum that can be obtained with them, and their inability to handle successfully sudden and violent fluctuations in boiler load.

Development in chain grate stokers in the past few years has been along the lines of changing from natural draft to forced blast stokers, and this development has reached the stage where it may be stated that in central station work where chain grates are to be installed, they will be of the blast rather than of the natural draft type.

In the method of feeding fuel, the progress of the fuel bed through the furnace, and the ash removal, the blast chain grate is similar to the natural draft grate. In construction, however, there is a difference. The Babcock & Wilcox blast chain-grate stoker, illustrated on page 222, is representative of this type and a brief description of this stoker will explain the class as a whole.

In construction, it differs from the natural draft chain grate in that the side frames consist of two side-wall blast boxes, which are divided into five or six compartments, depending upon the length of stoker. These two blast boxes are connected by transverse I-beams which divide the space between the upper and lower runs of the chain into compartments, corresponding to those of the side boxes. To the top of the transverse I-beams are attached several sets of skids, the space between the skids on the top of the beams being filled with spacers. The chain, sliding on the skids and these spacers, forms a seal at the top of the individual compartments. To the bottom of the transverse beams flexible seals are attached which, while allowing free movement of the chain, form an efficient seal at the bottom of the individual compartments. The return run of chain slides on plates which serve as a bottom plate for each compartment.

Each compartment is equipped with an individual damper, operated independently by a special damper-operating rig and so connected that the dampers on opposite sides of the compartments will act together and admit the air supply to both ends of a compartment equally. All of the side-wall blast boxes are connected to a common blast duct, the air pressure in the different compartments being regulated by adjustment of the individual compartment dampers.

8400 Horse-power Installation of Babcock & Wilcox Boilers and Superheaters at the Butler Street Plant of the Georgia Railway and Power Co., Atlanta, Ga. This Company Operates a Total of 15,200 Horse-power of Babcock & Wilcox Boilers

The air pressure, which rarely exceeds 2 inches in any compartment, is usually carried highest in the second and third compartments from the front, over which ignition has been completed and the maximum rate of combustion takes place.

As with natural-draft stokers, bridge-wall water boxes are installed, and in some cases side-wall water boxes at the level of the grate are also used to prevent the formation of clinker on the furnace walls at this point.

With blast chain grate stokers, the variation in length and height of arch for different fuels is not as great as in the case of natural-draft chain grates. The length is ordinarily such as to cover approximately 50 per cent of the grate, while the height is approximately 4 feet above the grate, sloping slightly upward toward the rear. An ignition arch from 12 to 18 inches long, approximately 21 inches above the grate, is placed immediately behind the stoker gate.

Satisfactory results may be obtained by operating blast chain grates as natural-draft stokers, though, because of the smaller air spaces in the links, as high a capacity cannot be obtained as with standard natural-draft stokers.

With the increased combustion rates made possible by the blast, the ability to control the blast under different portions of the fire, and the development of the stoker drive to give a wide range of stoker speeds, the blast chain grate stoker will handle any set of operating conditions as satisfactorily as will other types of blast stokers.

In addition to the class of fuel discussed, blast chain grates as built today are burning successfully small sizes of anthracite and coke breeze, fuels which can be burned on other types of stokers only with the greatest difficulty, if at all.

In chain grate stokers, both natural draft and blast, approximately only one-third of the total grate is in the furnace at one time, the particular third changing constantly, and this third, especially in the case of blast stokers, is amply ventilated. This freedom from exposure to high furnace temperature leads to low maintenance costs with this type of stoker.

Overfeed Stokers—The primary difference between the inclined front-feed and inclined side-feed stokers is in the direction in which the coal is fed to the grates.

In the front-feed class, the coal is fed from hoppers at the front end of the furnace to the upper portion of the grates, which are inclined downward toward the rear, ordinarily at an angle of approximately 45 degrees. The grates are reciprocated, taking alternately straight and inclined positions, the motion gradually carrying the fuel as it burns to the rear and bottom of the furnace. At the bottom of the inclined grates, flat dumping grates are provided for completing combustion and for cleaning. The fuel is coked on the upper portion of the grate and the volatile matter driven off in this process is ignited and burned as it passes over the burning carbon on the lower portion of the grate.

The method of cleaning fires by dumping with the front-feed stokers tends toward decreased capacity and efficiency at cleaning intervals. Further, this necessity for dumping fires tends more toward the production of smoke during cleaning periods than if the cleaning were continuous.

In the inclined side-feed stokers, which are ordinarily set in extension furnaces, the coal magazines or hoppers extend the full length of the grate at the top on both sides of the furnace, and the grates are inclined downward toward the horizontal center line of the furnace at the bottom. The inclined grates are actuated by rocking bars, the fuel being carried toward the bottom and center of the furnace as combustion progresses. At this point clinker grinders break and remove the refuse. Stokers of the side-feed type have arches extending over the full length of the grate.

Both front and side overfeed stokers satisfactorily handle coals of the caking class. The fires carried are ordinarily thin and the agitation of the grate bars keeps the fuel bed broken up and open, in this manner preventing caking.

Occasional efforts have been made to supply blast with overfeed stokers, but because of the fact that there was no possibility of regulating the blast under different portions of the grate such efforts were not successful. As natural-draft stokers they have the same objection to which the natural-draft chain grate is open, viz., the inability to develop high capacities or to handle satisfactorily sudden load fluctuations.

UNDERFEED STOKERS—The method in which coal is introduced into the furnace is approximately the same in all types of underfeed stokers, that is, from beneath by rams or plungers having a horizontal motion. The number of rams per retort varies with the different stokers, some being equipped with two or three while others have a single plunger for each retort, the coal being carried toward the rear of the furnace by means of auxiliary pushers. Coal is fed from hoppers at the stoker front by the action of the rams into retorts, the retort sides being cast-iron boxes connected to the blast duct. Some type of tuyeres or perforated cast-iron blocks forms the top of these boxes, the air for combustion being introduced through these openings. As the fresh coal is fed from beneath the coked and burning coal the volatile gases are driven off and are partially burned in their passage through the hottest portion of the fire at the top.

The methods of working the coal down the grates of inclined underfeed stokers vary somewhat in the different designs. In one class, it is brought about entirely by the action of the ram and auxiliary pushers; in a second class by gravity, the action of the plungers and a movement of extension overfeed grates at the rear of the tuyere boxes; and in a third design by gravity assisted by a movement of the tuyere boxes themselves.

At the rear of the tuyere boxes, or where auxiliary overfeed grates are used at the rear of such grates, there are either dumping plates, which are either hand or power operated, or some form of clinker grinder. Where dumping plates are used, there is the same objection as to lack of capacity and efficiency and smoke at dumping periods as in the case of the front overfeed stokers. Where clinker grinders are used, the cleaning is automatic though ordinarily not continuous.

In the horizontal type of underfeed stokers, the grates are horizontal lengthwise of the furnace, but slope downward across the furnace either from a single retort or a number of retorts to dumping plates. The coal, fed to the retort by plungers, rises in the retort and works downward on the sloping grates by gravity or in some designs by gravity assisted by alternate fixed and moving bars forming the sloping grate. Air for combustion is introduced through tuyeres forming the top and sides of the retorts, and, in the type utilizing movable grate bars, through both tuyeres and through the grate.

Underfeed stokers require no arch, and there is consequently a saving in furnace upkeep cost due to this item. On the other hand, furnace temperatures with underfeed stokers are ordinarily high, and this may result in a comparatively high upkeep cost of the other furnace brickwork, particularly if the draft is not properly regulated.

Underfeed stokers as a class are capable of handling heavy boiler overloads and rapidly fluctuating loads satisfactorily.

The fuel bed with underfeed stokers is ordinarily heavier than with any other type. This makes it possible to keep the excess air at a minimum, but it also makes necessary a blast appreciably higher than with blast chain grate stokers. For loads up to 300 per cent of rating, a blast of 5 inches is ordinarily required and when the loads are higher than this, 6 inches.

The fuel bed is kept agitated by the method of introducing the coal to the furnace, and this enables underfeed stokers to handle caking coals successfully.

SMOKE—The question of smoke and smokelessness in burning fuels has recently become a very important factor of the problem of combustion. Cities and communities throughout the country have passed ordinances relative to the quantities of smoke that may be emitted from a stack, and the failure of operators to live up to the requirements of such ordinances, resulting as it does in fines and annoyance, has brought their attention forcibly to the matter.

The whole question of smoke and smokelessness is to a large extent a comparative one. There are any number of plants burning a wide variety of fuels in ordinary hand-fired furnaces, in extension furnaces and on automatic stokers that are operating under service conditions, practically without smoke. It is safe to say, however, that no plant will operate smokelessly under any and all conditions of service, nor is there a plant in which the degree of smokelessness does not depend largely upon the intelligence of the operating force.

When a condition arises in a boiler room requiring the fires to be brought up quickly, the operatives in handling certain types of stokers will use their slice bars freely to break up the green portion of the fire over the bed of partially burned coal. In fact, when a load is suddenly thrown on a station, the steam pressure can often be maintained only in this way, and such use of the slice bar will cause smoke with the very best type of stoker. In a certain plant using a highly volatile coal and operating boilers equipped with ordinary hand-fired furnaces, extension hand-fired furnaces and stokers, in which the boilers with the different types of furnaces were on separate stacks, a difference in smoke from the different types of furnaces was apparent at light loads, but when a heavy load was thrown on the plant, all three stacks would smoke to the same extent and it was impossible to judge which type of furnace was on one or the other of the stacks.

In hand-fired furnaces much can be accomplished by proper firing. A combination of the alternate and spreading methods should be used, the coal being fired evenly, quickly, lightly and often, and the fires worked as little as possible. Smoke can be diminished by giving the gases a long travel under the action of heated brickwork before they strike the boiler heating surfaces. Air introduced over the fires and the use of heated arches, etc., for mingling the air with the gases distilled from the coal will also diminish smoke. Extension furnaces will undoubtedly lessen smoke where hand firing is used, due to the increase in length of gas travel and the fact that this travel is partially under heated brickwork. Where hand-fired grates are immediately under the boiler tubes and a high volatile coal is used, if sufficient combustion space is not provided the volatile gases, distilled as soon as the coal is thrown on the fire, strike the tube surfaces and are cooled below the burning point before they are wholly consumed and pass through the boiler as smoke. With an extension furnace, these volatile gases are acted upon by the radiant heat from the extension furnace arch, and this heat, with the added length of travel, causes their more complete combustion before striking the heating surfaces than in the former case.

Smoke may be diminished by employing a baffle arrangement which gives the gases a fairly long travel under heated brickwork and by introducing air above the fire. In many cases, however, special furnaces for smoke reduction are installed at the expense of capacity and economy.

From the standpoint of smokelessness, undoubtedly the best results are obtained with a good stoker, properly operated. As already stated, the best stoker will cause smoke under certain conditions. Intelligently handled, however, under ordinary operating conditions, stoker-fired furnaces are much more nearly smokeless than those which are hand fired, and are, to all intents and purposes, smokeless. In practically all stoker installations, there enters the element of time for combustion, the volatile gases as they are distilled being acted upon by ignition or other arches before they strike the heating surfaces. In many instances, too, stokers are installed with an extension beyond the boiler front, which gives an added length of travel, during which the gases are acted upon by the radiant heat from the ignition or supplementary arches, and here again we see the long travel giving time for the volatile gases to be properly consumed.

To repeat, it must be clearly borne in mind that the question of smokelessness is largely one of degree, and dependent to an extent much greater than is ordinarily appreciated upon the handling of the fuel and the furnaces by the operators, be these furnaces hand fired or automatically fired.

SMOKE AND EFFICIENCY OF COMBUSTION*—Though there is perhaps no phase of combustion that has been so fully discussed as that which results in the production of smoke, the common understanding of the loss from this source is at best vague and based in part, at least, on misconception. For this reason a brief consideration of smoke is included here, regardless of the amount of data on the subject available elsewhere.

Of the numerous and frequently unsatisfactory definitions of smoke that have been offered, that of the Chicago Association of Commerce Committee in its report on "Smoke Abatement and the Electrification of Railway Terminals in Chicago" is perhaps the best. This report defines smoke as "the gaseous and solid products of combustion, visible and invisible, including . . . . mineral and other substances carried into the atmosphere with the products of combustion."

From the standpoint of combustion loss, it is necessary to lay stress on the term "visible and invisible." The common conception of the extent of loss is based on the visible smoke, and such conception is so general that practically all, if not all, smoke ordinances are based on the visibility, density or color of escaping stack gases. As a matter of fact, the color of smoke, which is imparted to the gases by particles of carbon, cannot be taken as an indication of the stack loss. The invisible or practically colorless gases issuing from a stack may represent a combustion loss many times as great as that due to the actual carbon present in the gases, and but a small amount of

*From "Principles of Combustion in the Steam Boiler Furnace," by A. D. Pratt; published by The Babcock & Wilcox Company.

such carbon is sufficient to give color to large volumes of invisible gases which may or may not represent direct combustion losses. A certain amount of color may also be given to the gases by particles of flocculent ash and mineral matter. neither of which represents a combustion loss. The amount of such material in the escaping gases may be considerable where stokers of the forced draft type are used and heavy overloads are carried.

The carbon or soot particles in smoke from solid fuels are not due to the incomplete combustion of the fixed carbon content of the fuel. They result rather from the non-combustion or incomplete combustion of the volatile and heavy hydrocarbon constituents, and it is the wholly or partially incomplete combustion of these constituents that causes smoke from all fuels, solid, liquid or gaseous.

If the volatile hydrocarbons are not consumed in the furnace and there is no secondary combustion, there will, of course, be a direct loss resulting from the non-combustion of these constituents. While certain of these unconsumed gases may appear as visible smoke. the loss from this source cannot be measured with the ordinary flue-gas analysis apparatus and must of necessity be included with the unaccounted losses.

Where the combustion of the hydrocarbon constituents is incomplete a portion of the carbon component ordinarily appears as soot particles in the smoke. In the burning of hydrocarbons, the hydrogen constituent unites with oxygen before the carbon; for example, in the case of ethylene (C_2H_4).

$$C_2H_4 + 2O = 2H_2O + 2C$$

If, after the hydrogen is "satisfied," there is sufficient oxygen present with which that carbon component may unite, and temperature conditions are right, such combination will take place and combustion will be complete. If, on the other hand, sufficient oxygen is not present, or if the temperature is reduced below the combining temperature of carbon and oxygen, the carbon will pass off unconsumed as soot, regardless of the amount of oxygen present.

The direct loss from unconsumed carbon passing off in this manner is probably rarely in excess of one per cent of the total fuel burned, even in the case of the densest smoke. The loss due to unconsumed or partially consumed volatile hydrocarbons, on the other hand, though not indicated by the appearance of the gases issuing from a stack, may represent an appreciable percentage of the total fuel fired.

While the loss represented by the visible constituents of smoke leaving a chimney may ordinarily be considered negligible, there is a loss due to the presence of unconsumed carbon and tarry hydrocarbons in the products of combustion which, while not a direct combustion loss, may result in a much greater loss in efficiency than that due to visible smoke. These constituents adhere to the boiler heating surfaces and, acting as an insulating layer, greatly reduce the heat-absorbing ability of such surfaces. From the foregoing it is evident that the stack losses indicated by smoke, whether visible or invisible, result almost entirely from improper combustion. Assuming a furnace of proper design and fuel-burning apparatus of the best, there will be no objectionable smoke where there is good combustion. On the other hand, a smokeless chimney is not necessarily indicative of proper or even of good combustion. Large quantities of excess air in diluting the products of combustion naturally tend toward a smokeless stack, but the possible combustion losses corresponding to such an excess air supply have been shown.

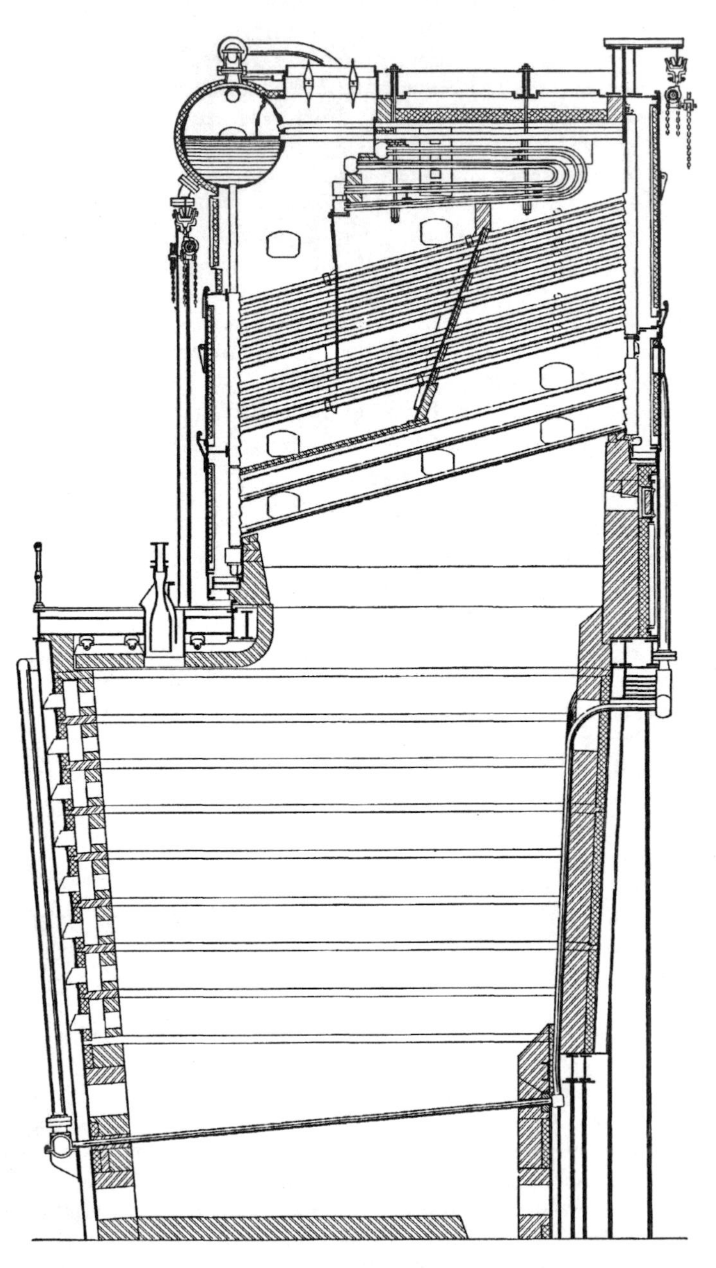

A Babcock & Wilcox Boiler Equipped with a Furnace for Burning Pulverized Coal

PULVERIZED COAL

COAL in pulverized form has been burned in industrial furnaces for many years. It is only within the past five years, however, that its use for steam-generating purposes can be considered to have passed the experimental stage. There is apparently a growing tendency toward the use of coal burned in a pulverized state, and such use will increase, in all probability, particularly in central station practice. The coals of better grades are becoming less and less available and insufficient for general central station demands. This fact, together with unsettled labor conditions and transportation difficulties, is forcing the power stations to the use of poorer grades of coal that, theoretically, at least, should be burned more efficiently in pulverized form than by other methods. Further, because of the size of the modern central station and the difficulty of maintaining a coal supply, the variation in the character of the coal available from day to day for steam generation may be great. Such variation might lead to appreciable losses in efficiency during the time the operators are determining the best methods of combustion for the coals of different characteristics burned on stokers, whereas if burned in pulverized form, the operation of burners for a change in coal, for the best efficiency, might be determined quickly.

As stated, there is apparently a growing tendency toward the use of pulverized coal in central station work and, with coals of certain characteristics, such use may become more or less general. On the other hand, there are many coals with which commercial efficiencies, or efficiencies based on operating costs, can be obtained with stokers that cannot be bettered, or in many instances equalled, when burning such coals in pulverized form.

The main objection to the use of pulverized coal from the standpoint of the boiler manufacturer and the boiler user has been the furnace volume ordinarily deemed necessary by the manufacturer of the pulverized coal-burning apparatus, and furnace brickwork trouble resulting from the use of these large furnaces and the temperatures developed with this class of fuel. This large furnace volume has been required largely because of the methods of introduction of the air for combustion and the "lazy" nature of the flame developed by such methods. It is possible that other methods of air introduction will be developed which may result in more rapid combustion and a shorter flame, in which case the necessity for what might appear to be an excessive furnace volume would be obviated.

Any reduction in the furnace volume would tend toward decreased furnace brickwork difficulties, in that the amount of brickwork would be less; on the other hand, because of the temperatures developed, such difficulties would probably be great regardless of furnace size. It is possible that some system of protecting the furnace walls by means of water heating surface connected to the circulation of the boiler will be developed, which will solve this phase of the problem.

DRYING—There are two types of dryers available for drying the coal before grinding, the single-shell and the double-shell. The latter, because of increased capacity and probably increased efficiency, is preferable where a minimum moisture content is desired.

The moisture content of pulverized coal is not particularly objectionable from the standpoint of efficiency of combustion. Tests have indicated that with coal containing slightly over 8 per cent of moisture, combustion is as complete and with as little excess air as with coal containing approximately 2 per cent of moisture, the difference in efficiency being that due to the difference in moisture loss in the flue gases. The moisture content has an effect on the grinding, the difficulty and cost of pulverization in general increasing with increased moisture. Surface moisture apparently has more effect in grinding than has inherent moisture.

Ideas as to the maximum allowable moisture for successful pulverization have undergone radical changes in the past few years. In the early installations, it was considered necessary to dry the coal to a point where the moisture did not exceed one per cent, and the capacity of grinding mills is ordinarily based on such a moisture content. Recent experience has indicated that coals containing an appreciably higher moisture content than one per cent can be satisfactorily pulverized without excessive cost. There are a number of installations handling coals containing 5 per cent moisture, and in some cases this figure is exceeded.

Grinding — In the earlier pulverizing plants, tube-and-ball grinding mills were used. These have in general been abandoned, however, and the mills in common use today are of the Raymond or the Fuller type.

As in the case of moisture content, ideas as to the necessary fineness to which coal should be ground for successful combustion have changed materially in the last few years. The specifications under which grinding mills are sold indicate the degree of pulverization deemed necessary in the earlier installations. These call for a degree of fineness for bituminous coal containing less than one per cent of moisture of 95 per cent passing a 100-mesh sieve and 82 per cent passing a 200-mesh sieve.

The present-day tendency is toward coarser grinding, and tests have shown that, with a properly designed furnace, the efficiency and the amount of combustible matter in the ash are approximately the same with coal ground to 98 per cent 40-mesh, 90 per cent 100-mesh and 12 per cent 200-mesh, as with the greater fineness given above.

The completeness of combustion is rather a function of burner and furnace design and of methods of air introduction than of fineness of grinding. The less the degree of pulverization, the longer the particles of coal must remain in suspension for complete combustion. This represents the necessity for a long length of flame travel and probably a greater flame velocity, which is as yet considered an objectionable feature from the standpoint of brickwork erosion. If some method of introducing air for combustion is satisfactorily developed which will result in more rapid combustion and hence a shorter flame, the necessity for length of flame travel might be decreased. On the other hand, even with a fineness of 200-mesh, the length of flame travel for fuel particles burned in suspension will probably be greater than that required where the fuel is burned in solid form, as on stokers.

Conveyors—The methods of moving the pulverized fuel from the grinding mill to the boiler supply bin were, in the early installations, almost entirely of the screw type. More recently, air-conveyor systems have been successfully used.

In the selection of apparatus for the preparation and delivery of pulverized coal to the boiler for any individual plant or set of operating conditions, the primary factors to be taken into consideration are the power consumption of such apparatus and the maintenance cost per unit of fuel handled.

Firing — Under this general heading must be included feeders, burners and methods of air introduction.

The feeder should be of a design that gives a positive and uniform delivery of fuel over the complete range of boiler capacity desired, and should have a method of control such as to allow immediate response to changes in operating conditions and load fluctuations. It should be of rugged construction to minimize difficulties arising from interrupted service.

Burners, while differing in design in different systems, may be generally classified under two types. In the first only a portion of the air for combustion is admitted through the burner with the pulverized coal, while in the second all of the air for combustion passes through the burner proper with the coal.

In the firing proper, there are two general methods used. In the first, the fuel is conveyed from the grinder to a storage bin adjacent to the point at which it is to be used, and from this bin taken to individual burners. This method has the advantage of probable uniformity of grinding, resulting from a continuity of grinding operations, and because the supply available gives a reserve fuel supply for suddenly increased load conditions. The disadvantages of such a system result from somewhat greater complication and the space occupied by the storage bins. In the second method, the grinder, conveyor and burner are one self-contained unit, the coal being pulverized and burned in a single step, there being no intermediate storage. Such a system has as an advantage a somewhat greater simplicity of installation and in its space requirements. With such a type of burner, however, the importance of interruption of service should be taken into consideration; without a reserve fuel supply, regardless of the ability of the burner, stoppage of the supply due to grinding trouble might not allow a fluctuating load to be handled successfully.

In the systems of pulverized fuel apparatus in most common use the air for combustion is ordinarily admitted in part through the burner proper with the coal and in part through secondary combustion air openings, varying in location, relative to the form of flame propagation, with different systems. Where secondary combustion air is used, such air is ordinarily drawn through ducts in the furnace walls, theoretically cooling these walls and adding to the efficiency of combustion through the use of preheated air.

More recent designs of burners introduce all of the air for combustion through the burner with the fuel. If this design is satisfactorily developed, and the mixture of pulverized fuel and air for combustion is completed before such mixture is delivered to the furnace, a more rapid ignition and probably a shorter flame might follow, which possibly would allow a reduction in the furnace volume below that apparently required where secondary air admission is used. This is a feature that development in the art will answer.

Furnace Design—The high efficiencies obtained in the properly designed pulverized fuel-burning plant are primarily the result of completing combustion in the furnace without carrying the flame into the boiler heating surfaces. Combustion can be completed in this manner only by the use of a large furnace volume and proper furnace form. With a suitable form of furnace, the capacity that can be developed is almost directly proportional to the furnace volume. General practice among the manufacturers is to recommend 2 cubic feet of furnace volume per boiler horse-power to be developed. Tests apparently indicate that a combustion rate of from 1 to 1½ pounds of coal per cubic foot of furnace volume per hour gives the best efficiency of combustion, though it is claimed that this rate may, with a proper design of furnace, be increased to 2 pounds per cubic foot without an appreciable decrease in efficiency of combustion. On the basis of 2 cubic feet of furnace volume per developed horse-power, a combustion rate of 1.75 pounds of coal per cubic foot would correspond to approximately 300 per cent of rating, while if the rate is increased to 2 pounds per cubic foot the corresponding capacity is 340 per cent.

SLAGGING—One of the greatest difficulties in the satisfactory burning of pulverized coal has resulted from the formation of slag. Such formation, while giving some trouble on the boiler heating surfaces, occurs principally in the furnace. Trouble from this source, which was excessive in the earlier installations, is gradually being reduced by improvements in furnace design and methods of air introduction, and today it may be considered to be largely eliminated except in the case of coals having a low fusion-point ash. The methods used in reducing slagging to a minimum are:

1st. The maintaining of a furnace temperature below that at which ash fusion will occur. By following this method, combustion is completed with an amount of excess air much higher than that with which it is possible to burn coal in pulverized form. From the standpoint of thermal efficiency of boiler and furnace, therefore, the method is not representative of the best practice: on the other hand, from the standpoint of commercial operating efficiency, this method must be given consideration under certain individual sets of conditions.

2nd. In a second method the furnace is designed to give three temperature zones. The upper is immediately under the furnace roof (where the fuel is introduced downwardly into the furnace), and in it the combustion occurs; this zone is maintained at a maximum temperature. Below the combustion zone is a neutral zone, and toward the bottom of the furnace there is a cooling zone in which, by the admission of auxiliary air, a temperature below the fusing temperature of the ash is maintained. In the downward passage of the gases, the ash, which is precipitated in the neutral zone, passes into the cooling zone and settles, unfused, on the floor of the furnace.

3rd. In the third method, a water screen toward the bottom of the furnace, and ordinarily connected to the circulation of the boiler, is used. It is claimed for this method that the ash particles, in passing through the water screen, are cooled and deposit on the furnace floor. Being in a comparatively low temperature zone, no fusion of the ash occurs and it may be removed by the ordinary methods of ash disposal.

ASH—The accurate determination of the ultimate settling point of the ash from a pulverized fuel burning installation is difficult, and statements bearing on this feature vary widely. One authority states that with a proper design of furnace, from 25 to 50 per cent of the ash remains in the combustion chamber, from 5 to 12 per cent settles in the boiler setting, from 25 to 35 per cent settles in the stack base, and from 12 to 25 per cent passes up the stack. With overhead stacks, the amount given as settling in the stack base must be added to that passing to the atmosphere. While it is frequently claimed that the ash issuing from the stack from pulverized fuel equipment is of a flocculent, non-carbon content, material that will not damage surrounding property, this question is a real problem in the successful commercial burning of pulverized coal and one to which the manufacturers and users of such equipment must and are giving a great amount of consideration and experimenting for the elimination of the ash discharge.

SOLID FUELS OTHER THAN COAL AND THEIR COMBUSTION

WOOD is vegetable tissue which has undergone no geological change. When newly cut, wood contains from 30 to 50 per cent of moisture. When dried in the atmosphere for approximately one year, the moisture content is reduced to 18 or 20 per cent.

Wood is ordinarily classified as hard wood, including oak, maple, hickory, birch, walnut and beech, and soft wood, including pine, fir, spruce, elm, chestnut, poplar and willow. While, theoretically, equal weights of wood substance should generate the same amount of heat, regardless of species, practically the varying form of wood tissue and the presence of rosins, gums, tannin, oils and pigments result in different heating values, and more particularly in a difference in the ease with which combustion can be accomplished. Rosin may increase the heating value as much as 12 per cent. Contrary to general opinion, the heat value per pound of soft wood is slightly greater than that of hard wood. Table 41 gives the chemical composition and the heat values of the common woods. The heat values of wood fuels are ordinarily reported on a dry basis. It is to be remembered, however, that because of the high moisture content, the ratio of the amount of heat available for steam generation to that of the dry fuel is much lower than that of practically all other solid fuels. Even woods that are air dried contain approximately 20 per cent of moisture, and this moisture must be evaporated and superheated to the temperature of the escaping gases before the heat evolved, for absorption by the boiler, can be determined.

TABLE 41

ULTIMATE ANALYSES AND CALORIFIC VALUES OF DRY WOODS

(GOTTLIEB)

Kind of Wood	C	H	N	O	Ash	B.t.u. per Pound
Oak	50.16	6.02	0.09	43.36	0.37	8316
Ash	49.18	6.27	0.07	43.91	0.57	8480
Elm	48.99	6.20	0.06	44.25	0.50	8510
Beech	49.06	6.11	0.09	44.17	0.57	8591
Birch	48.88	6.06	0.10	44.67	0.29	8586
Fir	50.36	5.92	0.05	43.39	0.28	9063
Pine	50.31	6.20	0.04	43.08	0.37	9153
Poplar	49.37	6.21	0.96	41.60	1.86	7834*
Willow	49.96	5.96	0.96	39.56	3.37	7926*

*B.t.u. calculated

While the term is ordinarily used to designate those compact substances familiarly known as tree trunks and limbs, wood, as used commercially for steam generation, is usually a waste product from some industrial process. At the present time, refuse from lumber and sawmills forms by far the greatest portion of this class of fuel. In such refuse the moisture content may run as high as 60 per cent and the composition of the fuel may vary over a wide range during different periods of mill operation. The fuel consists of sawdust, "hogged" wood and slabs, and the proportions of these may vary widely. Hogged wood is mill refuse and logs which have been passed through a "hog" machine or macerator which cuts or shreds the wood with rotating knives to a state in which it may be readily handled as fuel.

3520 Horse-power Installation of Babcock & Wilcox Boilers at the Portland Railway Light and Power Co., Portland, Ore. These Boilers Are Equipped with Wood Refuse Extension Furnaces at the Front and Oil-burning Furnaces at the Mud-drum End

Table 42 gives the heat value and moisture content of typical lumber mill refuse from various woods.

TABLE 42

MOISTURE AND CALORIFIC VALUE OF SAWMILL REFUSE

Kind of Wood	Nature of Refuse	Per Cent Moisture	B.t.u. per Pound Dry Fuel
Mexican White Pine	Sawdust and Hog Chips . . .	51.90	9020
Yosemite Sugar Pine	Sawdust and Hog Chips . .	62.85	9010
Redwood 75 per cent, Douglas Fir 25 per cent . . .	Sawdust, Box Mill Refuse and Hog	42.20	8977*
Redwood	Sawdust and Hog Chips . . .	52.98	9040*
Redwood	Sawdust and Hog Chips . . .	49.11	9204*
Fir, Hemlock, Spruce and Cedar	Sawdust	42.06	8949*

* Average of two samples.

FURNACE DESIGN—The principal features of furnace design for the satisfactory combustion of wood fuel are ample furnace volume and the presence of a large area of heated brickwork to radiate heat to the fuel bed. The latter factor is of particular importance in the case of wet wood, and ordinarily necessitates the use of an extension furnace. A furnace of this form not only gives the required amount of heated brickwork for proper combustion, but enables the fuel, in the case of hogged wood and sawdust, to be most readily fed to the furnace. With wet mill refuse, the furnace should be "bottled" at its exit to maintain as high a temperature as possible, the extent to which the bottling effect is carried being dependent primarily upon the moisture content of the fuel and being greater as the moisture content is higher. The bottling effect, which is ordinarily secured by a variation in the height of the extension furnace bridge wall, has, in several recent installations, been accomplished by the use of a "drop-nose" arch at the rear of the furnace combustion arch.

Secondary air for combustion is of assistance in securing proper results and may be admitted through the bridge wall to the furnace or, where there is a secondary combustion space behind the bridge wall, into that space.

In the case of hogged wood and sawdust, the fuel is fed through fuel chutes in the roof of the extension furnace, ordinarily being brought from the storage supply to the chutes by some type of conveyor system. With this class of wood fuel, in-swinging fire doors are placed at the furnace front for fire-inspection purposes. Where slabs are burned in addition to hogged wood and sawdust, large side-hinged slab firing doors are usually installed above the in-swinging doors.

Fuel chutes should be circular on the inside and square outside, such design enabling them to be installed most readily in the furnace roof. For ordinary mill refuse, the chute should be 12 inches in diameter, though for shingle mill refuse the size should be made 18 inches.

Each fuel chute should handle a square unit of grate surface, the dimensions of such units varying from 4 by 4 to 8 by 8 feet, depending upon the moisture content and nature of the fuel. Where it is necessary in order to secure sufficient grate surface, units of grate and chutes may be placed in tandem.

Satisfactory results have been obtained with both sprung and flat suspended arches. In the case of the former, it is perhaps well to install individual arches over separate grate units in width. The fuel fed through the chutes will form a cone at an angle of approximately 45 degrees, and the height of the arch varies with the dimensions of the grate unit handled by each chute. This height, where dampers are used in the fuel chutes, should be approximately 12 or 14 inches above the top of the fuel cone. Such a distance is necessary, particularly in the case of tandem chutes and grate units, to allow the products of combustion to pass from the furnace.

While the ash from wood fuel is light, because of high combustion rates its volume is great, and ashpits are ordinarily made deeper than for coal firing. Ash removal with wood fuel can be satisfactorily accomplished by sluicing.

Experience has indicated that better combustion results are obtained with natural draft than with forced blast. The amount of draft required at the damper is dependent not only on the capacity desired but also upon the moisture content of the fuel, being greater for a given capacity as the moisture content increases.

Capacities with wood fuel are ordinarily considered in terms of the maximum horse-power that can be developed per square foot of grate surface. While the difference in the ease with which satisfactory combustion may be accomplished resulting from the presence of rosin, gums, etc., in the fuel, discussed above, makes it difficult to make any definite statement as to such maximum capacities, a conservative figure may be taken as from 9 horse-power per square foot of grate surface with wood fuels containing 30 per cent of moisture or less to 4 horse-power for fuels containing from 45 to 50 per cent of moisture.

Mill refuse, as ordinarily burned, has been considered almost entirely a waste product which would have to be disposed of in some other manner if not used for steam generation. The growing tendency toward all fuel conservation, however, is making the combustion of such fuel with maximum efficiency of greater importance, and while capacity will probably always be the primary result sought, efficiency, with capacity, is being given more and more consideration.

Dry Wood—Dry sawdust, chips, blocks and veneer are frequently burned in plants of the wood-working industry. With such fuel, as with wet wood refuse, an ample furnace volume is essential, though because of the lower moisture content, the presence of heated brickwork is not as necessary as with wet wood fuel.

As an example of the results that may be obtained with dry wood refuse where proper attention is given to furnace design, one plant burning this class of fuel may be cited. In this plant, where the fuel is hogged kiln-dried wood, largely maple, a furnace volume of 6.3 cubic feet per rated boiler horse-power was installed. This volume was secured through height of furnace rather than through depth, though a small extension furnace was used, through the roof of which the fuel is fed. The boilers in the plant are normally operated at from 150 to 200 per cent of rating, though at times capacities as high as 300 per cent of rating are carried. The efficiency under ordinary operating conditions is approximately 65 per cent. The grate surface installed in these boilers is such as to give a ratio of grate surface to heating surface of 1 to 70. Because of the height of the furnace, a large portion of the fuel is burned in suspension before reaching the fuel bed, and it is probably this fact that enables the high overloads to be carried with such a grate ratio. While these results are probably better than those obtained in the ordinary plant burning this class of fuel, they are indicative of what may be secured with a proper design of furnace.

In a few localities cord wood is burned. With this as with other classes of wood fuel, a large combustion space is an essential feature. The percentage of moisture in cord wood may make it necessary to use an extension furnace, but ordinarily this is not required. Ample combustion space is in most cases secured by dropping the grates to the floor line, large double-deck fire doors being supplied at the usual fire-door level through which the wood is thrown by hand. Air is admitted under the grates through an excavated ashpit. The side, front and rear walls of the furnace should be corbelled out to cover about two-thirds of the total grate surface. This prevents cold air from streaming up the sides of the furnace and also reduces the grate surface. Cord wood and slabs form an open fire through which the frictional loss of the air is much less than in the case of sawdust or hogged material. The combustion rate with cord wood is, therefore, higher, and the grate surface may be considerably reduced. Such wood is usually cut in lengths of 4 feet or 4 feet 6 inches, and the depth of the grates should be kept approximately 5 feet to get the best results.

BAGASSE — Bagasse is the refuse of sugar cane from which the juice has been extracted, and from the beginning of the sugar industry has been the natural fuel for sugar plantation power plants.

In the early days of sugar manufacture, the cane was passed through a single mill and the defecation and concentration of the saccharine juice took place in a series of vessels mounted over a common flue with a fire at one end and a stack at the other. This method required an enormous amount of fuel, and it was frequently necessary to sacrifice the degree of extraction to obtain the necessary amount of bagasse and a bagasse that could be burned. In the primitive furnaces of early practice, it was necessary to dry the bagasse before it could be burned, and the amount of labor involved in spreading and collecting it was great.

With the general abolition of slavery and resulting increased labor cost of production, and with growing competition from European beet sugar, it was necessary to increase the degree of extraction, the single mill being replaced by the double mill, and the open wall or Jamaica train method of extraction described being replaced by vacuum evaporating apparatus and centrifugal machines. Later a third grinding was introduced, and the maceration and dilution of the bagasse were carried to a point where the last trace of sugar in the bagasse was practically eliminated. The amount of juice to be treated was increased by these improved manufacturing methods from 20 to 30 per cent, but the amount of bagasse available for fuel and its calorific value as fuel were decreased to an extent which the combustion capacity of the furnaces available could not meet. In order that the steam-generation end of manufacture should keep pace with the process end, it was necessary to develop a more efficient method of burning the bagasse commercially than that employed in the drying of the fuel.

During the transition period of manufacture many furnaces were "invented" for burning green bagasse, the saving in labor by this method over that necessary in spreading, drying and collecting the fuel obviously being the primary factor in the reduction of the cost of steam generation. None of these furnaces, however, gave satisfactory results until the hot-blast bagasse furnace was introduced in 1888. While furnaces of this design operated satisfactorily, their construction was expensive and. because of the cost to the planters in changing to improved sugar-manufacture apparatus, was difficult of introduction. The necessity of decreasing the cost of furnaces for burning green bagasse led to a series of experiments by The Babcock & Wilcox Company that ultimately resulted in the cold-blast bagasse furnace described hereafter in connection with the combustion of this class of fuel.

COMPOSITION AND CALORIFIC VALUE OF BAGASSE—The proportion of fiber contained in the cane and the density of the juice are important factors in the relation the bagasse fuel will have to the total fuel necessary to generate the steam required in a mill's operation. A cane rich in wood fiber produces more bagasse than a poor one, and a thicker juice is subjected to a higher degree of dilution than one not so rich.

Besides the percentage of bagasse in the cane, its physical condition has a bearing on its calorific value. The factors here entering are the age at which the cane must be cut, the locality in which it is grown, etc. From the analysis of any sample of bagasse its approximate calorific value may be calculated from the formula,

$$\text{B.t.u. per pound bagasse} = \frac{8550F + 7119S + 6750G - 972W}{100} \qquad (27)$$

Where F = percentage of fiber in cane, S = percentage of sucrose, G = percentage of glucose and W = percentage of water.

This formula gives the total available heat per pound of bagasse, that is, the heat generated per pound less the heat required to evaporate its moisture and superheat the steam thus formed to the temperature of the stack gases.

Three samples of bagasse in which the ash is assumed to be 3 per cent give from the formula:

$F = 50$ S and $G = 4.5$ $W = 42.5$ B.t.u. = 4183
$F = 40$ S and $G = 6.0$ $W = 51.0$ B.t.u. = 3351
$F = 33.3$ S and $G = 7.0$ $W = 56.7$ B.t.u. = 2797

A sample of Java bagasse having $F = 46.5$, $S = 4.50$, $G = 0.5$, $W = 47.5$ gives B.t.u. 3868.

These figures show that the more nearly dry the bagasse is, the higher the calorific value, though this is accompanied by a decrease in sucrose. The explanation lies in the fact that the presence of sucrose in an analysis is accompanied by a definite amount of water, and that the residual juice contains sufficient organic substance to evaporate the water present when a fuel is burned in a furnace. For example, assume the residual juice (100 per cent) to contain 12 per cent of organic matter. From the constant in formula *(27)* $(12 \times 7119) \div 100 = 654.3$ and $[(100 - 12) \times 972] \div 100 = 653.4$. That is, the moisture in a juice containing 12 per cent of sugar will be evaporated by the heat developed by the combustion of the contained sugar. It would. therefore, appear that a bagasse containing such juice has a calorific value due only to its fiber content. This is, of course, true only where the highest products of oxidization are formed during the combustion of the organic matter. This is not strictly the case, especially with a bagasse of a high moisture content which will not burn properly but which smoulders and produces a large quantity of products of destructive distillation, chiefly heavy hydrocarbons, which escape unburnt. The reasoning. however, is sufficient to explain the steam-making properties of bagasse of a low sucrose content, such as are secured in Java, as when the sucrose content is lower, the heat value is increased by extracting more juice and hence more sugar from it. The sugar operations in Java exemplify this and show that with a high dilution by maceration and heavy pressure the bagasse meets all of the steam requirements of the mills without auxiliary fuel.

A high percentage of silica or salts in bagasse has sometimes been ascribed as the reason for the tendency to smoulder in certain cases of soft fiber bagasse. This, however, is due to the large moisture content of the sample resulting directly from the nature of the cane. Soluble salts in the bagasse have also been given as the explanation of such smouldering action of the fire, but here too the explanation lies solely in the high moisture content, this resulting in the development of only sufficient heat to evaporate the moisture.

Table 43 gives the analyses and heat values of bagasse from various localities.

TABLE 43

ANALYSES AND CALORIFIC VALUES OF BAGASSE

Source	Moisture	C	H	O	N	Ash	B.t.u. per Pound Dry Bagasse
Cuba . . .	51.50	43.15	6.00	47.95	. .	2.90	7985
Cuba .	49.10	43.74	6.08	48.61	. .	1.57	8300
Cuba . . .	42.50	43.61	6.06	48.45	. .	1.88	8240
Cuba . . .	51.61	46.80	5.34	46.35	. .	1.51	. .
Cuba . .	52.80	46.78	5.74	45.38	. .	2.10	. .
Porto Rico .	41.60	44.28	6.66	47.10	0.41	1.35	8359
Porto Rico .	43.50	44.21	6.31	47.72	0.41	1.35	8386
Porto Rico .	44.20	44.92	6.27	46.50	0.41	1.90	8380
Louisiana . .	52.10	.				2.27	8230
Louisiana . .	54.00	. .	. .	. .	.	. .	8370
Louisiana . .	51.80	. . .	. .	. .	.	. .	8371
Java	. . .	46.03	6.56	45.55	0.18	1.68	8681

Table 44 gives the value of mill bagasse at different extractions, which data may be of service in making approximations as to its fuel value as compared with that of other fuels.

Furnace Design and the Combustion of Bagasse—With the advance in sugar manufacture there came, as described, a decrease in the amount of bagasse available for fuel. As the general efficiency of a plant of this description is measured by the amount of auxiliary fuel required per ton of cane, the relative importance of the furnace design for the burning of this fuel is apparent.

In modern practice, under certain conditions of mill operation and with bagasse of certain physical properties, the bagasse available from the cane ground will meet the total steam requirements of the plant as a whole; such conditions prevail, as described, in Java. In the United States, Cuba, Porto Rico and like countries, however, auxiliary fuel is almost universally a necessity. The amount will vary, depending to a great extent upon the proportion of fiber in the cane, which varies widely with the locality and with the age at which it is cut, and to a less extent upon the degree of purity of the manufactured sugar, the use of the maceration water and the efficiency of the mill apparatus as a whole.

TABLE 44

VALUE OF ONE POUND OF MILL BAGASSE AT DIFFERENT EXTRACTIONS

Per Cent Extraction of Weight of Cane	Per Cent Moisture in Bagasse	Fiber		Sugar		Molasses		B.t.u. per Pound of Bagasse			Pounds of Bagasse Equivalent to One Pound of Coal of 14,000 B.t.u.
		Per Cent in Bagasse	Fuel Value, B.t.u.	Per Cent in Bagasse	Fuel Value, B.t.u.	Per Cent in Bagasse	Fuel Value, B.t.u.	Total Heat Developed per Pound of Bagasse	Heat Required to Evaporate Moisture*	Heat Available for Steam Generation	
BASED UPON CANE OF 12 PER CENT FIBER AND JUICE CONTAINING 18 PER CENT OF SOLID MATTER. REPRESENTING TROPICAL CONDITIONS											
75	42.64	48.00	3996	6.24	451	3.12	217	4664	525	4139	3.38
77	39.22	52.17	4343	5.74	414	2.87	200	4958	483	4475	3.13
79	35.15	57.14	4757	5.14	371	2.57	179	5307	433	4874	2.87
81	30.21	63.16	5258	4.42	319	2.21	154	5731	372	5359	2.61
83	24.12	70.59	5877	3.53	256	1.76	122	6255	297	5958	2.35
85	16.20	80.00	6660	2.40	173	1.20	83	6916	200	6716	2.08
BASED UPON CANE OF 10 PER CENT FIBER AND JUICE CONTAINING 15 PER CENT OF SOLID MATTER. REPRESENTING LOUISIANA CONDITIONS											
75	51.00	40.00	3330	6.00	433	3.00	209	3972	678	3294	4.25
77	48.07	43.45	3617	5.66	409	2.82	196	4222	592	3630	3.86
79	44.52	47.62	3964	5.24	378	2.62	182	4524	548	3976	3.52
81	40.18	52.63	4381	4.73	342	2.36	164	4887	495	4392	3.19
83	35.00	58.82	4897	4.12	298	2.06	143	5436	431	5005	2.80
85	28.33	66.67	5550	3.33	241	1.67	116	5907	349	5558	2.52

* Assuming bagasse temperature = 80 degrees Fahrenheit and exit-gas temperature = 500 degrees Fahrenheit.

In general it may be stated that this class of fuel may be best burned in large quantities. Due to this fact, and in order to obtain the efficient combustion resulting from the burning of a bulk of this fuel, a single large furnace is frequently installed between two boilers, serving both, though there is, of course, a limit to the size of boiler units that may be set in this manner. A disadvantage of this type of installation results from the necessity of having two boiler units out of service when it is necessary to take the furnace down for repairs, requiring a greater boiler capacity than where single furnaces are installed to assure continuity of service. On the other hand, the lower cost of one large furnace as against that of two individual smaller furnaces, and the increased efficiency of combustion with the former, may more than offset this disadvantage.

As in the case of wet wood refuse and, as a matter of fact, for all fuels containing an excessive moisture content, the essential features of furnace design for the proper combustion of green bagasse are ample combustion space, a large mass of furnace brickwork for maintaining furnace temperature, and a length of gas travel sufficient to enable combustion to be completed before the boiler heating surfaces are encountered. The fuel is burned either upon a hearth or on grates. The objection to the latter method, particularly where blast is used, is that the air for combustion enters largely around the edges of the fuel pile where the bed is thinnest. Further, where the fuel is burned on grates, the tendency of the ash and refuse to stop the air spaces does not allow a constant combustion rate for a given draft, and since there is a combustion rate which represents the best efficiency with this class of fuel, such efficiency cannot be maintained throughout the entire period between cleaning intervals. Where the bagasse is burned on a hearth, the ash and refuse form on the hearth, do not affect the air supply, and allow a constant combustion rate to be maintained. When burned on a hearth, the air for combustion is admitted through a series of tuyeres extending around the furnace and upward from the hearth. In some cases a combination of grates and tuyeres has been used. When air is admitted through tuyeres, it impinges on the fuel pile as a whole and gives a uniform combustion. The tuyeres are connected to an annular space in which, where blast is used, the pressure is controlled by a blower.

As stated, bagasse is best burned in large quantities, with corresponding high combustion rates. When burned on grates with a natural draft of 0.3 inch of water in the furnace, a combustion rate of from 250 to 300 pounds per square foot of grate surface per hour may be obtained, while with a blast of 0.5 inch this rate may be increased to approximately 450 pounds. When burned on a hearth with a blast of 0.75 inch a combustion rate of approximately 450 pounds per square foot of hearth per hour may be obtained, while with the blast increased to 1.6 inches, this rate may be increased to approximately 650 pounds. These rates apply to bagasse containing about 50 per cent of moisture. It would appear that when burned on grates the most efficient combustion rate is approximately 300 pounds per square foot of grate per hour, and, as stated, this rate is obtainable with natural draft. When burned on a hearth, and with blast, the most efficient rate is about 450 pounds per square foot of hearth per hour, which rate requires a blast of approximately 0.75 inch.

A green bagasse furnace which meets the general requirements of proper furnace design described above and with which eminently satisfactory results have been obtained is illustrated in Fig. 13. The hearth on which the bagasse is burned is ordinarily elliptical in form. Air for combustion is admitted through a series of tuyeres shown above the hearth line. The supply of air is controlled by the amount and pressure of the air within the annular space to which the tuyeres are connected. Secondary air for combustion is admitted at the rear of the bridge wall, as indicated. The roof of the furnace is ordinarily spherical in form, with its top from 11 to 13 feet above the grate or hearth. The products of combustion pass from the primary combustion chamber under an arch to a secondary combustion chamber, as shown. A furnace of this design embodies the essential features of ample combustion space, the mass of heated brickwork necessitated by the high moisture content of the fuel, and a long travel of gases before the boiler heating surfaces are encountered.

The fuel is fed through the roof of the furnace, preferably by some mechanical method which will assure a constant fuel supply and at the same time prevent the inrush of cold air into the furnace.

This class of fuel deposits an appreciable quantity of dust and ash which, if not removed promptly, fuses into a hard, glass-like clinker. Ample provision should be made for the removal of this material from the furnace, the gas ducts, and the boiler setting and heating surfaces.

It is common practice, where the bagasse available is not sufficient to meet the steam requirements of the mill, to equip the bagasse-fired boilers with some form of auxiliary fuel furnace. Fig. 13 shows a mechanical atomizing oil burner as the auxiliary fuel-burning apparatus. Satisfactory results cannot be obtained in burning the two fuels at the same time, the oil burners being operated only when the supply of bagasse is exhausted and for a period of sufficient duration to accumulate a supply of the primary fuel.

Table 45 gives the principal results of several tests of a Babcock & Wilcox boiler equipped with a bagasse furnace as illustrated in Fig. 13.

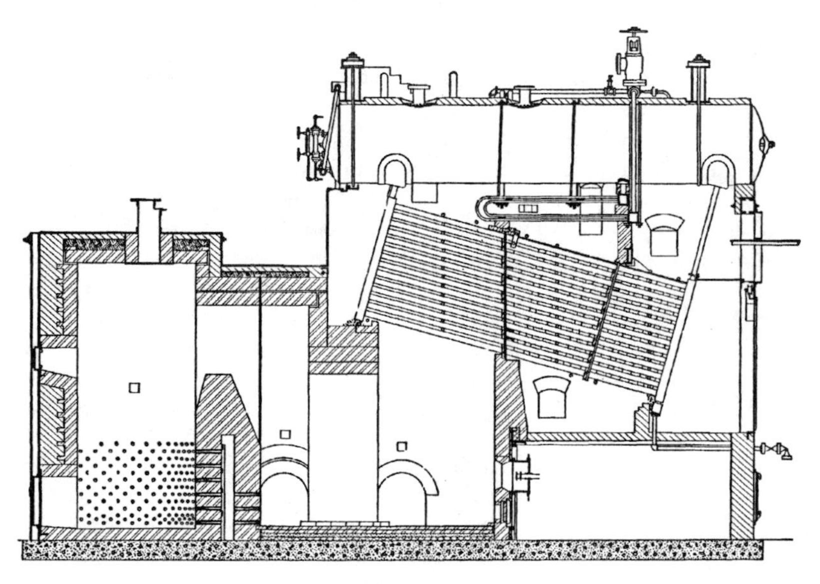

Fig. 13. A Babcock & Wilcox Boiler Equipped with a Babcock & Wilcox Bagasse Burner and a Babcock & Wilcox Mechanical Oil Burner

Tan Bark—Tan bark, or spent tan, is the fibrous portion of bark remaining after use in the tanning industry. It is usually very high in its moisture content, a number of samples giving an average of 65 per cent or about two-thirds of the total weight of the fuel. The weight of the spent tan is about 2.13 times as great as the weight of the bark ground. In calorific value, an average of 10 samples gives 9500 B.t.u. per pound dry.* The available heat per pound as fired, owing to the great percentage of moisture usually found, will be approximately 2700 B.t.u. Since the weight of the spent tan as fired is 2.13 as great as the weight of the bark as ground at the mill, one pound of ground bark produces an available heat of approximately 5700 B.t.u. Relative to bituminous coal, a ton of bark is equivalent to 0.4 ton of coal. An average chemical analysis of the bark is, carbon 51.8 per cent, hydrogen 6.04, oxygen 40.74, ash 1.42.

Tan bark is burned in isolated cases and in general the remarks on burning wet-wood fuel apply to its combustion. The essential features are a large combustion

*Dr. Henry C. Sherman, Columbia University.

TABLE 45

TESTS OF BABCOCK & WILCOX BOILERS WITH GREEN BAGASSE

Rated Horse-power	860	860	860
Duration of Test, hours	8.08	7.82	8.00
Steam Pressure by Gauge, pounds	110.9	112.7	111.8
Feed Temperature, degrees Fahrenheit	148.3	169.0	158.0
Degrees of Superheat, degrees Fahrenheit	105.5	140.9	121.2
Factor of Evaporation	1.1664	1.1639	1.1648
Pressure to Tuyeres, inches	.90	1.66	.76
Draft in Furnace, inches	.34	.24	.26
Draft at Damper, inches	.56	.88	.67
Exit-gas Temperature, degrees Fahrenheit	407	522	438
Flue-gas Analysis: CO_2, per cent	13.98	14.28	13.37
Flue-gas Analysis: O_2, per cent	6.51	6.00	7.19
Flue-gas Analysis: CO, per cent	.05	.30	.06
Bagasse Fired per Hour, pounds	10090	19210	13200
Moisture in Bagasse, per cent	49.5	48.5	49.7
Dry Bagasse Fired per Hour, pounds	5095	9893	6639
Bagasse Fired per Square Foot of Hearth per Hour, pounds	348	662	455
Water per Hour, actual, pounds	21704	39512	28663
Water per Hour from and at 212 degrees, pounds	25308	46005	33386
Evaporation from and at 212 degrees per Pound as Fired pounds	2.51	2.39	2.53
Evaporation from and at 212 degrees per Pound Dry Bagasse, pounds	4.97	4.65	5.03
Horse-power Developed	733.5	1333.5	967.7
Per Cent of Rated Horse-power Developed	85.4	155.2	112.7

space, large areas of heated brickwork radiating to the fuel bed, and draft sufficient for high combustion rates. The ratings obtainable with this class of fuel will not be as high as with wet-wood fuel, because of the heat value and the excessive moisture content. Mr. D. M. Myers found in a series of experiments that an average of from 1.5 to 2.08 horse-power could be developed per square foot of grate surface with horizontal return tubular boilers. This horse-power would vary considerably with the method in which the spent tan was fired.

City of San Francisco, Cal. Fire Fighting Station No. [illegible] 28[illegible] Horse-power of Babcock & Wilcox Boilers Equipped for Burning Oil Fuel

LIQUID FUELS AND THEIR COMBUSTION

FUEL oil is the only liquid fuel sufficiently abundant for use in steam generation. The term fuel oil is rather elastic in that it has had various meanings, not only for oils from different sources of supply, but also for an oil from a given source at different periods as improvements have come about in methods of refinement. In the early days of oil burning, when the demands for the refined products were of limited extent, much of the oil burned was "crude oil" or oil in its natural state as it came from the well. Today, with the marked advance in the methods of refining and in the demand for the products of such refining, the oil in use for fuel purposes is the residue from crude oil after it has been more or less refined. Dr. Day of the U. S. Bureau of Mines defines fuel oil as including "all oils which are not salable for some other special purpose at a higher price than that which prevails for oils to be sold as fuel oils or to be burned under boilers."

While geologists are not in entire agreement as to the origin of petroleum, whether it be of animal, vegetable or mineral origin, the generally accepted opinion is that such origin is of organic nature. Crude oil, while varying widely in physical characteristics, may be broadly divided into three classes in accordance with the predominant base, or the nature of the residue after distillation. These are: 1st, Paraffin base, which, in general, includes the lighter oils; 2nd, Asphaltic base, which ordinarily includes the heavier grade oils; 3rd, Mixed base, containing varying proportions of paraffin and asphaltic bases and including the broad class of intermediate grade oils.

While oil is found in varying quantities in practically every portion of the world, the largest supplies available today are from the United States, Mexico and Russia. The oil used in this country is entirely from our own fields and from Mexico.

The principal fields in this country from which oil is produced are:

1st. The Pennsylvania-Appalachian fields, which produce light-grade crude of paraffin base.

2nd. The Lima, Indiana and Illinois fields, which produce for the most part oils similar to the Pennsylvania fields.

3rd. The Mid-continent fields, including those of Oklahoma, Kansas and part of Texas, which produce medium grades of crude oils, largely of asphaltic base, although some are of paraffin base.

4th. The Gulf fields, including the states bordering the Gulf of Mexico. These produce somewhat heavier grades of crude oil of asphaltic base.

5th. The California fields, producing medium-grade oils of asphaltic and mixed base.

All of the oils available for fuel purposes may be said to fall within the following grades:

1st. "Topped" crude, from which the light oils such as gasoline and kerosene have been removed.

2nd. Residue oils, or oils left after complete distillation, but from which water, coke and other foreign matter have been removed.

3rd. Distillates, such as gas oil, which may be mixed with the heavier reduced crudes and residuums to make them sufficiently fluid.

4th. Mixtures of the above in varying proportions.

Where oil fuel is to be used, there are certain physical characteristics that must be taken into consideration in the determination of the suitability of the available supply. These characteristics, together with the manner in which they affect suitability for use in steam generation, may be summarized as follows:

Specific Gravity, or the ratio of weight of a given volume of oil to that of an equal volume of water. The specific gravity is of importance, particularly in the case of the lighter oils. since it has a bearing on the calorific value. In general. the higher the specific gravity, the lower the content of the lighter hydrocarbons and the lower the heat value of the oil.

Specific gravity of oil fuels is ordinarily determined by a direct reading on the Baumé hydrometer scale and is expressed in degrees Baumé. In 1922-1923, an agreement was reached between the U. S. Bureau of Standards, the U. S. Bureau of Mines. and the American Petroleum Institute to call the degrees on the hydrometer scale used with petroleum and its products "degrees A. P. I." instead of degrees Baumé, the latter term being restricted to degrees on a hydrometer scale used with other liquids than petroleum. As the oils in common use become heavier, the Baumé scale becomes more liable to error, but up to the present time its use has proven satisfactory for the purpose of comparing relative gravities. Table 46 gives the specific gravities corresponding to the Baumé scale, now also called the A. P. I. scale, together with the corresponding weights per gallon of oil.

Viscosity of oil is defined in a number of ways, but may be considered as the resistance of oil to flow or its resistance to internal movement. Since the viscosity of any oil will change with its temperature, it is necessary to set a standard temperature, usually 60 degrees Fahrenheit, for the determination of the comparative viscosity of different oils. While there are a number of viscosimeters available, The Babcock & Wilcox Company has adopted as standard in its oil-burning work the Engler apparatus. In this the viscosity is measured on a scale based on the ratio of the time required for 200 cubic centimeters of water to flow through an orifice to the time for the same volume of oil to pass through the same orifice.

Viscosity is perhaps the most important factor in determining the suitability of available fuel oils, particularly where these are heavy. Since all oils must be heated for satisfactory handling in pumping and atomizing, it is necessary to determine to what extent the viscosity must be reduced and to what temperature the oil must be heated to accomplish such reduction. In general, the heavy oils have a high viscosity and the lighter oils a low viscosity; on the other hand, it is of interest to note that for satisfactory handling in pumping and atomization, the oils having the lower viscosities require a reduction in such viscosity to an extent greater than in the case of the heavier oils having high viscosities. Oils of the same specific gravity do not necessarily have the same viscosity.

Flash Point. The temperature at which inflammable gas is driven from fuel oil is called its flash point. The oil is heated in a vessel, either open or closed to the atmosphere, and the flash point determined with a spark or flame as the distilled vapor ignites. The flash point as determined with the closed vessel, or the "closed cup" flash point, is lower than that as determined by the "open-cup" method, and is ordinarily accepted as standard.

The flash point is of importance in connection with the storage of oil, and the minimum allowed by most insurance regulations is 150 degrees "closed cup."

TABLE 46

SPECIFIC GRAVITY AND WEIGHT IN POUNDS PER GALLON OF FUEL OIL HAVING THE GIVEN BAUMÉ (A P I.) READINGS AT 60 DEGREES FAHRENHEIT

Degrees Baumé (A. P. I.)	Specific Gravity at 60 Degrees Fahrenheit	Pounds Per Gallon	Degrees Baumé (A. P. I.)	Specific Gravity at 60 Degrees Fahrenheit	Pounds Per Gallon
10.0	1.0000	8.331	22.5	0.9188	7.655
10.5	0.9965	8.302	23.0	0.9159	7.630
11.0	0.9930	8.273	23.5	0.9129	7.605
11.5	0.9895	8.244	24.0	0.9100	7.581
12.0	0.9861	8.215	24.5	0.9071	7.557
12.5	0.9826	8.186	25.0	0.9042	7.533
13.0	0.9792	8.158	26.0	0.8984	7.485
13.5	0.9759	8.130	27.0	0.8927	7.437
14.0	0.9725	8.102	28.0	0.8871	7.390
14.5	0.9692	8.074	29.0	0.8816	7.345
15.0	0.9659	8.047	30.0	0.8762	7.300
15.5	0.9626	8.019	31.0	0.8708	7.255
16.0	0.9593	7.992	32.0	0.8654	7.210
16.5	0.9561	7.965	33.0	0.8602	7.166
17.0	0.9529	7.939	34.0	0.8550	7.123
17.5	0.9497	7.912	35.0	0.8498	7.080
18.0	0.9465	7.885	36.0	0.8448	7.038
18.5	0.9433	7.859	37.0	0.8398	6.996
19.0	0.9402	7.833	38.0	0.8348	6.955
19.5	0.9371	7.807	39.0	0.8299	6.914
20.0	0.9340	7.781	40.0	0.8251	6.874
20.5	0.9309	7.755	41.0	0.8203	6.834
21.0	0.9279	7.730	42.0	0.8156	6.795
21.5	0.9248	7.705	43.0	0.8109	6.756
22.0	0.9218	7.680	44.0	0.8063	6.717

Specific Heat. The specific heat of fuel oil, which is of importance in the design of oil heaters, varies with the chemical composition, being higher as the hydrogen content is greater and lower as the carbon content is greater. The range of specific heat is from 0.4 in the case of California oils to 0.5 for the lighter Pennsylvania oils.

Chemical Composition. Liquid fuel is composed of carbon, hydrogen, oxygen, nitrogen, sulphur, silt, and moisture in varying proportions, the carbon and hydrogen being the main constituents. The heat value varies from 17,000 to 20,000 B.t.u. per pound, though the ordinary range of oil available for fuel purposes is from 17,500 to 19,000 B.t.u.

Typical analyses of fuel oils together with their heat values are given in Table 47.

From a consideration of the foregoing characteristics of oil, it is obvious that in making a selection of available oils for fuel purposes the primary factors are: 1st, maximum heat value; 2nd, minimum water, sulphur, and sediment content; 3rd, viscosity such as to require no undue heating for pumping and atomizing; 4th, flash point such as will meet local ordinances and insurance regulations.

OIL FUEL *vs.* COAL—In considering plant operation as a whole, oil fuel has many advantages over coal in the generation of steam. On the other hand, because of the developments and improvements during recent years in stoker and furnace design, some of the advantages have been greatly reduced and, in certain instances, the changes have given coal the advantage.

TABLE 47

COMPOSITION AND CALORIFIC VALUE OF VARIOUS OILS

Kind of Oil	Per Cent Carbon	Per Cent Hydrogen	Per Cent Sulphur	Per Cent Oxygen	Specific Gravity	Degrees Flash Point	Per Cent Moisture	B.t.u. Per Pound	Authority
California, Coaling	. . .	. . .	.	. . .	.927	134	. . .	17117	Babcock & Wilcox Co.
California, Bakersfield .	. .	. . .	.	. . .	.975	. .	. . .	17600	Wade
California, Bakersfield	.	. . .	1.30	. .	.992	. .	. . .	18257	Wade
California, Kern River .	. . .	. . .	. .	. .	.950	140	. . .	18845	Babcock & Wilcox Co.
California, Los Angeles .	. . .	. .	2.56	. . .	. .	. .	. .	18328	Babcock & Wilcox Co.
California, Los Angeles .	. . .	. . .	. .	. .	.957	196	. . .	18855	Babcock & Wilcox Co.
California, Los Angeles .	. . .	. . .	. .	. . .	.977	. .	.40	18280	Babcock & Wilcox Co.
California, Monte Christo	. . .		. .	. .	.966	205	. . .	18878	Babcock & Wilcox Co.
California, Whittier . .	. . .	. . .	.98	. .	.944	. .	1.06	18507	Wade
California, Whittier . .	. . .	. . .	.72	. .	.936	. .	1.06	18240	Wade
California	85.04	11.52	2.45	.99*	.	. .	1.40	17871	Babcock & Wilcox Co.
California	81.52	11.51	.55	6.92*	. .	230	. . .	18667	U. S. N. Liquid Fuel Board
California	. . .	.	.87	. .	.	. .	.95	18533	Blasdale
California	. . .	. . .	. .	. .	.891	257	. .	18655	Babcock & Wilcox Co.
California . .	.	. . .	2.45	. .	973	.	1.50†	17976	O'Neill
California	. . .	. . .	2.46	. .	.975	. .	1.32	18104	Shepherd
Texas, Beaumont . .	84.6	10.9	1.63	2.87	.924	180	. . .	19060	U. S. N. Liquid Fuel Board
Texas, Beaumont . . .	83.3	12.4	.50	3.83	.926	216	. . .	19481	U. S. N. Liquid Fuel Board
Texas, Beaumont . .	85.0	12.3	1.75	.92*	.	. .	. . .	19060	Denton
Texas, Beaumont .	86.1	12.3	1.60	. .	.942	.	. .	20152	Sparkes
Texas, Beaumont . .	. . .	. . .	.	. .	.903	222	. .	19349	Babcock & Wilcox Co.
Texas, Sabine	. .		.	. .	.937	143	.	18662	Babcock & Wilcox Co.
Texas	87.15	12.33	0.32	. . .	.908	370	. .	19338	U. S. N.
Texas	87.29	12.32	0.43	. . .	.910	375	. .	19659	U. S. N.
Ohio	83.4	14.7	0.6	1.3	.	. .	. . .	19580	
Pennsylvania . . .	84.9	13.7	.	1.4	.886	. .	. . .	19210	Booth
West Virginia	84.3	14.1		1.6	.841	.	. . .	21240	
Mexico	. . .	. . .	.	. . .	.921	162	. .	18840	Babcock & Wilcox Co.
Russia, Baku . . .	86.7	12.9	.	. .	.884	. .	.	20691	Booth
Russia, Novorossick . .	84.9	11.6	.	3.46	. .	. .	. . .	19452	Booth
Russia, Caucasus . . .	86.6	12.3	.	1.10	.938	. .	. . .	20138	
Java	87.1	12.0	.	.9	.923		. . .	21163	
Austria, Galicia . . .	82.2	12.1	5.7	. .	.870	. .	. . .	18416	
Italy, Parma	84.0	13.4	1.8	. .	.786	. .	. . .	. .	
Borneo	85.7	11.0	.	3.31	. .	. .	. . .	19240	Orde

*Includes N. †Includes silt.

EFFICIENCY—In the average small plant using oil as fuel, the boiler efficiencies obtained in every-day operation are, without question, higher than those obtained, or perhaps obtainable, with hand-fired coal. In the same class of plant, where stokers are used, if a difference in efficiency exists, it is probably in favor of the oil-fired installation, though with the grade of operators and load conditions ordinarily found in this class of plant, this difference is not great.

In the larger plants, where only stoker-fired boilers need be considered, the method of burning the oil is a factor. If steam atomizing oil burners are used, the efficiencies at low ratings are somewhat higher than are ordinarily found in stoker-fired plants. The difference, however, is small, and with a slight change in the personnel of the operating crew might readily be overthrown in favor of coal firing, for oil fuel presents to the careless operator possibilities of wastefulness much greater than in the case of coal-fired plants. At high ratings, with modern stoker equipment, the efficiency with coal is higher than is obtainable with steam atomizing oil burners. At what point in the efficiency-capacity curve the two cross is dependent upon a number of factors, but is probably slightly above 150 per cent of the boiler rating.

With mechanical atomizing oil burners, the efficiencies at different ratings with oil and stoker-fired coal parallel each other up to the maximum rating, being in favor of oil at all capacities, though the difference is not great. In plants where economizers are installed, the difference in combined efficiency in favor of the oil-fired units is less than where such apparatus is not used, for with a given amount of economizer surface, the greater ratio of gas to water weight in the case of coal gives a higher additive efficiency due to the economizer than in the case of oil fuel.

Considered from the standpoint of efficiency alone, recent results obtained when burning pulverized coal in a properly designed furnace would indicate that efficiencies with this method of firing coal are as high as are obtainable with oil fuel and mechanical atomizing burners, though the range of capacities with the former fuel is not as yet so great as has been covered with the latter.

In discussing the comparative efficiency to be obtained with oil and coal, stress is commonly laid on the absence of ashpit loss with the former, while the difference, which is appreciable, in loss due to the burning of the hydrogen content of the two fuels, is frequently overlooked.

Capacity—Oil-fired boilers can carry loads that cannot be approached with hand-fired coal. When comparing capacities of such units with stoker-fired boilers, however, as in the case of efficiency, the method of burning the oil is a factor. With modern stoker equipment, boiler capacities are obtainable greatly in excess of those possible with boilers equipped with steam atomizing burners.

The question of maximum capacity of boilers equipped with mechanical atomizing oil burners as compared with that of stoker-fired units is dependent upon the number of burners installed, the size of the stoker, and the blast available. While it is true that in marine practice, boiler capacities, expressed in terms of evaporation per square foot of boiler heating surface, have been carried greatly in excess of any carried by stoker-fired units in stationary practice, it is also true that in stationary practice stoker-fired units have developed capacities higher than any as yet developed by oil-fired mechanical burner boilers. Provided proper furnaces and burner and stoker equipment are installed, it may be stated that for a given available draft and blast, the maximum capacity obtainable with stokers and with mechanical atomizing oil burners will be approximately the same.

Labor—The saving in labor in the oil-fired plant as compared with the hand-fired coal-burning plant is great, because of the elimination of coal passers and ash handlers. This saving in favor of the oil-burning installation as compared with the stoker-fired plant is not so great, though it still exists, for the operation of the stokers and of the coal and ash-handling apparatus will require a greater amount of labor than will properly designed oil-burning equipment.

CONTROL AND FLEXIBILITY—In oil-fired plants, where regulation of combustion for varying load conditions is by hand, the degree of control and the flexibility are not greater than in the modern stoker-fired plant. There are, however, regulating devices available that enable a wide range of capacity and a regulation of combustion conditions to meet such range, to be satisfactorily handled automatically with oil, which give an ease of control that cannot be obtained in the stoker-fired plant, even where the automatic features of the latter are carried to the limits of the art. This possible method of better combustion control gives to the oil-fired plant a somewhat greater flexibility than is possible in the stoker-fired plant.

CLEANLINESS—The cleanliness in handling and using oil as compared to that in handling coal and ashes is too obvious to need comment. The tendency of boiler heating surfaces to foul with properly operated oil-fired furnaces is, in general, not so great as with stoker firing, regardless of operation. This leads to a decrease in the cost of labor for blowing or cleaning boilers, and to a somewhat better efficiency of heat absorption due to relatively clean exterior heating surface.

STORAGE AND HANDLING—Oil can be stored without danger of deterioration or of spontaneous combustion. While the latter may be prevented with coal by proper storage methods, all coal tends to disintegrate and deteriorate in heat value when stored, almost regardless of the methods used. The handling of oil from storage into the plant and to the oil-burning equipment is much more simple than the methods necessary in coal handling, and there is no ash or refuse resulting from combustion of the former fuel to be removed.

COST—Many tables have been prepared to show comparative costs of generating steam with oil and with coal. Such tables, however, are based either on the relative heating values of the two fuels without reference to the difference in efficiency, or take into consideration the variation in efficiency while maintaining a fixed calorific value of the fuels. There are so many factors entering into the cost of the generation of a given amount of steam aside from the heat value of the fuel or the boiler efficiency shown that no table of comparative costs can be anything but the most approximate guide, and in comparing costs each installation, with its individual operating characteristics, should be considered as a problem by itself.

BURNING OIL FUEL—The requirements for the successful burning of oil fuel and the methods of meeting such requirements may be summarized as follows:

1st. The atomization of the oil must be thorough, this requirement being met by the selection of the proper type of burner.

2nd. When atomized, the oil must be brought into contact with a sufficient amount of air for combustion in a manner that assures a thorough mixture of oil and air, and at the same time the amount of air so supplied must be kept at a minimum. Proper methods of introducing air into the furnace and means of controlling this air enable this requirement to be met.

3rd. The furnace must be of such form as to maintain the temperatures essential for high efficiency, and built of a grade of refractory material that will stand up under high temperatures. Further, the form must be such as to assure complete combustion of the gases before they strike the boiler heating surfaces. This requirement is met by providing ample combustion space and length for flame travel.

4th. The burners must be so located relative to the boiler surfaces and must be so operated that there will be no localization of heat on certain portions of the surfaces or trouble from overheating and blistering will result.

OIL BURNERS—Oil burners in use today are of two general classes, steam and mechanical atomizing burners, though a third class, having certain characteristics of both, has been recently developed.

STEAM ATOMIZING BURNERS—In stationary oil-fired practice the burners in almost universal use until recently were of the steam atomizing type. Compressed air was sometimes used as the atomizing agent, but air burners were never developed to give the commercially satisfactory results obtainable with steam.

Steam atomizing burners may be subdivided into flat-flame and round-flame designs, the former being in most common use and being generally accepted as giving more efficient results. Such burners are of the outside mixing type, the steam passing through a slot or row of holes below a similar slot through which the oil leaves the burner, the oil being picked up by the steam and atomized as it leaves the burner tip. Air for combustion with flat-flame burners is introduced through checkerwork beneath the burner, this checkerwork forming a portion of the furnace floor and designed to admit the proper amount of air under the flame as it spreads and as combustion progresses.

For ease of pumping and proper atomization, the oil must be heated, and such heating should be done as close to the burner as possible. The temperature necessary depends upon the oil used, and varies from 150 degrees with the better grades of lighter oils to 190 degrees with the heavy grades of Mexican oils. If the oil is heated to a point where any appreciable vaporization occurs, there is a tendency toward an uneven oil flow and a spluttering action of the flame. Any undue moisture in the oil or in the steam used for atomization also leads to a spluttering flame, and while atomization with superheated steam is no better than that with dry saturated steam, the presence of superheat assures an absence of moisture in the steam at the burner.

The pressure at which oil is brought to steam atomizing burners varies from 40 to 60 pounds.

The steam consumption of steam atomizing burners, aside from the steam for pumping and heating, *i. e.*, the steam actually used in atomization, varies with the design of burner, the boiler capacity, and the class of operation. While the average consumption in all plants using this type of burner is probably between 2 and 4 per cent, there is no reason why, with a proper design of burner and good operation, such consumption should be in excess of 2 per cent, and in certain plants under the most careful operation, a burner consumption of less than one per cent has been shown. In general, the steam consumption in terms of the total steam output of the boiler will decrease as the capacity increases. The use of one of the available regulating devices by which oil pressure, steam to burners, and draft are automatically controlled, leads not only to high efficiency at varying loads but also to low steam consumption.

Steam atomizing burners of whatever design should be equipped with strainers for removing any accumulation of sediment carried by the oil. These should

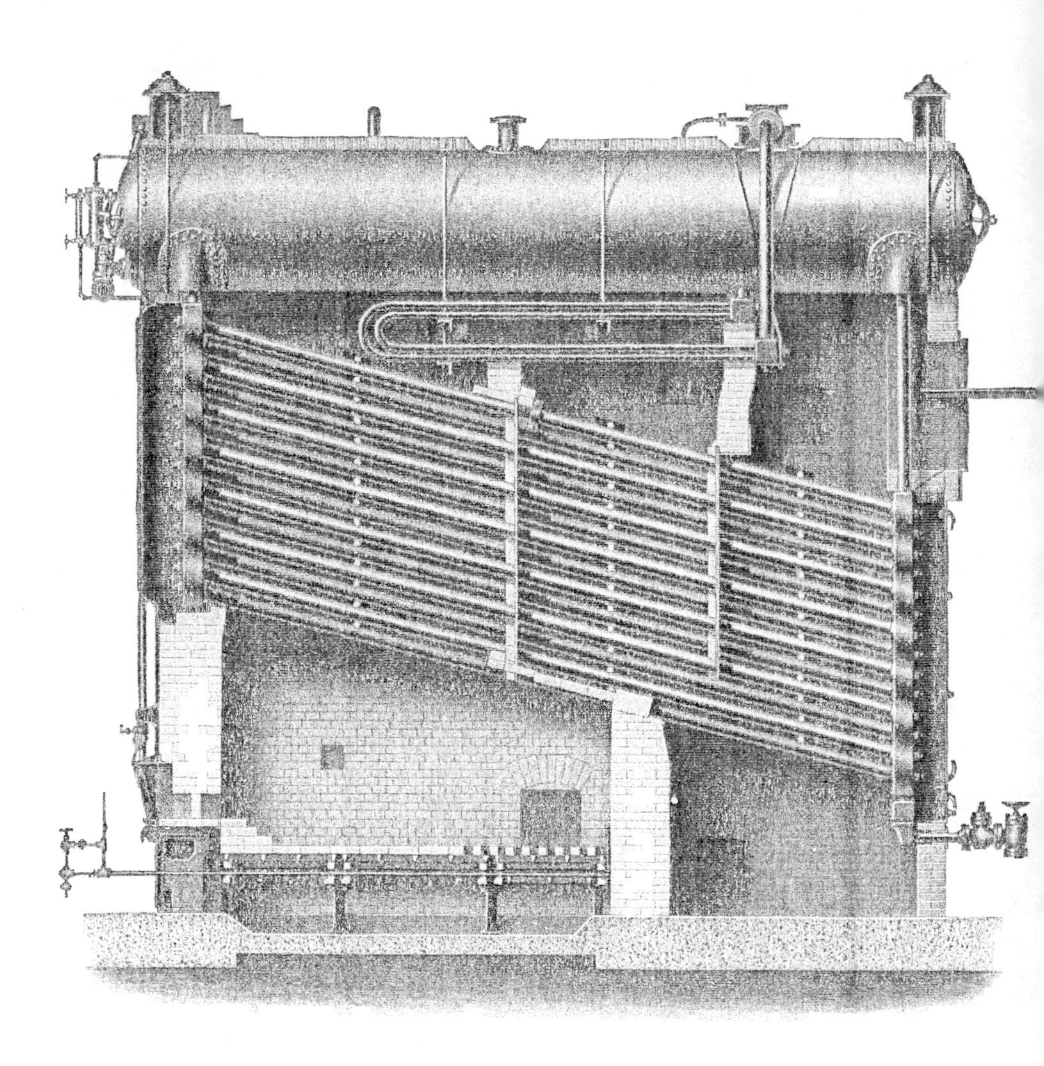

A Babcock & Wilcox Boiler with a Babcock & Wilcox Steam Atomizing Oil Burner

be in duplicate and located so as to allow rapid removal and cleaning without affecting boiler operation. By-pass connections should be provided between oil and steam ducts of the burner to allow blowing out of oil lines and burner.

The Babcock & Wilcox steam atomizing burner is typical of the flat-flame outside mixing type and is shown in Fig. 14, the construction being clearly indicated in the illustration.

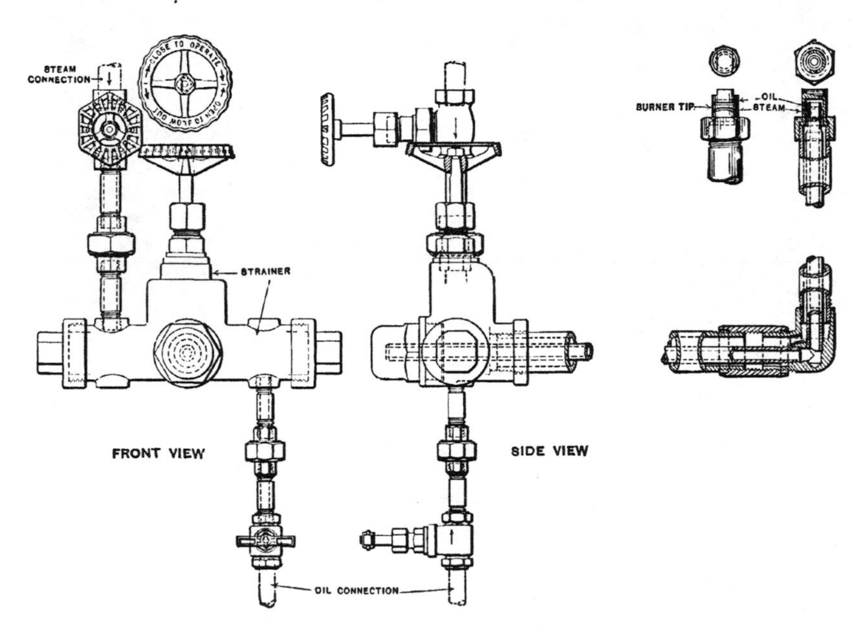

FIG. 14. THE BABCOCK & WILCOX STEAM ATOMIZING OIL BURNER

The advantages of the steam atomizing oil burner may be summarized as follows: simplicity of design and ease of manufacture; relatively low first cost and cost of installation; the low temperature to which it is necessary to heat the oil and the low pressure at which it is brought to the burner lead to low first cost of heater and pump equipment; high efficiencies obtainable at low and moderate ratings; low labor costs in operation, particularly in plants equipped with automatic regulating devices.

The disadvantages of the steam atomizing burner, which are the factors that led to the development of the mechanical burner for stationary practice, are the relatively high steam consumption of burners, the limitation in boiler capacity obtainable, and the decrease in boiler efficiency with an increase in capacity.

While the steam consumption for atomization should, as indicated, be below 2 per cent, this is a direct loss and is chargeable against the boiler efficiency.

Because of the manner in which air for combustion is admitted to the furnace with practically all designs of flat-flame burners, the number that can be installed is limited to a single row across the boiler furnace. While, through the use of special furnace design, two rows are possible in certain instances, this defect may

Boiler Room of the Singer Building, New York, Where Five Babcock & Wilcox Boilers, 1042 Horse-power, Equipped with Babcock & Wilcox Mechanical Oil Burners and Superheaters, Are Operated

be considered as inherent for the class of burners as a whole. Individual burners must be located at such distance apart that there is no appreciable impingement of the fan-shaped flames of adjacent burners or on the furnace side walls. When such impingement occurs, and when a capacity has been reached where the length of flame is such that it extends into the boiler heating surfaces, the limit in the amount of oil that can be burned except at the expense of a great loss in efficiency has been reached, and such limit in the combustion rate represents the practicable limit of boiler capacity. This limit varies with the boiler under which the burners are installed, but with the amount of heating surface per foot of furnace width now in common use, the maximum capacity obtainable is not in excess of 250 per cent, while the maximum practicable limit is ordinarily accepted as approximately 200 per cent.

The gross efficiencies shown with steam atomizing oil burners in carefully conducted tests are among the highest authentic efficiencies reported. The high efficiencies, however, were obtained at comparatively low ratings, and that such efficiency falls off much more rapidly with an increase in capacity with steam than with mechanical atomizing burners is clearly indicated by the figures given in Table 48.

TABLE 48

TESTS OF BABCOCK & WILCOX BOILERS WITH OIL FUEL

Plant / Location	Pacific Light and Power Company, Redondo, Cal.		Savannah Electric Company, Savannah, Ga		Narragansett Electric Company, Providence, R. I.	
Type of Burner	Steam Atomizing		Mechanical		Mechanical	
Rated Horse-power	604	604	665	665	600	600
Duration of Test, hours	7	7	4	4	10	10
Steam Pressure by Gauge, pounds	184.7	184.9	190	200	207.9	209.1
Feed Water Temperature, degrees Fahrenheit	93.4	101.2	213.5	212.5	166.6	171.5
Superheat, degrees Fahrenheit	83.7	144.3	43.0	74.1	66.2	104.2
Factor of Evaporation	1.2227	1.2475	1.0776	1.0965	1.1071	1.1224
Draft in Furnace, inches	.014	.19	.11	.08	.05	.05
Draft at Damper, inches	.046	.47	.08	.28	.14	.71
Blast at Burners, inches	. . .	. . .	.37	1.56	.35	2.30
Exit Gas Temperature, degrees Fahrenheit	406	537	438	518	510	666
Flue Gas Analysis: CO_2, per cent	14.3	12.1	14.3	14.3	14.75	14.90
Flue Gas Analysis: O_2, per cent	3.8	6.8	1.6	1.6	1.38	1.46
Flue Gas Analysis: CO, per cent	0.0	0.0	0.0	0.0	.04	.03
Oil Burned per Hour, pounds	1439	2869	1661	3176	2022	4511
Evaporation from and at 212 degrees per Hour, pounds	22639	40375	26077	48935	30903	66782
Evaporation from and at 212 degrees per Pound of Oil, pounds	15.73	14.07	15.70	15.41	15.28	14.80
Horse-power Developed	659	1171	756	1418	896	1936
Percentage of Rated Horse-power Developed	108.6	193.8	113.7	213.3	149.3	322.6
Heat Value per Pound of Oil, B.t.u.	18326	18096	18471	18420	17934	18119
Efficiency, per cent	83.29	76.02	82.43	81.18	82.69	79.26

Babcock & Wilcox Boilers Equipped with Babcock & Wilcox Mechanical Oil Burners and Babcock & Wilcox Superheaters in One of the Stations of the San Diego Consolidated Gas & Electric Company. This Company Operates [illegible] Horse Power of Babcock & Wilcox Boilers

Mechanical Atomizing Oil Burners — Mechanical burners were originally developed in marine practice. Here the steam consumption of steam atomizing burners, being a direct loss and necessitating the evaporation of an amount of makeup equal to such consumption, made this type impracticable. Further, steam atomizers were not adapted to the rates of evaporation at which marine boilers, particularly in the Navy, are at times operated

Mechanical burners for atomization of the oil employ only the pressure of the oil as it comes to the burner. In practically all of the successful burners of this type, the oil is given a rapid whirling motion in a chamber at the burner tip and is liberated suddenly from a small orifice concentric with the axis of rotation. The oil is sprayed by centrifugal force in the form of a hollow cone, the individual oil particles flying off radially. There is no rotating motion of the spray itself aside from that which may be given by the method of introducing air for combustion around the burner tip.

Fig. 15. The San Diego Type of Babcock & Wilcox Mechanical Atomizing Oil Burner

The Babcock & Wilcox mechanical burner of the San Diego design, illustrated in Fig. 15 is typical of mechanical burners as a class and a brief description of this burner will indicate their general features.

A conical casting is introduced through the boiler front wall and serves as an orifice for the introduction of oil and air into the furnace. To this is attached the register casting which is fitted with adjustable automatic air register doors. To the front of the register casting is attached a front plate through the center of which passes a distance piece equipped at the inner end with a slightly conical impeller plate. The distance piece with the impeller plate and burner may be moved in or out on its axis relative to the cone, decreasing or increasing the area of the air passage around the impeller plate. The burner introduces the atomized oil through a central opening in the impeller plate, which has blades that give a whirling motion to the air entering the furnace around the burner tip. The cone through which the oil and air enter the furnace is also equipped with blades. These blades and the walls of the truncated cone direct such air as enters around the edge of the impeller plate along the axial lines of the burner. Where forced blast is used with mechanical burners, the register casting is within the blast duct, the front plate forming a portion of the double front.

Combustion conditions are controlled by adjustment of the boiler damper, by adjustment of the air registers, the position of the burner end relative to the cone, and by oil pressure and temperature.

Burner Capacity—Mechanical burners as originally developed in marine work were of relatively small size and the greatest quantity of oil that could be burned was approximately 300 pounds per hour per burner, even under the fire-room pressure available in naval work, which is as high as 8 inches. It was desirable to increase the capacity of individual burners and to increase such capacity with a draft or blast lower than that necessary with the early burners, and it was along these lines that the development of mechanical oil burners has progressed.

Mechanical atomizers now used in marine practice have burned a maximum of 4040 pounds of oil per burner per hour. This rate, however, was obtained under the very high blast available in Navy practice and with oil brought to the burner under very high pressure. For this maximum rate, the blast pressure at the burners was 9 inches and the pressure of oil 340 pounds, neither of which would be practicable in stationary practice.

As developed for stationary work, the modern mechanical burner is capable of handling approximately 1500 pounds of oil per burner per hour with a blast pressure of less than 3 inches, which is no higher, if as high, as the pressures necessary in stoker-fired work, and an oil pressure at the burner that cannot be considered excessive. In stationary practice it is probable that this maximum rate of 1500 pounds per hour per burner will not be exceeded, at least for some time to come. This capacity, together with the possibility of installing in the stationary boiler furnace a greater number of burners than can be installed in the ordinary marine boiler furnace, enables boiler capacities to be developed which will meet any peak-load conditions that may occur, or that could satisfactorily be handled with stoker-fired boilers.

While the figures above represent capacities obtainable with mechanical burners using blast, this type of burner operates satisfactorily under natural-draft conditions, boiler capacities up to 250 per cent of rating being readily obtained with a draft of 1.5 inches available at the boiler damper. Mechanical burners under natural-draft conditions will, however, require a greater draft at the boiler damper for a given boiler capacity than will steam atomizing burners, this being due to the necessity of pulling all air for combustion through the rather limited areas of the air registers.

Oil Heating and Oil Pressures—To give the best results in atomization with mechanical burners, oil should be heated to a point where the viscosity is reduced to from 3 to 5 degrees Engler. With the oils in common use, this corresponds to a temperature of from 220 to 250 degrees Fahrenheit. The pressure at the burner varies between 100 and 200 pounds, though in special cases this is increased to 250 pounds.

STEAM CONSUMPTION OF BURNERS—While with mechanical oil burners there is no direct steam consumption due to atomization, there is, as in the case of steam atomizers, a certain amount used in pumping and heating, which is somewhat greater than in the case of steam atomizing burners, because of higher oil temperatures and pressures. This gross consumption, however, is small, being less than one per cent, and since a large portion of the heat in the steam used for heating and pumping may be returned to the system, the net consumption is appreciably less than the figure given.

The advantages of the mechanical oil burner, in addition to the advantages listed for the steam atomizer, are the high efficiency obtainable and the maintaining of such efficiency over a wide range of boiler capacity, capacities obtainable comparable with those that may be secured with any fuel, simplicity of furnace design, and ability to handle sudden and wide swings in boiler load without appreciable loss in efficiency.

The first cost of mechanical burners is higher than that of steam atomizing burners because of greater complication in manufacture; and, because of oil temperatures and pressures higher than those required in steam atomizing, the cost of heaters and pumps is greater than for the latter type. These increased costs, however, are negligible as compared to the increased efficiencies and capacities obtainable with mechanical atomizing burners.

The efficiencies that may be obtained with mechanical burners and the extent to which these efficiencies hold up at high boiler capacities are indicated in the figures in Table 48.

The third type of oil burner, which is of recent development, may be described as a conical flame, steam atomizing burner, utilizing the methods of mechanical atomizers in the manner in which air for combustion is brought into contact with the atomized oil. The field for this type of burner lies somewhere between that of the steam and mechanical atomizing types. The main factors leading to the development of this burner were a reduction in the temperature to which the oil must be heated and in the pressure to which it must be raised from those necessary with mechanical burners, with a corresponding decrease in heater and pump costs, and the desirability of obtaining with the methods of air introduction in use with mechanical burners a higher boiler capacity for a given available draft at the boiler damper.

The use of steam as an atomizing agent reduces the temperature and pressure to those necessary with the standard type of steam atomizing burners, while the aspirating effect of the steam used in atomization makes it unnecessary to have a high draft available in the boiler furnace for drawing air for combustion through the air registers, as in the case of mechanical burners under natural draft conditions.

Under such conditions, the capacities obtainable with burners of this type will be higher than with the other types, because of the possibility of the installation of a greater number of burners in a given furnace than is possible with the flat-flame steam atomizing burner, and because of lower furnace draft requirements

Tamarack Power Plant No. 2 Jenckes Spinning Company, Pawtucket, Operating Ten 400 Horse-power [illegible] Babcock & Wilcox Boilers Equipped with Babcock & Wilcox Superheaters

than in the case of the natural-draft mechanical burner, though capacities comparable with those obtainable with blast mechanical burners will never be approached. The efficiency with this class of burner is somewhat higher than that obtainable with the flat-flame type at higher ratings, but appreciably lower than that secured with mechanical burners. The objection as to steam consumption in atomization is the same as in all steam atomizing burners. The form of flame is such as to lead to less furnace brickwork and tube difficulties due to impingement than with the ordinary flat-flame steam atomizing burner.

While in the average small steam plant, steam atomizing burners will continue to be used with satisfactory results, it is probable that in the central power station, mechanical burners will supplant the steam atomizing burner in the future, because of load conditions.

Furnace Design—In common with practice with all fuels, the tendency in furnace design for the combustion of oil has been primarily toward increased furnace volume, and with oil as with coal this has been the result of the demand for higher boiler capacities.

The problem of furnace design is somewhat different for steam and mechanical atomizing burners, though volume is essential for both where high capacities are desired.

While the ignition with steam atomizing burners is more rapid than with most fuels, on account of the methods ordinarily used in introducing air for combustion with flat-flame burners, the flame has a definite fan-like shape. If combustion space is installed sufficient to take care of what may be called flame thickness, the important factor in steam atomizing oil-fired furnaces is the provision for length of flame travel rather than of actual volume. The length for flame travel should be such that combustion may be completed before the flame enters the boiler heating surfaces. A furnace in which the volume increases in the direction of flame propagation is of assistance in accomplishing this result.

Mechanical burners atomize oil in such manner that ignition is more rapid than with steam atomizers, with a resulting shorter flame. Further, the form of flame with mechanical burners is such that the furnace volume may be more fully utilized than is possible with steam atomizing burners. With mechanical burners, then, the volume is dependent upon the number of burners installed and the capacity to be developed rather than on the length of flame travel, although, as with steam atomizers, length should be provided such as will give complete combustion in the furnace at the maximum rates of burning.

Because of the high temperatures developed, the refractory material entering into the furnace construction should be of the best. With steam atomizing, because of the methods of introducing air for combustion, the furnace floor is well protected. With mechanical burners, however, where the floor is subjected to maximum temperatures care must be taken in the floor construction in order to protect foundations. A construction of furnace floor that has given good results is 2 inches Nonpareil brick, ¼-inch millboard, 2 inches calcined Sil-o-cel, and the whole covered with 5 inches of fire brick. Where this construction is used on a concrete foundation, it is well to place 4-inch hollow tile immediately above the concrete.

From the standpoint of furnace brickwork upkeep, the location of the burners relative to the furnace walls and, in the case of mechanical burners, to the furnace floor is of importance. Burners should be located in such manner that there will be no impingement of flames on the side walls or furnace floor or erosion of brickwork will result. While, because of the softer flame with mechanical burners, such erosion will not be as rapid as with steam atomizers, nevertheless it will take place. Reference has been previously made to the necessity of locating burners in such manner as to obviate any localization of heat on boiler surfaces.

OPERATION OF BURNERS—In starting up boilers equipped with oil burners of whatever type, all burners and strainers should be thoroughly cleaned in order that there be no improper atomization resulting from an irregular flow of oil. Burners should be lighted individually with oil-soaked waste. No oil should be allowed to accumulate on the furnace floor as a result of poor atomization or explosions may occur when proper ignition is secured.

When the burners are lighted, the supply of steam to the burners, in the case of steam atomizers, must be reduced to the minimum required for proper atomization. Where insufficient steam is supplied, particles of burned oil drop to the furnace floor giving a scintillating appearance to the flame. Upon ignition of mechanical burners, the air supplied should be reduced to a minimum, the limit to which such reduction may be extended being indicated by a pulsating action in the furnace or excessive smoke.

Good combustion conditions in the furnace are probably most readily indicated by the stack. A smokeless stack is not necessarily a sign of proper combustion, and the highest efficiency of combustion ordinarily exists if a slight, light gray haze issues from the stack.

STATION OF THE TAMPA ELECTRIC COMPANY, OPERATING 2200 HORSE-POWER OF BABCOCK & WILCOX BOILERS

GASEOUS FUELS AND THEIR COMBUSTION

OF the gaseous fuels available for steam-generating purposes those in most common use are blast-furnace gas, natural gas and by-product coke oven gas.

BLAST-FURNACE GAS, as implied by the name it bears, is a by-product of the blast furnace of the iron and steel industry. The gasification of the solid fuel in the blast furnace comes about in a number of steps. As the air (blast) comes into contact with the bed of incandescent fuel at the tuyeres, it combines with the carbon of the coke to form carbon monoxide ($2C + O_2 = 2CO$). While this reaction may be indirect and carbon dioxide formed first, the final result is not affected. The water vapor of the blast also reacts with the carbon at this point ($C + H_2O = CO + H_2$), though the amount of carbon monoxide and hydrogen from this source is small. These reactions result in a high temperature, this being in excess of 3100 degrees Fahrenheit. The hearth reactions in which the oxides of the metalloids are reduced and the sulphur eliminated produce carbon monoxide as follows:

(*a*) $SO_2 + 2C = S + 2CO$
(*b*) $MnO + C = Mn + CO$
(*c*) $FeS + CaO + C = Fe + CaS + CO$
(*d*) $P_2O_5 + 5C = 2P + 5CO$

Such iron oxide or ore as may reach the region of the hearth will be reduced directly by the carbon to give carbon monoxide as follows:

(*a*) $FeO + C = Fe + CO$
(*b*) $Fe_2O_3 + 3C = 2Fe + 3CO$
(*c*) $Fe_3O_4 + 4C = 3Fe + 4CO$

The gases as they leave the hearth contain the inert gases introduced in the blast, carbon monoxide, and a small amount of hydrogen, and remain in this condition in their passage upward through the furnace until their temperature is reduced to approximately 1650 degrees Fahrenheit, which will be at a point some 15 or 20 feet above the bosh. At this point calcination of the limestone takes place, resulting in the production of carbon dioxide ($CaCO_3 = CaO + CO_2$). From this point to the top of the furnace, the condition and composition of the gases undergo continual changes, the temperatures being sufficiently low for carbon dioxide to remain in equilibrium with solid carbon, and the carbon monoxide content of the gases acting as a reducing agent. The reactions which bring about these changes are between the iron and carbon oxides, and are those resulting from the reduction of the ore. While through these latter reactions a portion of the carbon monoxide is oxidized to carbon dioxide, the gases as they reach the blast furnace " top " contain a sufficient quantity of carbon monoxide to make them a valuable asset as a means of power production.

The gases will contain a certain amount of moisture picked up in the progress through the furnace in drying the stock, such moisture content usually being between 25 and 50 grains per cubic foot. The temperature of the gas leaving the furnace will vary appreciably with the speed of combustion at the tuyeres, the quality of the coke used, the class of ore treated and other factors, and will be between 390 and 510 degrees Fahrenheit.

The heat value of blast-furnace gas will vary with the same factors as affect the temperature, such value being from 85 to 100 B.t.u. per cubic foot. Representative analyses and heat values of this class of gaseous fuel, which vary considerably, are given in Table 49.

TABLE 49

TYPICAL ANALYSES OF BLAST-FURNACE GAS

Carbon Dioxide	12.0	10.0	13.0	15.0	14.0	15.4
Carbon Monoxide . .	26.0	27.4	25.8	24.3	25.1	23.6
Hydrogen.	3.0	3.6	3.8	3.0	3.0	3.0
Nitrogen	59.0	59.0	57.4	57.7	57.9	58.0

The volume of gas produced depends upon a number of factors and varies between 133,000 and 150,000 cubic feet per ton of iron produced. The amount of coke consumed per ton of pig produced also varies with the nature of the ore, the quality of the coke, the quality of the product, and the equipment and management of the plant, but an average of 2000 pounds of coke per ton of iron represents ordinary practice. If the heat value of the coke is 12,800 B.t.u. per pound and the rate is 2000 pounds per ton of iron, the total heat value per ton is 25,600,000 B.t.u. If the volume of the gas produced is 140,000 cubic feet per ton, and this gas has a heat value of 95 B.t.u. per cubic foot, the total heat value of the gas is 13,300,000 B.t.u. per ton, or somewhat over 50 per cent of that of the value of the coke fired.

If an availability factor of 85 per cent is assumed to cover loss of gas when tapping, poor gas and irregular gas supply. the total heat value of the gas is 11,305,000 B.t.u. per ton of iron produced. Of this quantity approximately 30 per cent is required for the stoves, leaving 70 per cent or 7,913,500 B.t.u. per ton available for power production. If this gas is burned under boilers with an efficiency equal to that which could be obtained when burning coal containing 13,500 B.t.u. per pound, the amount of coal equivalent to the net gas available for power production, in a 500-ton furnace plant, is 146 tons per day. In view of these figures the enormity of the waste of early blast-furnace practice, where the gas was allowed to escape to the atmosphere, is obvious.

The volume of gas per ton of iron and its heat value per cubic foot both increase with an increase in the coke rate, but such increase is accompanied by a decrease in iron output. Furnace operators naturally look upon a blast furnace as a producer of iron and not of gas, and have endeavored to keep the fuel consumption at a minimum while keeping their product up to the desired standard without reference to the quantity or quality of the gas produced. With the increasing supply of coke available from by-product ovens, it is possible that in the future the desirability of a constant and sufficient supply of high heat value gas may offset the factor of low fuel consumption. Such a method of furnace operation would tend toward the highly desirable balance between gas supply and power demand and would make unnecessary the installation of direct coal-fired boilers.

Following their first installation in the United States in 1903, many gas engines of fairly large size were installed utilizing blast-furnace gas in the production of power. The trend of recent development, however, has been along the lines of better cleaning of the gas and its combustion in the stoves and under boilers. The more recent power installations in steel plants have comprised turbine generating sets with their boiler equipment, and in several instances gas-engine installations have been replaced by steam units.

Reference has been made to the wastefulness of early practice in allowing the gas to escape to the atmosphere. Hardly less wasteful were the early attempts to burn the gas under boilers. Steel-mill operators, even after this use of the fuel was common, were inclined until recently to look upon the gas as a waste product, and in generating steam were interested in the quantity of steam produced without reference to efficiency of combustion. It is only recently that it has been generally appreciated to what extent any increase in the efficiency with which blast-furnace gas is burned affected the amount of coal fired directly under the boiler, and it was this fact that led to the development of the modern blast-furnace gas-fired boiler unit. As an indication of the lack of interest of steel-mill operators in the efficiency of the combustion of this fuel as late as 1915, two quotations from a paper by Mr. Ambrose N. Diehl, presented in that year before the American Iron and Steel Institute, are of interest. Mr. Diehl states: "The average efficiency of a blast-furnace plant using common burners and operating without the aid of technical supervision is not over 50 per cent and frequently much lower," and " of a total of 60 blast furnaces in the corporation under observance, the tests and information from the operators indicate a practice of not exceeding 55 per cent."

These low efficiencies were not so much due to the lack of ability of the boiler heating surfaces to absorb heat efficiently as to the methods of burning the fuel. The furnaces installed were in part responsible for the low efficiency of combustion, in that they were of insufficient volume and did not allow sufficient length of flame travel, but the type of burner in common use was primarily responsible. These early burners were of the box type, the air for combustion being drawn through an opening around the rectangular box through which the gas was introduced into the furnace. With this class of burner there was no possibility of securing any proper mixture of air and gas, and practically no method of regulating the air supply to varying gas conditions in the main. With burners of this design, the amount of excess air used in combustion was probably rarely less than 100 per cent, and even with such excess, secondary combustion extending throughout the whole boiler setting was common. Blast-furnace gas is inherently slow of ignition, and this, together with improper mixture of gas and air, low furnace temperature, and lack of furnace volume and provision for sufficient length of gas travel, naturally resulted in this secondary combustion.

The development of the present-day blast-furnace gas-fired boiler unit has been along three lines: burners, furnace design, and arrangement of boiler heating surface.

Burners—The efficiency of the present blast-furnace gas burner is due primarily to the methods of introducing air for combustion. The gas is broken up into thin streams and the air so introduced as to assure a thorough mixture before ignition. While the majority of such burners have until very recently been of the aspirating type, a design gaining in favor and with which exceptionally efficient results are being obtained is one in which both the gas and the air for combustion are brought to the burner under pressure. The main advantage of this type of burner is in the

Babcock & Wilcox Boilers Fired with Blast-furnace Gas at the Bethlehem Steel Co. Bethlehem, Pa. This Company Operated 40,000 Horse-power of Babcock & Wilcox Boilers at the Date When it Acquired the Cambria and Lackawanna Steel Works

possibility of regulating the air supply for varying gas conditions, and such supply may be maintained at a minimum regardless of the gas pressure in the main. It is of interest to note that with burners of this design blast-furnace gas may be burned with as little as 15 per cent excess air as compared with 100 per cent excess in the early box types. In a number of plants using this design of burner there are systems of automatic control by which the induced and forced draft fan speeds are regulated from the pressure in the gas main. Such systems minimize the amount of fire-room labor required and result in efficient combustion for all gas conditions.

FURNACE DESIGN—The primary factors to be considered in the design of furnace for the efficient combustion of blast-furnace gas are furnace volume, furnace form, length of gas travel and means for maintaining furnace temperatures.

In the early furnaces for this class of fuel furnace volumes installed were from 0.5 to 0.8 cubic foot per rated boiler horse-power, while today this volume is made from 2.5 to 3 cubic feet per rated horse-power.

While a large volume is essential for efficient combustion, this volume should be so arranged as to eliminate to as great an extent as possible any "dead" spaces, which tends to minimize pulsation. The flames should be kept in contact with heated brick surfaces for as great a length of time as possible. While, because of the better mixture of air and gas, the flame from the modern burner is appreciably shorter than with the early types, complete combustion before the flames reach the boiler heating surfaces, necessary for efficient heat absorption by the boiler, can be secured only by giving the flames a long travel in the furnace.

Furnace temperatures with blast-furnace gas are low, probably with washed gas and the best of combustion conditions not exceeding an average of 2100 to 2200 degrees Fahrenheit. While with unwashed gas this temperature would be somewhat higher, due to the sensible heat in the gas as it comes to the burner, and possibly higher efficiencies might be expected because of such higher furnace temperature, in all probability the greater tendency of the boiler surfaces to become fouled with unwashed gas would more than offset the possibility of better efficiency due to higher temperature. Since the maximum furnace temperature with this fuel is low, the furnace brickwork must be so arranged as to assure this maximum being maintained.

BOILER DESIGN—The improvements in burner and furnace design discussed were such that when they were installed with any properly designed boiler they led to efficiencies very much higher than were obtainable from the earlier installations. Since high furnace temperature is essential for high boiler efficiency, in the case of blast-furnace gas, when such temperatures are at best low, it is necessary to use some means of offsetting such lack of furnace temperature in order to show efficiencies comparable to those obtainable with boilers fired by other fuels. For equal amounts of boiler-heating surface exposed to the direct radiant heat of the furnace, the amount of heat absorbed through direct radiation with blast-furnace gas as the fuel is only about 40 per cent of the amount so absorbed in stoker-fired boilers. To give a total absorption for the former comparable to that of the stoker-fired unit, it is necessary, then, to increase the absorption by convection over that common in stoker-fired practice, and this is accomplished only by the use of high gas velocities over that portion of the boiler surfaces absorbing heat solely by convection, as has been done so successfully in the development of the modern waste heat boiler.* Such high

*See page 279.

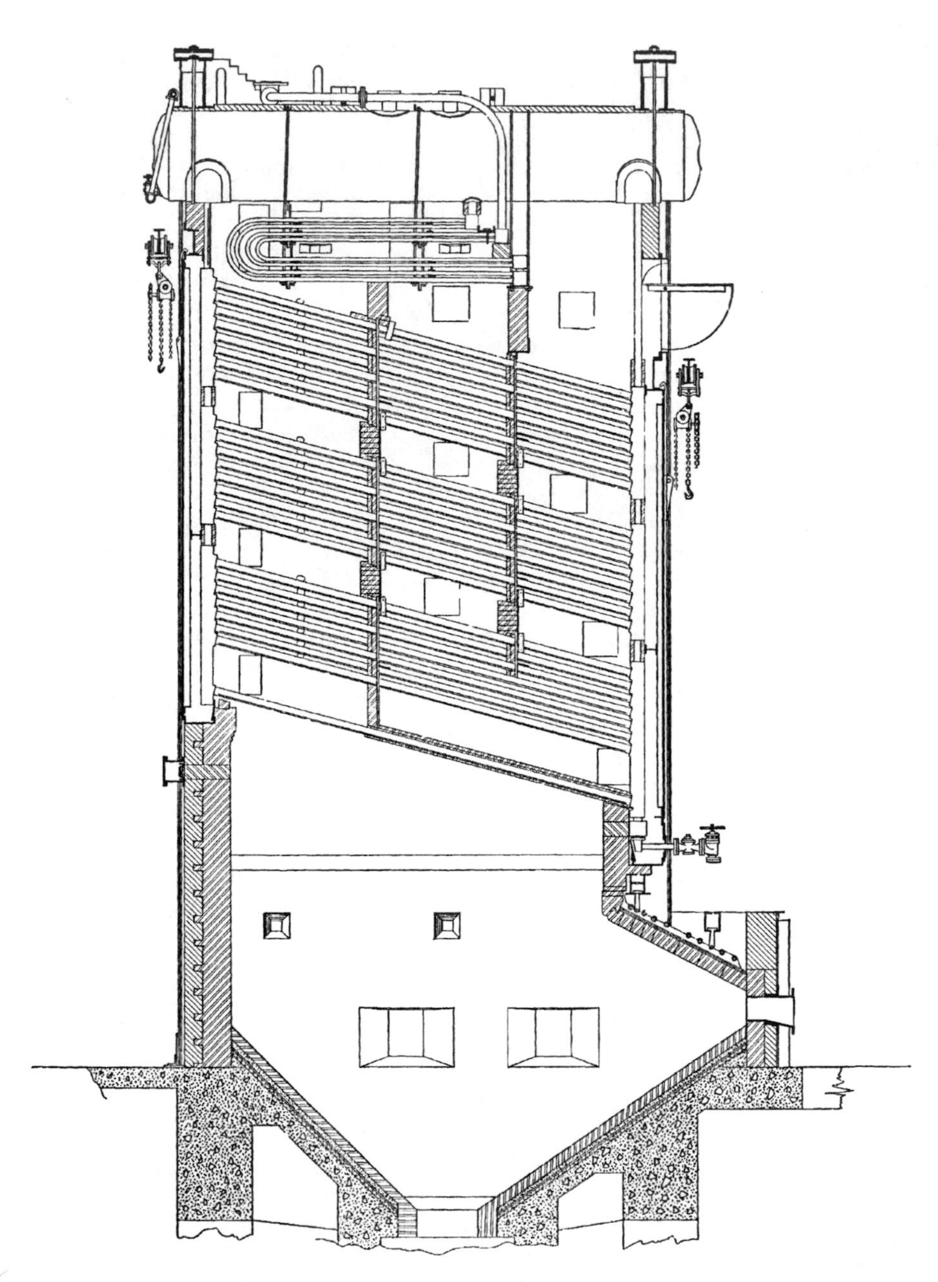

A Babcock & Wilcox Boiler with a Furnace for Burning Blast-furnace Gas

velocities which result in high rates of heat transfer come about to an extent naturally with blast-furnace gas, since even with the best of combustion conditions the weight of products of combustion per horse-power developed is high as compared with stoker-fired practice, being with gas approximately 70 pounds as against some 45 pounds in the case of stokers. In the modern blast-furnace gas-fired boiler the velocity is further increased by decreasing the ratio of gas passage area through the boiler to that of the boiler heating surface.

The increased gas velocity over the surfaces of this design of boiler results in a frictional resistance which in the greater number of installations necessitates an induced draft fan, but if such fan be turbine-driven and the exhaust from the turbine utilized in feed-water heating the gross and net steam output from the unit as a whole are very nearly the same.

The design of a boiler utilizing the high gas velocity principle set over a furnace meeting the requirements for efficient combustion of blast-furnace gas is illustrated on page 272. The efficiencies and capacities obtainable with such a combined unit are indicated in Table 50 and it is of interest to compare these with the figures given by Mr. Diehl as representing the practice of 1915.

TABLE 50

TESTS OF HIGH GAS VELOCITY BLAST-FURNACE GAS-FIRED BABCOCK & WILCOX BOILER

Boiler Heating Surface, square feet	12900	12900	12900
Duration of Test, hours	168	168	96
Steam Pressure by Gauge, pounds	151	147	148
Feed-water Temperature, degrees Fahrenheit	150	169	174
Superheat, degrees Fahrenheit	213	199	185
Factor of Evaporation	1.228	1.200	1.188
Draft at Damper, inches	1.7	2.2	2.4
Gas Pressure at Burners, inches	2.8	2.9	3.0
Air Pressure at Burners, inches	1.5	1.7	1.6
Temperature Exit Gases, degrees Fahrenheit	451	468	461
Flue-gas Analysis: CO_2, per cent	23.3	22.8	24.0
Flue-gas Analysis: O_2, per cent	2.0	2.4	1.2
Flue-gas Analysis: CO, per cent	0.0	0.0	0.0
Gas Burned per Hour, cubic feet	1042500	1050200	1045200
Gas Burned per Hour, pounds	86528	87167	86762
Actual Evaporation per Hour, pounds	63166	67982	58661
Evaporation from and at 212 degrees per hour, pounds	77569	81578	69689
B.t.u. per Cubic Foot Dry Gas	91.2	94.6	90.0
Percentage of Moisture in Gas	8.0	8.0	8.0
Efficiency,* per cent	75.71	75.63	75.93
Horse-power Developed	2248	2365	2020
Percentage of Rated Horse power	174	183	157

*Computed by indirect heat balance, assuming 5 per cent radiation and unaccounted loss.

Before leaving the subject of blast-furnace gas as fuel, there is one further feature to which reference should perhaps be made, viz., the use of economizers with boilers burning this fuel. As has been stated, the weight of products of combustion per horse-power produced is appreciably higher than such weight with stoker-fired

1600 Horse-power Installation of Babcock & Wilcox Boilers and Superheaters at the Carnegie Natural Gas Co., Underwood, W. Va. Natural Gas is the Fuel Burned under these Boilers

units. This means that even regardless of the low exit temperatures from the blast-furnace gas-fired boiler utilizing the high gas velocity principle, the question of economizer installation should be given careful consideration, for with such an increase in ratio of gas weight to weight of water handled in the economizer, the rate of heat transfer will be sufficiently higher than in stoker-fired practice to enable a given amount of work to be done in the economizer with a smaller amount of economizer surface.

NATURAL GAS—While natural gas is burned for power generation, in a number of communities where such fuel is available, it is being more and more conserved for domestic purposes. It will, however, in all probability continue to be burned in some localities, particularly at certain seasons of the year, and for that reason a brief description of the methods of its combustion is included here.

The analyses and calorific value of natural gas vary over a considerable range in different sections of the country. Typical analyses are given in Table 51.

TABLE 51

TYPICAL ANALYSES (BY VOLUME) AND CALORIFIC VALUES OF NATURAL GAS FROM VARIOUS LOCALITIES

Locality of Well	H	CH_4	CO	CO_2	N	O	Heavy Hydro-carbons	H_2S	B.t.u. per Cubic Foot Calculated*
Anderson, Ind.	1.86	93.07	0.73	0.26	3.02	0.42	0.47	0.15	1017
Marion, Ind.	1.20	93.16	0.60	0.30	3.43	0.55	0.15	0.20	1009
Muncie, Ind.	2.35	92.67	0.45	0.25	3.53	0.35	0.25	0.15	1004
Olean, N. Y.	. . .	96.50	0.50	. .	. . .	2.00	1.00	. .	1018
Findlay, O.	1.64	93.35	0.41	0.25	3.41	0.39	0.35	0.20	1011
St. Ive, Pa.	6.10	75.54	Trace	0.34			18.12	.	1117
Cherry Tree, Pa.	22.50	60.27	. .	2.28	7.32	0.83	6.80	. .	842
Grapeville, Pa.	24.56	14.93	Trace	Trace	18.69	1.22	40.60	. .	925
Harvey Well, Butler Co., Pa.	13.50	80.00	Trace	0.66	. . .	. .	5.72	. .	998
Pittsburgh, Pa.	9.64	57.85	1.00	. .	23.41	2.10	6.00	. .	748
Pittsburgh, Pa.	20.02	72.18	1.00	0.80	. .	1.10	4.30	.	917
Pittsburgh, Pa.	26.16	65.25	0.80	0.60	. . .	0.80	6.30	. .	899

*B.t.u. approximate. For method of calculation, see page 196.

The heat value of this gas is much higher than that of blast-furnace gas because of the hydrocarbon content, and consequently the speed of ignition is more rapid and the furnace temperatures developed are higher.

As with other classes of gaseous fuels, the furnace volumes supplied in the early installations were too small, and frequent unsatisfactory results were due to this cause. Furnace volumes for this fuel as offered today contain approximately 2 cubic feet per rated boiler horse-power. Because of the higher furnace temperatures developed, the necessity for large amounts of heated furnace brickwork is not as great as with blast-furnace gas, although such brickwork should be sufficient and so located as to maintain this temperature at a maximum.

Until the past few years natural gas was burned under boilers almost universally with burners that may be called of the Bunsen type. The number of burners installed was based on about 30 rated horse-power per burner, and overloads of 200 per cent could on this basis be satisfactorily handled. It was felt that by using such a number of burners, dangers of stratification of air and gas could be obviated, and the variation in boiler load handled satisfactorily by the cutting in and out of individual burners. While good results were obtained with the Bunsen type of burner. provided a suitable furnace was installed, this design has certain disadvantages when compared with modern burners for this fuel. The gas pressure at the burner must be high to give a proper mixture of gas and air for combustion. This, together with the relatively large size of burner units, tends toward a blow-pipe action of the flame at high ratings which frequently leads to excessive tube losses. Further, since in the Bunsen type of burner the air is drawn in by aspiration, it is difficult to control the ratio of air to gas at different rates of burning to give the highest efficiency.

A design of natural-gas burner known as the "multiple unit" type has been recently developed with which particularly good results have been obtained. One design of such burners consists of a cast-iron gas box, from which lead a number of small nozzles to the boiler side of the unit, where there is located a mixing plate provided with a number of holes corresponding to the number of gas nozzles. The whole is surrounded by an air box. except for a space directly behind the gas box, which extends around the burner and is equipped with a telescopic slide by which the area for air admission may be reduced or increased. Gas pressure gauges and air indicators calibrated for various gas pressures are attached to each unit.

The gas supply with the multiple unit type of burner is divided into much finer subdivisions than in the Bunsen type, each individual gas stream having its own mixing chamber, all controlled by a single gas and a single air regulator. Because of the great number of small units, the gas pressure rarely exceeds 6 inches and averages approximately 3 inches as compared with a range of from 8 to 12 inches used with the Bunsen type of burner. Inasmuch as the capacity of each multiple unit burner may be increased or decreased independently of other units by a change in gas pressure controlled by a gas valve on each unit, with a corresponding adjustment of the air supply slide, the flexibility of load obtained with this design is greater than with other types of natural-gas burners.

A central station in which both Bunsen and multiple unit burners are in use under boilers of the same size and furnace design gives some figures of comparative results that are of interest. With the Bunsen type of burners the maximum capacity obtainable was equivalent to 270 per cent of the boilers' rated capacity, while the maximum efficiency shown was 76.0 per cent. With the multiple unit type. the maximum capacity and maximum efficiency were 305 and 80.0 per cent, respectively. With the Bunsen type of burner the carbon dioxide content of the gases was 8.5 per cent as a maximum, while with the multiple unit type this was 10.0 per cent. These percentages of carbon dioxide in the flue gases correspond respectively to approximately 34 and 15 per cent of excess air used in combustion.

By-product Coke-oven Gas is the product of the destructive distillation of coal in the by-product oven. The design of by-product ovens varies, but in all classes the volatile products of the coal coked. instead of burning at the source as in the case of beehive or retort ovens, are carried through a standpipe to a collecting main. from which they are taken and cooled and the by-products removed. The principal by-products recovered by the complete by-product plant are gas, tar, crude naphthalene, ammonia and benzols, though the secondary by-products would extend this list greatly.

The extent of by-product recovery varies in different plants, depending upon cost of recovery and the quality of gas desired, these factors in turn depending upon the nature and location of the plant. There are today many by-product coke ovens installed primarily for gas making, the "straight"* gas, after the benzols have been removed, being sold for illuminating and domestic purposes.

In the by-product plant of the complete steel mill, and it is to this class of plant that the use of by-product gas for power generating purposes is almost entirely limited, there is use for all the surplus gas, both from blast furnaces and by-product ovens, in the manufacturing and heating of steel and for the generation of power.

The analysis of by-product coke-oven gas and its heat value vary with the amount of by-product recovery. Typical analyses and calorific values are given in Table 52.

TABLE 52

TYPICAL ANALYSES OF BY-PRODUCT COKE-OVEN GAS

CO_2	O	CO	CH_4	H	N	B.t.u. per Cubic Foot
0.75	Trace	6.0	28.15	53.0	12.1	505
2.00	Trace	3.2	18.80	57.2	18.0	399
3.20	0.4	6.3	29.60	41.6	16.1	551
0.80	1.6	4.9	28.40	54.2	10.1	460

In the modern regenerative steel mill oven, a portion of the gas produced is burned in the ovens for coking purposes, this amount with the ordinary coals representing approximately 40 per cent of the heat in the total gas generated, leaving 60 per cent as surplus gas. While the heat of this surplus gas varies with by-product recovery, its volume will be approximately 6000 cubic feet per ton of coal coked.

In heat value this fuel is more nearly like natural gas than is blast-furnace gas, and, in general, the features to be followed in furnace design for natural gas will hold for coke-oven gas. It is usually burned with burners of the Bunsen type, and because of the large tar and hydrocarbon content, which tends to deposit on and clog the burners, and for which provision as to cleaning should be made, other designs of burners such as have been developed for natural gas could probably not be used successfully. By-product gas comes to the burners at a pressure lower than natural gas where the Bunsen type of burners is used, and this, together with the lower heat value, necessitates a somewhat longer gas opening for the former to develop equal boiler capacities from an equal number of burners. Provision should be made in burners for this fuel for blowing out the condensation of the moisture content of the gas, which comes to the burners saturated.

* As distinguished from the "rich" gas of the earlier installations, in which a system of gas separation was used. In this practice, the rich gas of the early portion of the coking process was sold, and the "lean" gas of the latter portions of the process was used in the ovens for coking purposes

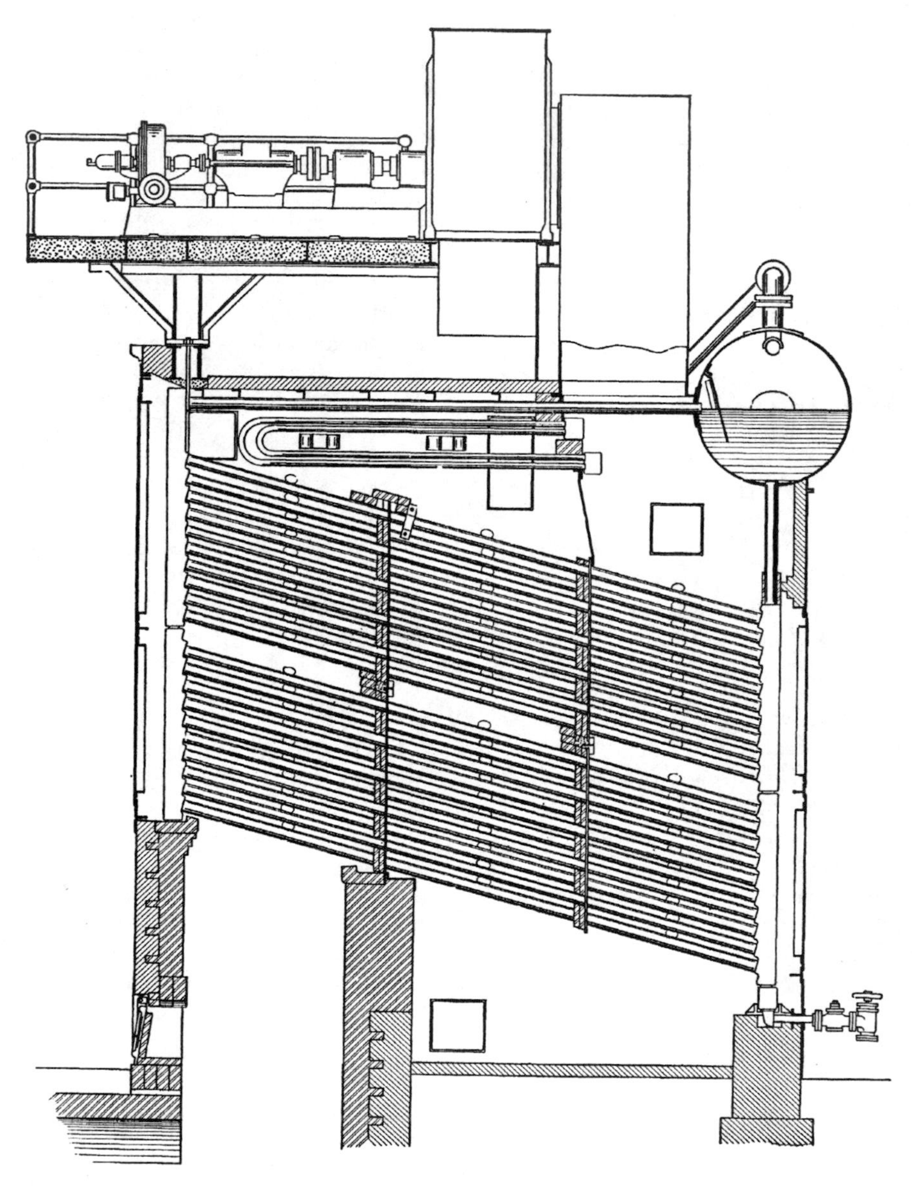

A BABCOCK & WILCOX WASTE-HEAT BOILER WITH AN OVERHEAD FAN, DEVELOPED FIRST FOR USE WITH OPEN-HEARTH STEEL FURNACES

THE UTILIZATION OF WASTE HEAT

THE utilization for the generation of steam of the available heat in the waste gases from various industrial process furnaces cannot be considered new, for records show that Babcock & Wilcox water-tube boilers were used for this purpose as early as 1874 and cylinder and flue boilers prior to that date. The methods of early utilization, however, as compared with present-day practice, were essentially different, and it may be stated that prior to 1915 all waste-heat boiler installations were made with industrial furnaces yielding gases at temperatures which today would be considered high in waste-heat work.

Non-interference with the operation of the primary furnace was the chief factor affecting early waste-heat installations, and from the standpoint of the operator, ample draft at the furnace outlet was the indication of such non-interference. In order that the draft should be ample, it was necessary with the stacks in use that the boilers installed should give low gas velocities with consequent minimum draft resistance, and such boilers were either entirely without baffles or were but partially baffled, giving the gases a straight path over the boiler heating surfaces from the furnace to the stack. Such lack of resistance to the gases led to high exit gas temperatures from the boiler, but these were considered desirable as assisting draft. No particular attention was given in the early installations to the amount and arrangement of heating surface to meet the requirements of varying types of industrial furnaces, and operators were satisfied with whatever power was developed, without reference to such amount as compared with the maximum obtainable.

The modern methods of waste-heat utilization are of particular interest because of the diametrically opposite standpoint from which the problem has been approached.

The theory upon which the design of the modern waste-heat boiler is based has been developed through a more thorough understanding of the laws governing the rate of heat transfer from hot gas to boiler heating surfaces, the principal factor involved being an appreciation of the effect of gas velocity on such transfer rate. Experiments have shown that the heat-transfer rate in itself, while increasing somewhat with increasing difference in temperature between the gas and the absorbing surface, is primarily dependent upon the gas velocity over such surface, the temperature difference having its main effect in total heat absorption. Carefully conducted experiments by The Babcock & Wilcox Company,* while showing the influence of increased gas velocity on the heat transfer rate for but a single form of gas-passage channel, are indicative of the extent of such increase in the transfer rate regardless of the form of channel.

The necessity for high transfer rates in waste-heat boilers results from the general low temperatures of the gases as they enter the boilers.

In direct-fired coal or oil-fired practice, where furnace temperatures of from 2400 to 2800 degrees are commonly developed, the amount of heat absorbed by direct radiation from the furnace represents a very large percentage of the total heat absorbed by the boiler. This is particularly true at the lower ratings—and it is only at such low capacities that waste-heat and direct-fired boilers can be compared—and with the increasing tendency in direct-fired practice toward the exposure of a greater amount of boiler heating surface to direct radiation.

*"Experiments on the Rate of Heat Transfer from a Hot Gas to a Cooler Metallic Surface," The Babcock & Wilcox Company, 1919.

The amount of heat absorbed through radiation by unit surface exposed to such radiation varies approximately as the fourth power of the absolute temperature of the radiating medium. With equal amounts of heating surface exposed to furnace temperature, in the case of a direct-fired boiler with such temperature 2700 degrees and a waste-heat boiler with such temperature 1200 degrees, the amount of heat absorbed through direct radiation approximately 14 times as great in the direct-fired as in the waste-heat unit, and with the greater amount of surface that would ordinarily be exposed in direct-fired practice, the ratio of absorption through radiation would be even greater.

If the results with the two boilers are to be comparable from the standpoint of total heat absorption as measured by steam output, the increased absorption through radiation in the direct-fired boiler must be offset by a corresponding increase in absorption by convection in the waste-heat unit over that absorbed by convection in the direct-fired boiler.

With a definite arrangement of a given amount of boiler heating surface, gas velocities and, hence, transfer rates would, at the same boiler output, be higher in the case of the waste-heat than in that of the direct-fired boiler, this resulting from the higher ratio of gas to steam weight in the former unit. Any increase in convection absorption due to this fact would not, however, be sufficient to offset the greater radiation absorption in the direct-fired boiler, and for comparable rates of steam output, the heating surface of the waste-heat unit must be arranged to give gas velocities greatly in excess of those common in direct-fired practice. It may be well to emphasize again that in comparing the operation and results of waste-heat and direct-fired boilers, only low capacities, and the gas velocities found at such capacities, should be considered in the case of the latter. This is due to the fact that the average waste-heat boiler utilizing low and moderate temperature waste gas ordinarily develops capacities approximating rating.

The velocity necessary in waste-heat as compared with direct-fired practice varies with the temperature of the waste gases available and with the draft requirements of the industrial furnace supplying the gases, as well as with the arrangement of direct-fired heating surface. For boilers developing approximately rating in the two classes of work, with waste-gas temperature of 1200 to 1300 degrees entering, and the arrangement of direct-fired boiler surface ordinarily found in stoker-fired work, the velocity of gas through the waste-heat unit is probably slightly more than three times that through the direct-fired boiler.

As the temperature of the waste gas available increases, the ratio of the necessary gas velocity for comparable rates of steam output in waste-heat and direct-fired boilers will increase, and the capacity at which operation and results are comparable will increase.

The high gas velocity necessary for high heat transfer rates results naturally in high draft resistance in the modern waste-heat boiler. In addition to the greatly increased draft loss in the boiler, as compared with early practice, the reduction in gas temperature leaving the boiler also has an important bearing on the draft.

Open-hearth steel furnace practice without waste-heat boilers as compared with present-day waste-heat installations in this class of work clearly illustrates the effect of high gas resistance in the boiler and low gas temperature on draft. Where no boilers are installed it is customary in open-hearth practice to use stacks approximately 160 feet in height, which give a draft of from 1.6 to 1.8 inches depending upon gas temperature, this draft being sufficient for the satisfactory operation of the furnace. The installation of a modern waste-heat boiler would reduce the temperature of the gases from 1100 or 1200 degrees to 450 or 500 degrees, at which temperature, even without taking into consideration any flue connection or boiler loss, a 160-foot stack would give only approximately 0.9 inch draft, which would not be sufficient for furnace operation. If the draft loss through the boiler is 2.0 inches, that due to resistance and turns in the flues 0.75 inch, and that required at the furnace checkers 1.5 inches, the total draft requirement for the operation of the boiler and furnace is 4.25 inches. To give this draft with gas temperatures of 450 or 500 degrees would necessitate a stack height in excess of 700 feet, the impracticability of which is obvious.

While, because of the draft requirements of the primary furnace, the above illustration may represent an extreme case, it may be broadly stated that in any class of waste-heat work where boilers using in their design the principle of high gas velocity are installed, the use of a natural draft stack, except for by-pass purposes, is impracticable, and in all such work induced draft fans must be installed. It is the necessity for the use of such fans that was the primary factor against which the modern waste-heat boiler in its development and introduction had to contend, for furnace operators almost universally took the stand that no waste-heat boiler installation requiring an induced draft fan could be made that would not lead directly to interference with furnace operation. Experience in every class of waste-heat work, however, has shown that the better control of draft at the furnace or that made possible by the use of fans has led to better furnace operation without in any way interfering with such operation, and in many classes of work has led to an increased furnace output that is directly traceable to the fan installation.

The maximum allowable draft loss through a waste-heat boiler to give the best commercial return varies appreciably in different classes of industrial furnaces. Where, for the proper operation of the primary furnace, a draft at its outlet of from 1.75 to 2.00 inches is required, the loss allowable through the boiler is obviously not as great as in the case of a boiler set with a furnace requiring at its outlet a draft of 0.15 or 0.25 inch. With a given weight of gas, the power required to overcome the resistance through a given boiler would be the same in the two instances. The additional power required to give the higher draft at the furnace outlet, however, would probably more than offset the gross return due to the same draft loss through the boiler. Conditions in different classes of waste-heat work and in different installations in the same class vary widely, and while experience has shown the approximate maximum allowable draft loss for all classes of work for securing the best commercial return, each installation should be considered as a problem by itself.

Any fans used in waste-heat work should be of ample size and capacity to handle the weight of gas that will be discharged from the furnace at its maximum rate of operation, and at such rate give a draft sufficient to overcome the resistance through the boiler and connecting flues and still give ample draft at the furnace outlet for proper operation. Fans should be figured liberally, and it is to be remembered that a fan of ample size may be operated as economically as a smaller fan, whereas the smaller fan, if overloaded, has an appreciably lower efficiency.

Because of the possibility of recovering the greater portion of the power used for the fan drive in the heat of the exhaust, turbine-driven fans are in general use in waste-heat boiler practice. The amount of heat regained in this manner depends upon a number of factors, but in ordinary practice is such as to bring the gross and the net boiler output relatively close. Where there is no source of steam available other than the waste-heat boilers, it is necessary, for starting up, to have at least one motor-driven fan unit.

The types of industrial furnace with which the modern design of waste-heat boiler is in satisfactory operation may be divided broadly into two classes, those operating at an approximately constant rate of furnace output, in which the gas temperature and gas weight are practically constant, and those which, because of the nature of the process, operate in fixed cycles, in which the temperature of the gas and its weight vary over a considerable range at different periods of such cycle. Another classification of industrial furnaces may be made between regenerative and non-regenerative furnaces, and while in general the gas temperatures from the former are higher than those from the latter, this distinction does not necessarily hold for all waste-heat work.

Table 53, which shows the different classes of industrial furnaces with which high gas velocity waste-heat boilers have been installed and are in successful operation and the temperature of the gases in each class of work, indicates the extent of waste-heat utilization by modern methods. The temperatures given represent averages for both constant and cycle furnace operation, and are approximately what may be expected at the boiler inlet. The temperatures at the furnace outlet may be appreciably higher than those shown, the loss between the furnace and the boiler varying considerably in different classes of work with the length and arrangement of flues, and particularly with the care used in the elimination of air infiltration at the furnace outlet and in the flues.

TABLE 53

WASTE-GAS TEMPERATURES*

Type of Primary Furnace	Temperature Degrees Fahrenheit
Nickel Refining Furnace	2500–3000
Beehive Coke Ovens	1950–2300
Zinc Refining Furnace	1700–2000
Heating Furnace	1700–1900†
Copper Refining Furnace	1450‡
Cement Kiln (Dry Process)	1150–1350
Cement Kiln (Wet Process)	800–1100
Open-hearth Steel Furnace (Producer Gas Fired)	1200–1300
Open-hearth Steel Furnace (Oil, Tar or Natural Gas)	800–1100
Gas Benches	1050–1150
Oil Stills	900–1000
Glass Tanks	800–1000

*As ordinarily found at inlet of waste-heat boiler and not necessarily the temperature leaving the primary furnace.

†During operating periods. With furnace kept hot, but heating no material, average temperature 1000 to 1100 degrees Fahrenheit.

‡Average over 36-hour cycle: range 500 to 2100 degrees Fahrenheit.

The securing of accurate gas temperatures in waste heat is difficult, as in all boiler work, and such temperature measurements must be taken at a point where they are not influenced by any leakage of outside air. Further, where there is any tendency toward a laneing action of the gases as they leave the furnace or come to the boiler, every effort should be made to secure average temperatures.

Air infiltration at all portions of the system should be kept at a minimum. Such leakage tends not only to a reduction in gas temperatures with a corresponding decrease in power developed, but also to an increase in the load on the fan, which in turn leads to a decrease in net output from the boiler.

Much of the waste gas available for power purposes carries a great amount of dust of varying characteristics. With proper methods of soot blowing, waste-heat boiler heating surface may be kept as clean as in direct-fired work and by the use of properly designed soot pockets in flues and boiler setting, much of the dust content may be precipitated. Where such dust has a reclaimable value, properly designed conveying systems below the soot pockets enable it to be carried to any desired point.

In some classes of waste-heat work, ordinarily with regenerative furnaces, if a mixture of combustible gas and air is allowed to reach the boiler setting, explosions may occur. Methods of furnace operation have been developed, however, largely through the manner of operating the valves controlling the regenerative system, which practically obviate any danger of such explosions.

The horse-power which can be developed by waste-heat boilers depends upon the weight and temperature of the gases available, the temperature to which the particular amount and arrangement of boiler heating surface installed will cool the gases, and the specific heat of the gases. The boiler horse-power that may be expected may be expressed:

$$\text{Horse-power} = \frac{W\,(T - t) \times c}{33479} \qquad (28)$$

where W = Weight of gas in pounds per hour,
T = Temperature of entering gas in degrees Fahrenheit,
t = Temperature of exit gas in degrees Fahrenheit,
c = Mean specific heat of gas over the range $T - t$.

The specific heat of the gas* is a factor which is perhaps of more importance in waste heat than in any other class of boiler work. From formula *(28)* it is clear that the power to be expected is directly proportional to the specific heat of the gas, and while this varies with the temperature range, the extent of such variation will be greater with the moisture content of the gas, for example, the moisture content of waste gases where oil as against coal is the fuel burned in the primary furnace, or in cement waste-heat practice where the process is wet as against dry.

Table 54 indicates the amount of power that may be developed with the modern design of waste-heat boiler set with a number of classes of industrial furnaces. It is of interest to note the horse-power developed with gases of the inlet temperatures shown, and also the low exit gas temperatures resulting from the use of the high gas velocity principle as compared with early waste-heat practice.

While economizers were not ordinarily used with the early installation of the modern waste-heat boilers, primarily because of the low gas temperatures leaving such boilers, they are in the present state of the art receiving more consideration and are coming into more general use. It is now recognized that, because of the high ratio of gas to water weights in waste-heat economizer practice, economizer transfer

* See page 146.

TABLE 54

TESTS OF BABCOCK & WILCOX WASTE-HEAT BOILERS

Type of Primary Furnace	Beehive Coke Ovens	Heating Furnace	Cement Kilns	Open Hearth Steel Furnace	Glass Tank
Plant		Bethlehem Steel Co.	Bath Portland Cement Co.	Alan Wood Iron & Steel Co.	Owens Bottle Machine Co.
Location		South Bethlehem, Pa.	Bath, Pa.	Ivy Rock, Pa.	Fairmont, W. Va.
Boiler Heating Surface, square feet	10200	5840	14800*	5830	2860
Horse-power	1020	584	1480*	583	286
Steam Pressure, gauge, pounds	165	152	139	113	119
Feed Water Temperature, degrees Fahrenheit	158	66	287	46	60
Superheat, degrees Fahrenheit	86	173	. .	163	53
Gas per hour, pounds	155100	87571	194735	61000	43660
Entering Gas Temperature, degrees Fahrenheit	2158	1745	1325	1436	808
Exit Gas Temperature, degrees Fahrenheit	477	436	506	464	401
Draft at Boiler Damper, inches	4.4	1.87	3.90	3.60	3.15
Draft at Boiler Inlet, inches	2.0	.68	.73	1.60	.96
Draft Loss through Boiler, inches	2.4	1.19	3.17	2.00	2.19
Horse-power Developed	1956	784	1280	461	133
Per cent of Rated Horse-power	192	134	87	79	46

* Two boilers of 720 horse-power each. Each boiler equipped with an economizer containing 2048 square feet of surface or 34 per cent. Temperature of feed to economizer, 209 degrees; temperature of gas from economizer, 405 degrees; power developed by economizer, 111 horse-power; total horse-power of boilers and economizers, 1391.

rates are obtained, at the boiler ratings common in such practice, greatly in excess of those at the same ratings in direct-fired work, and these high transfer rates more than offset the low temperature of gas leaving the boiler. In direct-fired boiler practice the rise in feed-water temperature through the ordinary economizer is roughly one degree for each drop in gas temperature of two degrees; in waste-heat practice, the rise in feed temperature is approximately one degree for each degree drop in gas temperature through the economizer. This fact makes possible an increase in feed-water temperature with economizers in waste-heat work greatly in excess of that which may be obtained with a given amount of economizer surface in direct-fired practice, regardless of the low exit gas temperatures from the boiler in the case of the former. Ordinary practice with waste-heat boilers where economizers are used is the installation of approximately 35 per cent of the boiler heating surface in the economizer.

CHIMNEYS AND DRAFT

THE height and diameter of a properly designed chimney depend upon the amount of fuel to be burned, its nature, the design of the flue, with its arrangement relative to the boiler or boilers, and the altitude of the plant above sea level. There are so many factors involved that as yet there has been produced no formula which is satisfactory in taking them all into consideration, and the methods used for determining stack sizes are largely empirical. In this chapter a method sufficiently comprehensive and accurate to cover all practical cases will be developed and illustrated.

DRAFT is the difference in pressure available for producing a flow of the gases. If the gases within a stack be heated, each cubic foot will expand, and the weight of the expanded gas per cubic foot will be less than that of a cubic foot of the cold air outside the chimney. Therefore, the unit pressure at the stack base due to the weight of the column of heated gas will be less than that due to a column of cold air. This difference in pressure, like the difference in head of water, will cause a flow of the gases into the base of the stack. In its passage to the stack the cold air must pass through the furnace or furnaces of the boilers connected to it, and it in turn becomes heated. This newly heated gas will also rise in the stack and the action will be continuous.

The intensity of the draft, or difference in pressure, is usually measured in inches of water. Assuming an atmospheric temperature of 62 degrees Fahrenheit and the temperature of the gases in the chimney as 500 degrees Fahrenheit, and, neglecting for the moment the difference in density between the chimney gases and the air, the difference between the weights of the external air and the internal flue gases per cubic foot is 0.0347 pound, obtained as follows:

Weight of a cubic foot of air at 62 degrees Fahrenheit	= 0.0761 pound
Weight of a cubic foot of air at 500 degrees Fahrenheit	= 0.0414 pound
Difference	= 0.0347 pound

Therefore, a chimney 100 feet high, assumed for the purpose of illustration to be suspended in the air, would have a pressure exerted on each square foot of its cross sectional area at its base of $0.0347 \times 100 = 3.47$ pounds. As a cubic foot of water at 62 degrees Fahrenheit weighs 62.32 pounds, an inch of water would exert a pressure of $62.32 \div 12 = 5.193$ pounds per square foot. The 100-foot stack would, therefore, under the above temperature conditions, show a draft of $3.47 \div 5.193$ or approximately 0.67 inch of water.

The method best suited for determining the proper proportion of stacks and flues is dependent upon the principle that if the cross-sectional area of the stack is sufficiently large for the volume of gases to be handled, the intensity of the draft will depend directly upon the height; therefore, the method of procedure is as follows:

1st. Select a stack of such height as will produce the draft required by the particular character of the fuel and the amount to be burned per square foot of grate surface.

2nd. Determine the cross-sectional area necessary to handle the gases without undue frictional losses.

The application of these rules follows:

DRAFT FORMULA—The force or intensity of the draft, not allowing for the difference in the density of the air and of the flue gases, is given by the formula:

$$D = 0.52\, H \times P \left(\frac{1}{T} - \frac{1}{T_1}\right) \qquad (29)$$

in which

D = draft produced, measured in inches of water,
H = height of top of stack above grate bars, in feet,
P = atmospheric pressure in pounds per square inch,
T = absolute atmospheric temperature,
T_1 = absolute temperature of stack gases.

In this formula no account is taken of the density of the flue gases, it being assumed that it is the same as that of air. Any error arising from this assumption is negligible in practice, as a factor of correction is applied in using the formula to cover the difference between the theoretical figures and those corresponding to actual operating conditions.

The force of draft at sea level (which corresponds to an atmospheric pressure of 14.7 pounds per square inch) produced by a chimney 100 feet high with the temperature of the air at 60 degrees Fahrenheit and that of the flue gases at 500 degrees Fahrenheit is,

$$D = 0.52 \times 100 \times 14.7 \left(\frac{1}{521} - \frac{1}{961}\right) = 0.67$$

Under the same temperature conditions this chimney at an atmospheric pressure of 10 pounds per square inch (which corresponds to an altitude of about 10,000 feet above sea level) would produce a draft of,

$$D = 0.52 \times 100 \times 10. \left(\frac{1}{521} - \frac{1}{961}\right) = 0.45$$

For use in applying this formula it is convenient to tabulate values of the product

$$0.52 \times 14.7 \left(\frac{1}{T} - \frac{1}{T_1}\right)$$

which we will call K, for various values of T_1. With these values calculated for assumed atmospheric temperature and pressure, *(29)* becomes $D = K H$. *(30)*

For average conditions the atmospheric pressure may be considered 14.7 pounds per square inch, and the temperature 60 degrees Fahrenheit. For these values and various stack temperatures K becomes:

Temperature Stack Gases	*Constant K*
750	0.0084
700	0.0081
650	0.0078
600	0.0075
550	0.0071
500	0.0067
450	0.0063
400	0.0058
350	0.0053

DRAFT LOSSES—The intensity of the draft as determined by the above formula is theoretical and can never be observed with a draft gauge or any recording device.

However, if the ashpit doors of the boiler are closed and there is no perceptible leakage of air through the boiler setting or flue, the draft measured at the stack base will be approximately the same as the theoretical draft. The difference existing at other times represents the pressure necessary to force the gases through the stack against their own inertia and the friction against the sides. This difference will increase with the velocity of the gases. With the ashpit doors closed, the volume of gases passing to the stack is a minimum and the maximum force of draft will be shown by a gauge.

As draft measurements are taken along the path of the gases, the readings grow less as the points at which they are taken are farther from the stack, until in the boiler ashpit, with the ashpit doors open for freely admitting the air, there is little or no perceptible rise in the water of the gauge. The breeching, the boiler damper, the baffles and the tubes, and the coal on the grates all retard the passage of the gases, and the draft from the chimney is required to overcome the resistance offered by the various factors. The draft at the rear of the boiler setting where connection is made to the stack or flue may be 0.5 inch, while in the furnace directly over the fire it may not be over, say, 0.15 inch, the difference being the draft required to overcome the resistance offered in forcing the gases through the tubes and around the baffling.

One of the most important factors to be considered in designing a stack is the pressure required to force the air for combustion through the bed of fuel on the grates. This pressure will vary with the nature of the fuel used, and in many instances will be a large percentage of the total draft. In the case of natural draft, its measure is found directly by noting the draft in the furnace, for with properly designed ashpit doors it is evident that the pressure under the grates will not differ sensibly from atmospheric pressure.

Loss in Stack—The difference between the theoretical draft as determined by formula *(29)* and the amount lost by friction in the stack proper is the available draft, or that which the draft gauge indicates when connected to the base of the stack. The sum of the losses of draft in the flue, boiler and furnace must be equivalent to the available draft, and as these quantities can be determined from records of experiments, the problem of designing a stack becomes one of proportioning it to produce a certain available draft.

The loss in the stack due to friction of the gases can be calculated from the following formula:

$$\Delta D = \frac{fW^2CH}{A^3} \qquad (31)$$

in which

ΔD = draft loss in inches of water,

W = weight of gas in pounds passing per second,

C = perimeter of stack in feet,

H = height of stack in feet,

f = a constant with the following values at sea level:

0.0015 for steel stacks, temperature of gases 600 degrees Fahrenheit.

0.0011 for steel stacks, temperature of gases 350 degrees Fahrenheit.

0.0020 for brick or brick-lined stacks, temperature of gases 600 degrees Fahrenheit.

0.0015 for brick or brick-lined stacks, temperature of gases 350 degrees Fahrenheit.

A = area of stack in square feet.

This formula can also be used for calculating the frictional losses for flues, in which case, $C =$ the perimeter of the flue in feet, $H =$ the length of the flue in feet, the other values being the same as for stacks.

The available draft is equal to the difference between the theoretical draft from formula *(30)* and the loss from formula *(31)*, hence:

$$d' = \text{available draft} = K\,H - \frac{fW^2CH}{A^3} \qquad (32)$$

Table 55 gives the available draft in inches that a stack 100 feet high will produce when serving different horse-powers of boilers, with the methods of calculation

TABLE 55

AVAILABLE DRAFT

CALCULATED FOR 100-FOOT STACK OF DIFFERENT DIAMETERS ASSUMING STACK TEMPERATURE OF 500 DEGREES FAHRENHEIT AND 100 POUNDS OF GAS PER HORSE-POWER

FOR OTHER HEIGHTS OF STACK MULTIPLY DRAFT BY HEIGHT ÷ 100

Horse-power	Diameter of Stack in Inches														
	36	42	48	54	60	66	72	78	84	90	96	102	108	114	120
100	.64														
200	.55	.62													
300	.41	.55	.61												
400	.21	.46	.56	.61											
500		.34	.50	.57	.61										
600		.19	.42	.53	.59										
700			.34	.48	56	.60	.63								
800			.23	.43	.52	.58	.61	.63							
900				.36	.49	.56	60	.62	.64						
1000				.29	.45	.53	.58	.61	63	.64					
1100					.40	.50	.56	.60	.62	.63	.64				
1200					.35	.47	.54	.58	.61	.63	.64	.65			
1300					.29	.44	.52	.57	.60	.62	.63	.64	.65		
1400						.40	.49	.55	.59	.61	.63	.64	.65	.65	
1500						.36	.47	.53	.58	.60	.62	.63	.64	.65	.65
1600						.31	.43	.52	.56	.59	.62	.63	.64	.65	.65
1700							.41	.50	.55	58	.61	.62	.64	.64	65
1800							.37	.47	.54	.57	.60	.62	.63	.64	.65
1900							34	.45	.52	56	.59	.61	.63	.64	64
2000								.43	.50	.55	.59	.61	.62	.63	64
2100								.40	.49	.54	.58	.60	.62	.63	.64
2200								.38	.47	.53	.57	.59	.61	.62	.64
2300								.35	.45	.52	.56	.59	.61	.62	.63
2400								.32	.43	50	.55	58	.60	.62	.63
2500									.41	.49	.54	.57	.60	.61	.63

Horse-power	Diameter of Stack in Inches							
	90	96	102	108	114	120	132	144
2600	.47	.53	.56	.59	61	.62	.64	.65
2700	.45	.52	.55	.58	.60	.62	64	.65
2800	.44	50	.55	.58	.60	.61	.64	.65
2900	.42	.49	.54	.57	.59	.61	.63	.65
3000	.40	.48	.53	.56	.59	.61	.63	.64
3100	.38	.47	.52	.56	.58	.60	.63	.64
3200		.45	.51	.55	58	.60	.63	.64
3300		.44	.50	.54	.57	.59	.62	.64
3400		.42	.49	.53	.56	.59	.62	.64
3500		.40	.48	.52	.56	.58	.62	.64
3600			.47	.52	.55	.58	.61	.63
3700			.45	.51	.55	.57	.61	.63
3800			.44	.50	.54	.57	.61	.63
3900			.43	.49	.53	.56	.60	.63
4000			.42	.48	.52	.56	.60	.62
4100			.40	.47	.52	.55	.60	.62
4200			.39	.46	.51	.55	.59	.62
4300				.45	.50	.54	.59	.62
4400				.44	49	.53	.59	.62
4500				.43	.49	.53	.58	.61
4600				.42	.48	.52	.58	.61
4700				.41	.47	.51	.57	.61
4800				40	46	.51	.57	.60
4900					.45	.50	.57	.60
5000					.44	.49	.56	.60

FOR OTHER STACK TEMPERATURES ADD OR DEDUCT BEFORE MULTIPLYING BY HEIGHT ÷ 100 AS FOLLOWS*

For 750 Degrees F. Add 17 inch	For 650 Degrees F. Add .11 inch	For 550 Degrees F. Add .04 inch.	For 400 Degrees F. Deduct .09 inch.
For 700 Degrees F. Add .14 inch.	For 600 Degrees F. Add .08 inch.	For 450 Degrees F. Deduct .04 inch.	For 350 Degrees F. Deduct .14 inch.

Results secured by this method will be approximately correct for other heights. While the weight of gas per developed horse-power used in the computation of the values in this table is higher than is ordinarily found under good combustion conditions, the use of such weight gives a margin of safety, and in determining stack sizes it is strongly recommended that the stack sizes be amply safe.

HEIGHT AND DIAMETER OF STACKS—From this formula *(32)* it becomes evident that a stack of certain diameter, if it be increased in height, will produce the same available draft as one of larger diameter, the additional height being required to overcome the added frictional loss. It follows that among the various stacks that would meet the requirements of a particular case there must be one which can be constructed more cheaply than the others. In selecting a stack for any given set of conditions, this fact should be kept in mind, and the builder or manufacturer of stacks or chimneys should be consulted as to cost with reference to height and diameter.

A method of determining stack diameters for stoker-fired coal-burning plants that has been found to give satisfactory results is to use a diameter which will give a cross-sectional stack area of 35 square feet per rated 1000 boiler horse-power. Such an area will allow all but exceptional overloads to be carried. With fuels such as oil, where the weight of the products of combustion per horse-power developed is less than where coal is burned, this area may be reduced to approximately 30 square feet per rated 1000 horse-power. In districts where oil is not the natural fuel, however, and may be burned only temporarily, as where there is a possibility that at some time coal will be used, the possible future boiler capacity should not be handicapped and the larger area should be used.

LOSSES IN FLUES—The loss of draft in straight flues due to friction and inertia can be calculated approximately from formula *(31)*, which was given for loss in stacks. It is to be borne in mind that C in this formula is the actual perimeter of the flue and is least, relative to the cross-sectional area, when the section is a circle, is greater for a square section, and greatest for a rectangular section. The retarding effect of a square flue is 12 per cent greater than that of a circular flue of the same area, and that of a rectangular with sides as 1 and 1½, 15 per cent greater. The greater resistance of the more or less uneven brick or concrete flue is provided for in the value of the constants given for formula *(31)*. Both steel and brick flues should be short and should have as near a circular or square cross section as possible. Abrupt turns are to be avoided, but as long easy sweeps require valuable space, it is often desirable to increase the height of the stack rather than to take up more space in the boiler room. Short right-angle turns reduce the draft by an amount which can be roughly approximated as equal to 0.05 inch for each turn. The turns which the gases make in leaving the damper box of a boiler, in entering a horizontal flue and in turning up into a stack should always be considered. The cross-sectional areas of the passages leading from the boilers to the stack should be of ample size to provide against undue frictional loss. It is poor economy to restrict the size of the flue and thus make additional stack height necessary to overcome the added friction. The general practice is to make flue areas the same or slightly larger than that of the stack; these should be, preferably, at least 20 per cent greater, and a safe rule to follow in figuring flue areas is to allow 42 square feet per 1000 horse-power. It is unnecessary to maintain the same size of flue the entire distance behind a row of boilers, and the areas at any point may be made proportional to the volume of gases

L Street Station, Edison Electric Illuminating Company of Boston, Operating Babcock & Wilcox Boilers of an Aggregate of 23,000 Horse Power, Equipped with Babcock & Wilcox Superheaters

that will pass that point. That is, the areas may be reduced as connections to various boilers are passed.

With circular steel flues of approximately the same size as the stacks, or reduced proportionally to the volume of gases they will handle, a convenient rule is to allow 0.1 inch draft loss per 100 feet of flue length and 0.05 inch for each right-angle turn. These figures are also good for square or rectangular steel flues with areas sufficiently large to provide against excessive frictional loss. For losses in brick or concrete flues, these figures should be doubled.

Underground flues are less desirable than overhead or rear flues for the reason that in most instances the gases will have to make more turns where underground flues are used and because the cross-sectional area of such flues will often be decreased on account of an accumulation of dirt or water which it may be impossible to remove.

In tall buildings, such as office buildings, it is frequently necessary in order to carry spent gases above the roofs, to install a stack the height of which is out of all proportion to the requirements of the boilers. In such cases it is permissible to decrease the diameter of a stack, but care must be taken that this decrease is not sufficient to cause a frictional loss in the stack as great as the added draft intensity due to the increase in height which local conditions make necessary.

In such cases, also, the fact that the stack diameter is permissibly decreased is no reason why flue sizes connecting to the stack should be decreased. These should still be figured in proportion to the area of the stack that would be furnished under ordinary conditions or with an allowance of 42 square feet per 1000 horse-power, even though the cross-sectional area appears out of proportion to the stack area.

Loss in Boilers—The draft loss through the boiler proper depends upon the arrangement of the boiler heating surface and baffles, the class of fuel burned, the quantity of fuel burned, and the combustion conditions under which it is burned. A variation in any one of these factors with the others remaining constant might cause an appreciable variation in the draft loss through the boiler. These factors can enter into so many combinations that it is obvious that no statement of draft loss through boilers can be made that will be in any way comprehensive, and each combination of boiler, fuel and combustion conditions must be considered independently.

Loss in Furnace—The draft loss in the furnace or through the fuel bed varies between wide limits. The air necessary for combustion must pass through the interstices of the coal on the grate. Where these are large, as is the case with broken coal, but little pressure is required to force the air through the bed; but if they are small, as with bituminous slack or small sizes of anthracite, a much greater pressure is needed. If the draft is insufficient the coal will accumulate on the grates and a dead smoky fire will result with the accompanying poor combustion; if the draft is too great, the coal may be rapidly consumed on certain portions of the grate, leaving the fire thin in spots and a portion of the grates uncovered with the resulting losses due to an excessive amount of air.

Draft Required for Different Fuels—For every kind of fuel and rate of combustion there is a certain draft with which the best general results are obtained. A comparatively light draft is best with the free burning bituminous coals, and the amount to use increases as the percentage of volatile matter diminishes and the fixed carbon increases, being highest for the small sizes of anthracites. Numerous other factors, such as the thickness of fires, the percentage of ash and the air spaces in the grates, bear directly on this question of the draft best suited to a given combustion rate. The effect of these factors can be found only by experiment. It is almost impossible, therefore, to show by one set of curves the furnace draft required at various rates of combustion for all of the different conditions of fuel, etc., that may be met.

RATE OF COMBUSTION—The amount of coal which can be burned per hour per square foot of grate surface is governed by the character of the coal and the draft available. The area of the grate and the ratio of this area to the boiler heating surface will depend upon the nature of the fuel to be burned, and the stack should be so designed as to give a draft sufficient to burn the maximum amount of fuel per square foot of grate surface corresponding to the maximum evaporative requirements of the boiler.

SOLUTION OF A PROBLEM—Select a size of brick-lined steel stack suitable for a 2000 rated horse-power power plant operating under the following conditions:

Boilers are equipped with a blast stoker and are to be operated at a maximum of 5000 total horse-power. Coal containing 13,500 B.t.u. per pound is burned in such a manner that the weight of the products of combustion is 15 pounds of gas per pound of coal. The efficiency at 250 per cent of rating is 74.5 per cent, and the exit gas temperature is 600 degrees. The boiler heating surface is arranged so that the draft loss through the boiler proper at this rating is 0.8 inch. The length of the flue between the stack and the most distant boiler is 100 feet, containing three right-angle turns, including those from the boiler to the flue and from the flue up into the stack.

The total draft required under these conditions is:

Draft in furnace	0.15 inch
Draft loss, boiler	0.80 inch
Draft loss, flue	0.10 inch
Draft loss, turns	0.15 inch
Total draft required	1.20 inches

An efficiency of 74.5 per cent with 13,500 B.t.u. coal represents an evaporation from and at 212 degrees of 10,364 pounds of water per pound of coal burned, and the total weight of coal is

$$5000 \times 34.5 \div 10{,}364 = 16{,}644 \text{ pounds per hour.}$$

The total gas weight is, then,

$$16{,}644 \times 15 = 249{,}660 \text{ pounds per hour.}$$

Formula *(32)* as given is used in the computation of the draft that a given stack will produce. Transformed to give the height necessary to produce a given draft, this formula becomes

$$H = \frac{d'}{K - \dfrac{fW^2C}{A^3}} \qquad (32a)$$

If in accordance with the recommendations as to stack areas, approximately 35 square feet of cross-sectional area is provided per rated 1000 horse-power, a stack 9 feet 6 inches in diameter should be selected. We then have as the values for use in formula *(32a)*

$d' = 1.20$ $\quad K = 0.0075$ $\quad f = 0.002$

$W = 249{,}660 \div 3{,}600 = 69.4$ $\quad W^2 = 4{,}816$

$C = 29.8$ $\quad A = 70.9$ $\quad A^3 = 356.414$

$$H = \frac{120}{0.0075 - \frac{0.002 \times 4{,}816 \times 29.8}{356{,}414}} = 180$$

Thus to meet the particular conditions assumed, it would be necessary to install a stack 180 feet high above the point where the flue enters. As a matter of fact, the efficiency assumed is fairly high and the weight of the products of combustion represents good combustion conditions, and while a 180-foot stack would serve under such conditions, in an actual installation to take care of a 250 per cent load, the stack would probably be made 200 or 225 feet in order that such a load might be carried under poor operating conditions. It is probably best in computing stack heights to assume the lowest efficiency that is likely to be found and a gas weight corresponding to the poorest combustion conditions liable to occur.

CORRECTION IN STACK SIZES FOR ALTITUDES—It has ordinarily been assumed that a stack height for altitude will be increased inversely as the ratio of the barometric pressure at the altitude to that at sea level, and that the stack diameter will increase inversely as the two-fifths power of this ratio. Such a relation has been based on the assumption of constant draft measured in inches of water at the base of the stack for a given rate of operation of the boilers, regardless of altitude.

If the assumption be made that boilers, flues and furnace remain the same, and further that the increased velocity of a given weight of air passing through the furnace at a higher altitude would have no effect on the combustion, the theory has been advanced* that a different law applies.

Under the above assumptions, whenever a stack is working at its maximum capacity at any altitude, the entire draft is utilized in overcoming the various resistances, each of which is proportional to the square of the velocity of the gases. Since boiler areas are fixed, all velocities may be related to a common velocity, say, that within the stack, and all resistances may, therefore, be expressed as proportional to the square of the chimney velocity. The total resistance to flow, in terms of velocity head, may be expressed in terms of weight of a column of external air, the numerical value of such head being independent of the barometric pressure. Likewise the draft of a stack, expressed in height of column of external air, will be numerically independent of the barometric pressure. It is evident, therefore, that if a given boiler plant, with its stack operated with a fixed fuel, be transplanted from sea level to an altitude, assuming the temperatures remain constant, the total draft head measured in height of column of external air will be numerically constant. The velocity of chimney gases will, therefore, remain the same at altitude as at sea level, and the weight of gases flowing per second with a fixed velocity will be proportional to the atmospheric density or inversely proportional to the normal barometric pressure.

To develop a given horse-power requires a constant weight of chimney gas and air for combustion. Hence, as the altitude is increased, the density is decreased and, for the assumptions given above, the velocity through the furnace, the boiler passes, breeching and flues must be correspondingly greater at altitude than at sea level. The mean velocity, therefore, for a given boiler horse-power and constant weight of gases will be inversely proportional to the barometric pressure, and the velocity head

* See "Chimneys for Crude Oil," C. R. Weymouth, Trans. A. S. M. E., Dec. 1912.

McAlpin Hotel, New York City, Operating 2300 Horse-power of Babcock & Wilcox Boilers

measured in column of external air will be inversely proportional to the square of the barometric pressure.

For stacks operating at altitude it is necessary to increase not only the height but also the diameter, as there is an added resistance within the stack due to the added friction from the additional height. This frictional loss can be compensated by a suitable increase in the diameter, and when so compensated it is evident that, on the assumptions as given, the chimney height would have to be increased at a ratio inversely proportional to the square of the normal barometric pressure.

In designing a boiler for high altitudes, as already stated, the assumption is usually made that a given grade of fuel will require the same draft measured in inches of water at the boiler damper as at sea level, and this leads to making the stack height inversely as the barometric pressures, instead of inversely as the square of the barometric pressures. The correct height, no doubt, falls somewhere between the two values, as larger flues are usually used at the higher altitudes, whereas to obtain the ratio of the squares, the flues must be the same size in each case, and again the effect of an increased velocity of a given weight of air through the fire at a high altitude, on the combustion, must be neglected. In making capacity tests with coal fuel, no difference has been noted in the rates of combustion for a given draft suction measured by a water column at high and low altitudes, and this would make it appear that the correct height to use is more nearly that obtained by the inverse ratio of the barometric readings than by the inverse ratio of the squares of the barometric readings. If the assumption is made that the value falls midway between the two formulæ, the error in using a stack figured in the ordinary way by making the height inversely proportional to the barometric readings would differ about 10 per cent in capacity at an altitude of 10,000 feet, which difference is well within the probable variation of the size determined by different methods. It would, therefore, appear that ample accuracy is obtained in all cases by simply making the height inversely proportional to the barometric readings and increasing the diameter so that the stacks used at high altitudes have the same frictional resistance as those used at low altitudes, although, if desired, the stack may be made somewhat higher at high altitudes than this rule calls for in order to be on the safe side.

The increase of stack diameter necessary to maintain the same friction loss is inversely as the two-fifths power of the barometric pressure.

Table 56 gives ratio of barometric readings at various altitudes to sea-level readings, values for the square of this ratio and values of the two-fifths power.

TABLE 56

STACK CAPACITIES, CORRECTION FACTORS FOR ALTITUDES

Altitude Height in Feet Above Sea Level	Normal Barometer	R Ratio Barometer Reading Sea Level to Altitude	R^2	$R^{\frac{2}{5}}$ Ratio Increase in Stack Diameter
0	30.00	1.000	1.000	1.000
1000	28.88	1.039	1.079	1.015
2000	27.80	1.079	1.164	1.030
3000	26.76	1.121	1.257	1.047
4000	25.76	1.165	1.356	1.063
5000	24.79	1.210	1.464	1.079
6000	23.87	1.257	1.580	1.096
7000	22.97	1.306	1.706	1.113
8000	22.11	1.357	1.841	1.130
9000	21.28	1.410	1.988	1.147
10000	20.49	1.464	2 144	1.165

These figures show that the altitude affects the height to a much greater extent than the diameter and that practically no increase in diameter is necessary for altitudes up to 3000 feet.

For high altitudes the increase in stack height necessary is, in some cases, such as to make the proportion of height to diameter impracticable. The method to be recommended in overcoming, at least partially, the great increase in height necessary at high altitudes is an increase in the grate surface of the boilers which the stack serves, in this way reducing the combustion rate necessary to develop a given power and hence the draft required for such combustion rate.

STACKS FOR OIL FUEL—The requirements of stacks connected to boilers under which oil fuel is burned are entirely different from those where coal is used. While more attention has been paid to the matter of stack sizes for oil fuel in recent years, there has not as yet been gathered the large amount of experimental data available for use in designing coal stacks.

In the case of oil-fired boilers, the loss of draft through the fuel bed is partially eliminated. While there may be practically no loss through any checkerwork admitting air to the furnace when a boiler is new, the areas for the air passage in this checkerwork will be decreased in a short time, due to the silt which is present in practically all fuel oil. The loss in draft through the boiler proper at a given rating will be less than in the case of coal-fired boilers, this being due to a decrease in the volume of the gases. Further, the action of the oil burner itself is to a certain extent that of a forced draft. To offset this decrease in draft requirement, the temperature of the gases entering the stack will be somewhat lower where oil is used than where coal is used, and the draft that a stack of a given height would give, therefore, decreases. The factors as given above, affecting as they do the intensity of the draft, affect directly the height of the stack to be used.

As already stated, the volume of gases from oil-fired boilers being less than in the case of coal makes it evident that the area of stacks for oil fuel will be less than for coal. It is assumed that these areas will vary directly as the volume of the gases to be handled.

In designing stacks for oil fuel there are two features which must not be overlooked. In coal-firing practice there is rarely danger of too much draft. In the burning of oil, however, this may play an important part in the reduction of plant economy, the influence of excessive draft being more apparent where the load on the plant may be reduced at intervals. The reason for this is that, aside from a slight decrease in temperature at reduced loads, the tendency, due to careless firing, is toward a constant gas flow through the boiler regardless of the rate of operation, with the corresponding increase of excess air at light loads. With excessive stack height, economical operation at varying loads is almost impossible with hand control. With automatic control, however, where stacks are necessarily high to take care of known peaks, under lighter loads this economical operation becomes less difficult. For this reason the question of designing a stack for a plant where the load is known to be nearly a constant is easier than for a plant where the load will vary over a wide range. While great care must be taken to avoid excessive draft, still more care must be taken to assure a draft suction within all parts of the setting under any and all conditions of operation. It is very easily possible to more than offset the economy gained through low draft, by the losses due to setting deterioration, resulting from such lack of suction. Under conditions where the suction is not sufficient to carry off the products of combustion, the action of the heat on the setting brickwork will cause its rapid failure.

It becomes evident, therefore, that the question of stack height for oil-fired boilers is one which must be considered with the greatest of care. The designer must guard, on the one hand, against the evils of excessive draft with the view to plant economy, and, on the other, against the evils of lack of draft from the viewpoint of upkeep cost. Stacks for this work should be proportioned to give ample draft for the maximum overload that a plant will be called upon to carry, all conditions of overload carefully considered. At the same time, where this maximum overload is figured liberally enough to insure a draft suction within the setting under all conditions, care must be taken against the installation of a stack which would give more than this maximum draft.

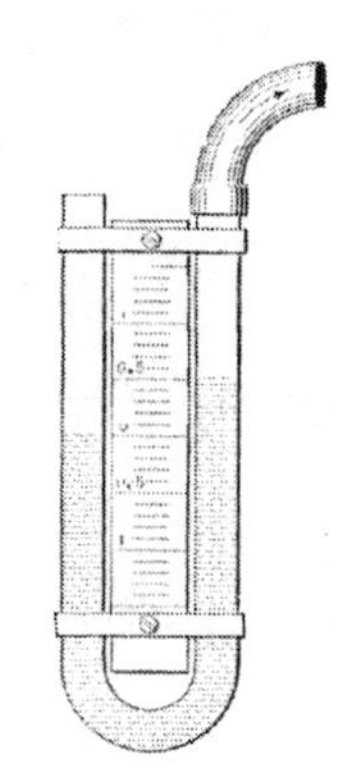

Fig. 15a. U-tube Draft Gauge

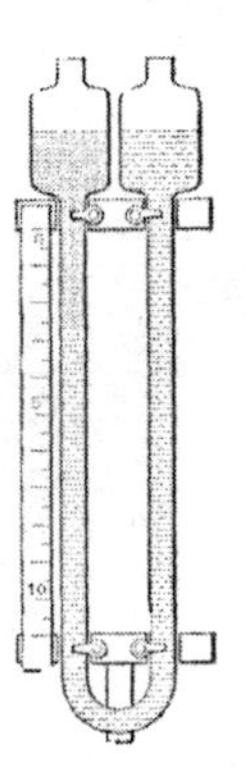

Fig. 16. Barrus Draft Gauge

Draft Gauges—The ordinary form of draft gauge, Fig. 15a, which consists of a U-tube, containing water, lacks sensitiveness in measuring such slight pressure differences as usually exist, and for that reason gauges which multiply the draft indications are more convenient and are much used.

An instrument which has given excellent results is one introduced by Mr. G. H. Barrus, which multiplies the ordinary indications as many times as desired. This is illustrated in Fig. 16, and consists of a U-tube made of one-half inch glass, surmounted by two larger tubes, or chambers, each having a diameter of 2½ inches. Two different liquids which will not mix, and which are of different color, are used, usually alcohol colored red and a certain grade of lubricating oil. The movement of the line of demarcation is proportional to the difference in the areas of the chambers and the U-tube connecting them. The instrument is calibrated by comparison with the ordinary U-tube gauge.

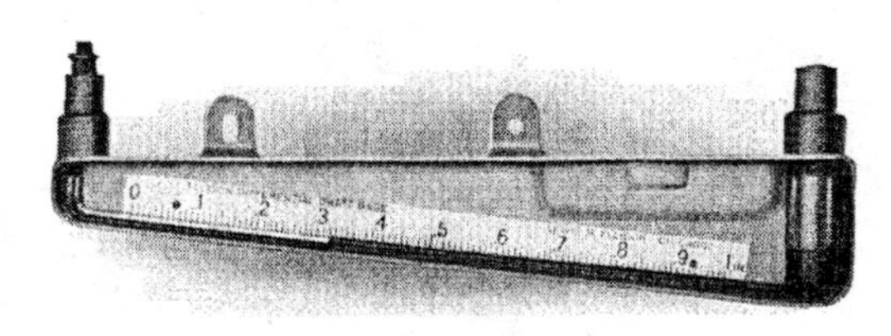

Fig. 17 Ellison Draft Gauge

In the Ellison form of gauge the lower portion of the ordinary U-tube has been replaced by a tube slightly inclined to the horizontal as shown in Fig. 17. By this arrangement any vertical motion in the right-hand upright tube causes a very much greater travel of the liquid in the inclined tube, thus permitting extremely small variations in the intensity of the draft to be read with facility.

The gauge is first leveled by means of the small level attached to it, both legs being open to the atmosphere. The liquid is then adjusted until its meniscus rests at the zero point on the left. The right-hand leg is then connected to the source of draft by means of a piece of rubber tubing. Under these circumstances, a rise of level of one inch in the right-hand vertical tube causes the meniscus in the inclined tube to pass from the point 0 to 1.0. The scale is divided into tenths of an inch and the subdivisions are hundredths of an inch.

The makers furnish a non-drying oil for the liquid, usually a 300 degrees test refined petroleum.

A very convenient form of the ordinary U-tube gauge is shown in Fig. 18. This is a small modified U-tube with a sliding scale between the two legs of the U and with connections such that either a draft suction or a draft pressure may be taken. The tops of the sliding pieces extending across the tubes are placed at the bottom of the meniscus and accurate readings in hundredths of an inch are obtained by a vernier.

Fig. 18 Vernier Draft Gauge

Two 619 Horse-power Babcock & Wilcox Boilers at the Pharmaceutical Works of Parke, Davis & Company, Detroit

EFFICIENCY AND CAPACITY OF BOILERS

ASIDE from proper design and construction, the two factors of primary importance entering into the consideration of what constitutes a satisfactory boiler are its efficiency and that characteristic which is, perhaps ineptly, known as its "capacity." The present chapter deals with these factors only with a view to giving the terms a definite meaning as applied to steam-generating apparatus, together with the method of determining such values by actual test. The relation existing between efficiency and capacity is discussed in the following chapter on the "Selection of Boilers."

In the present consideration of the terms efficiency and capacity, the steam-generating apparatus as a whole is assumed, and may include a boiler alone, a boiler equipped with a superheater or a boiler and superheater with which an economizer is installed. In the following, a boiler is used as representing any combination of heat-absorbing surfaces, the methods of computation for the different combinations being the same.

EFFICIENCY—The term efficiency as specifically applied to a steam boiler is directly represented by the ratio of the total heat absorbed to the total heat in the fuel burned. Where the nature of the fuel is such that ash and refuse remain after combustion, such refuse, because of the limitations of fuel-burning apparatus, contains a certain amount of unconsumed combustible matter. Obviously, the boiler cannot properly be charged with the lack of absorption of the heat that such combustible matter would have evolved had it been burned; on the other hand, since the boiler user pays, in purchasing fuel, for the heat represented by this unconsumed combustible matter, it must be charged against the combined boiler, furnace and grate. The efficiency of the boiler as ordinarily given thus becomes in reality the efficiency of the boiler furnace, and grate or stoker, and is expressed:

$$\left.\begin{matrix}\text{Efficiency of boiler,}\\ \text{furnace, and grate}\end{matrix}\right\} = \frac{\text{Heat absorbed per pound of fuel}}{\text{Heat value per pound of fuel}} \qquad (33)$$

The efficiency as thus expressed is the same whether based on fuel as fired or on dry fuel. Any correction that would be made in changing from one basis to the other would appear in both numerator and denominator of *(33)* as the same value and would cancel.

A second efficiency basis, which is originally assumed to be a measure of the efficiency of the boiler alone as distinguished from that of the boiler and grate, is the so-called efficiency based on combustible. While the difficulties arising from any attempt to separate efficiencies, discussed hereafter, must be recognized, common practice has established the use of the combustible basis as the ordinary efficiency that is assumed to make comparable the relative performance of boilers, irrespective of the grates or stokers installed, to an extent that makes advisable its inclusion here.

The heat in the products of combustion passed over boiler heating surface for absorption is due to the combustible portion of the fuel that is burned, regardless of what proportion of the total combustible constituent of the fuel this may represent. In determining the heat absorbed per pound of combustible burned, in order that any loss due to unconsumed combustible matter discharged with the ash and refuse shall not be charged against the boiler, the total combustible constituent of

PORTION OF 14,300 HORSE-POWER INSTALLATION OF BABCOCK & WILCOX BOILERS AND SUPERHEATERS EQUIPPED WITH BABCOCK & WILCOX CHAIN GRATE STOKERS AT THE BLUE ISLAND PLANT OF THE PUBLIC SERVICE CO. OF NORTHERN ILLINOIS. THIS COMPANY OPERATES 31,000 HORSE POWER OF BABCOCK & WILCOX BOILERS AND SUPERHEATERS IN ITS VARIOUS STATIONS

the fuel must be corrected for such unconsumed combustible matter. On such a basis, the efficiency based on combustible is expressed:

$$\left.\begin{array}{c}\text{Efficiency of boiler}\\ \text{and furnace}\end{array}\right\} = \frac{\text{Heat absorbed per pound combustible burned}}{\text{Heat value per pound combustible}} \qquad (34)$$

If the loss in unconsumed combustible in the ash could be eliminated, the efficiencies given by formulæ *(33)* and *(34)* would be the same. Thus, with liquid and gaseous fuels, if combustion is complete, there is no difference in the efficiencies as given on either basis.

Boiler performance is so directly related to grate or stoker performance that it is difficult and, where absolute accuracy is desired, impossible to separate the efficiency of the boiler from that of the combined unit of boiler and grate, even where proper correction is made for unconsumed combustible matter appearing in the ash and refuse.

Many methods have been proposed for a separation of boiler efficiency from that of the combined unit, including furnace and grate or stoker, and it is probable that this ground has been more thoroughly covered than any other feature of boiler testing. No method offered, however, has been able to make a separation of efficiencies in a manner such that the value for the efficiency of the boiler depends solely upon its design and is not influenced by the design of furnace and grate or stoker. In practically all of the proposed methods of separating efficiencies, that of the furnace and grate or stoker is based on the analysis of furnace or exit gas and the amount of unconsumed combustible in the ash. As has been indicated, one of the principal sources of loss in the operation of a steam-generating unit is in the amount of excess air supplied for combustion. There is no method of properly proportioning the loss from such excess air between boiler and stoker, though presumably it is largely due to the stoker. What is a more important factor in the difficulty of separation of efficiencies is that resulting from delayed or secondary combustion. This occurs in the passage of the gases through the heating surface of the boiler that cannot be considered as a part of the furnace, and which will occur under conditions of improper furnace operation regardless of the design of boiler. The effect of secondary combustion cannot be indicated by any analysis of furnace or exit-flue gas, or by an analysis of the ash, and shows directly against the boiler efficiency, while resulting from improper furnace or stoker design or operation. The factor of delayed combustion may affect the combined efficiency results as much as 10 per cent for the same analysis of flue gases leaving the boiler and the difficulty of separating the responsibility for such loss between boiler and furnace or grate is clear. On the subject of the difficulty of differentiation between boiler and combined steam-generation unit efficiency, Dr. D. S. Jacobus says, in a paper presented before The American Society of Mechanical Engineers in 1922:

> "Should it be possible to express accurately the furnace and stoker efficiency, a purchaser should obtain the same overall economy from his boilers with any arrangement of stokers and furnaces having a given efficiency. There is no method whereby this result could be insured and whereby the purchaser would be protected should he assume that the boiler and furnace efficiency would be an exact measure of the relative advantages of using one or another of the various stokers and furnaces in connection with his boilers."

The differentiation between the efficiency of boiler and furnace is of greater interest to the operator than to the boiler manufacturer, but from the standpoint of

24,420 Horse-power Installation of Babcock & Wilcox Boilers and Superheaters Equipped with Babcock & Wilcox Chain Grate Stokers in the Quarry Street Station of the Commonwealth Edison Co., Chicago, Ill.

both it emphasizes the importance of and necessity for coordination between furnace, grate or stoker, and boiler design, if the best results are to be obtained. If the boiler manufacturer is called upon to make a guarantee of performance of a combined unit, the design and operation of an element of which are not his responsibility, he must, as a matter of protection, make certain qualifications as to the performance of those elements of the combined unit which he does not furnish. Such qualifications are ordinarily expressed in terms of completeness of combustion in the boiler furnace, a minimum percentage of carbon dioxide and a maximum percentage of carbon monoxide in the flue gases, and, in the case of solid fuels, a minimum of unconsumed combustible matter in the discharged ash and refuse.

In connection with any attempt to allocate specifically losses in boiler-test work, either in the case of a combined unit or where it is endeavored to make a further segregation, the following quotation from The American Society of Mechanical Engineers' Test Code for Stationary Steam Generating Units is of interest:

> "Boiler testing should not be lightly undertaken by anyone who has not had some training under an experienced testing engineer, if reliable results are to be expected. The whole matter should be thoroughly understood, both theoretically and practically. Accurate tests depend very largely upon the care and faithfulness of the observers. It is much easier to make mistakes than is realized by those who are not familiar with practical testing.
>
> "The absolute accuracy of the results of a boiler test, even when conducted with the greatest care, is doubtful, but there is yet no possible basis upon which to determine what the probable limits of error might be. It is generally conceded, however, that there are several sources of indeterminate error, the more important of which are discussed below. The limits of accuracy of a test may very reasonably be taken to be within plus or minus 3 per cent.
>
> "One of the sources of probable error is the sampling of coal. Even when the greatest care is taken to obtain a representative sample, there may be an indeterminate error in ascertaining the heat value of the coal, even though the laboratory analysis is most reliable. With modern apparatus, these laboratory determinations should be substantially correct as regards the sample tested; but the question as to how truly the same represents the whole is always present and cannot be answered indubitably.
>
> "Another source of error is the moisture contained in the coal. As explained in the preceding paragraph, the sampling is more or less uncertain. It is contended by some that if the attempt is made to determine the moisture during the test, the methods of drying and weighing are unreliable; while others contend that though the moisture as determined in the laboratory is accurate so far as the sample delivered to the laboratory is concerned, this sample probably does not represent the bulk of the coal actually burned, since there must inevitably have been more or less loss of moisture during the collection, preparation and handling of the sample.
>
> "Similarly, it is problematical whether the samples collected for the determination of the moisture in steam and for gas analysis are representative of the bulk, although the testing of the samples obtained may be quite accurate.
>
> "It is not unusual for heat balances to be reported to the nearest B.t.u. and to the nearest one-tenth of 1 per cent. But the present state of the art of boiler testing does not provide means for attaining anything like this accuracy. In general, results should be reported only to the nearest significant figure. Reporting results of any kind in small units is likely to convey an erroneous idea as to the real accuracy of the figures.
>
> "It is, therefore, quite logical in the case of guarantee tests that a substantial compliance with the guarantee be accepted as full compliance therewith. A limit of tolerance should be agreed upon beforehand by the parties to the test. The amount of this tolerance might well bear some relation to the care exercised in arranging the details and in the conducting of the test."

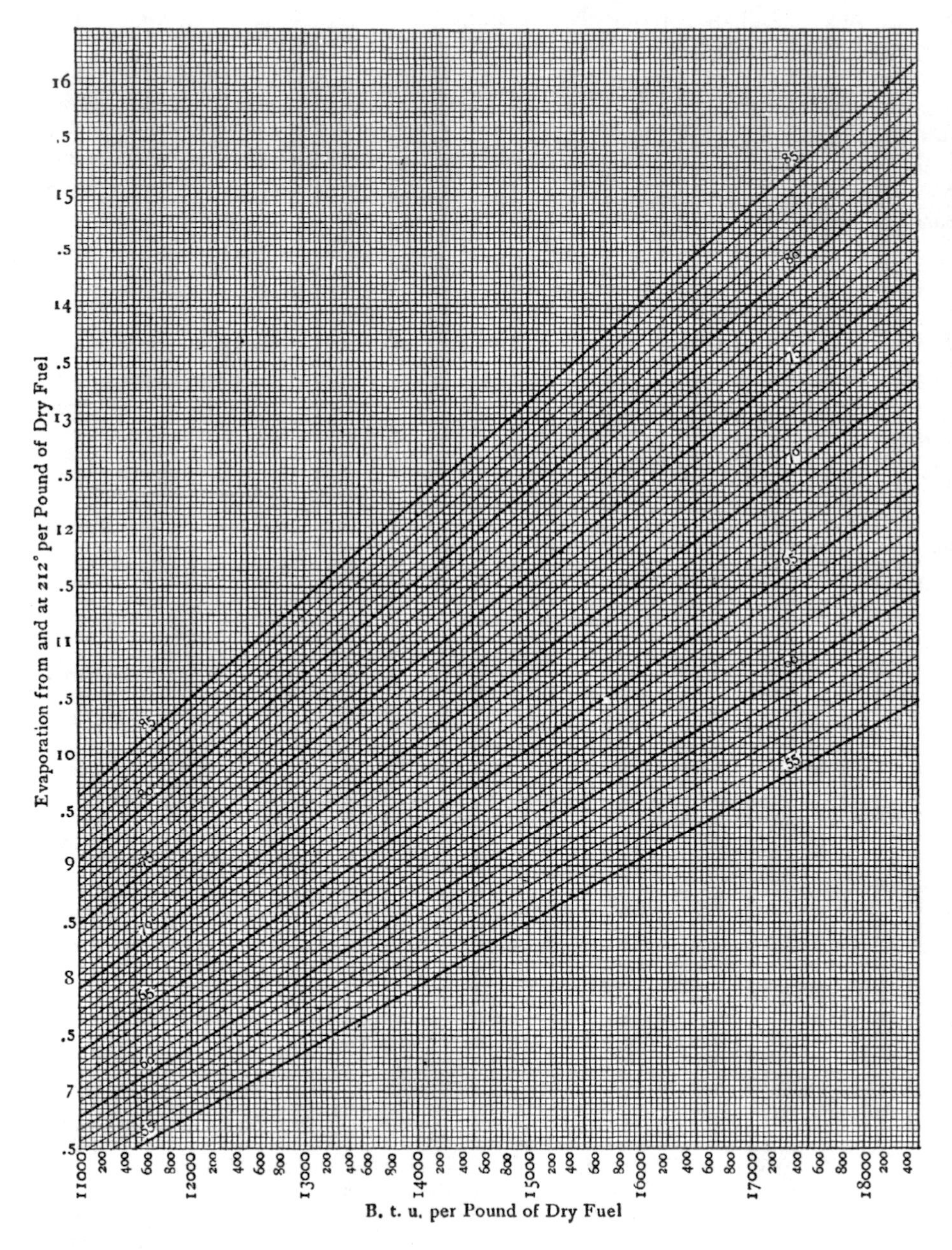

FIG. 19. EFFICIENCY CHART. CALCULATED FROM MARKS AND DAVIS' TABLES
DIAGONAL LINES REPRESENT PER CENT EFFICIENCY

The computation of the efficiencies described may best be indicated by an example.

Assume the following data obtained from an actual boiler test:

(*a*) Steam pressure by gauge, 200 pounds.
(*b*) Feed temperature, 180 degrees.
(*c*) Superheat, 200 degrees.
(*d*) Total weight of fuel fired per hour, 3,761 pounds.
(*e*) Moisture in fuel, 5 per cent.
(*f*) Total ash and refuse per hour, 447 pounds.
(*g*) Total evaporation per hour, actual, 30,000 pounds.
(*h*) Heat value per pound of dry coal, 13,000 B.t.u.
(*i*) Heat value per pound of combustible, 14,444 B.t.u.

The factor of evaporation is, from Tables 22 and 20, and by formula *(4)*, 1.1972. The equivalent evaporation from and at 212 degrees is, therefore, 35,916 pounds per hour.

The total weight of dry fuel is $3,761 \times 0.95 = 3,573$ pounds per hour, and the evaporation from and at 212 degrees per pound of dry fuel is $35,916 \div 3,573 = 10.05$ pounds. The heat absorbed per pound of dry fuel is therefore $10.05 \times 970.4 = 9,752$ B.t.u. Hence the efficiency, by *(33)*, is $9,752 \div 13,000 = 75.00$ per cent.

The total combustible burned is $3,573 - 447 = 3,126$ pounds per hour, and the evaporation from and at 212 degrees per pound of combustible is $35,916 \div 3,126 = 11.49$ pounds. Hence the efficiency based on combustible is from *(34)*, $(11.49 \times 970.4) \div 14,444 = 77.19$ per cent.

If item (*h*) were expressed in terms of fuel as fired, and the evaporation from and at 212 degrees computed on such basis, the efficiency per pound of fuel would be found to be the same, whether the wet or dry basis were used.

For approximate results, the efficiency may be computed from a chart. Fig. 19 shows such a chart based on the evaporation and heat value per pound of dry fuel, and efficiencies may be read directly to within one-half of one per cent. The same chart may be used for efficiency determinations on the fuel as fired or the combustible basis, provided evaporations and heat values corresponding to such basis are used in the graphic computations.

The difference in the efficiency shown in a test of any combination of steam generating units, and 100 per cent efficiency, includes the unavoidable and avoidable losses arising in the conversion of the heat energy of the fuel to that transmitted to the steam as delivered by the generating unit.

Unavoidable losses result directly from the necessity of discharging the products of combustion at a temperature above that of the temperature of the fuel, regardless of the combination of heat-absorbing apparatus, and from the moisture and hydrogen content of the fuel. The avoidable losses, or rather those that to an extent can be controlled, result from the amount of air in excess of that necessary for perfect combustion that is heated to the temperature of the exit gases; the moisture accompanying such excess; unconsumed solid combustible matter discharged with the ash and refuse; unconsumed gaseous combustible matter discharged with the exit gases; and heat radiated from the setting of the unit.

These losses, segregated as far as available data and methods of determination permit, added to the boiler efficiency will total less than 100 per cent, the difference

2400 Horse Power Installation of Babcock & Wilcox Boilers in the Union Station Power House of the Pennsylvania Railroad Co. Pittsburgh, Pa. This Company has a Total of 33,000 Horse Power of Babcock & Wilcox Boilers Installed

between such sum and 100 representing losses that cannot be charged against any listed items. The amount of such unaccounted loss is a measure of the accuracy of the test as a whole, and for this reason the tabulation of the efficiency of the unit and the determinable losses is known as a "heat balance." The items given in the ordinary heat balance may be listed as follows:

1st. Loss due to the moisture content of the fuel. Such moisture must be evaporated and superheated to a temperature represented in the computation of the loss by that of the exit gases. The loss is represented by the heat which would be developed by the cooling of the water vapor to the initial temperature of the *fuel* and condensing it to liquid water at that temperature. The amount of heat that would be so developed is the total heat of the water vapor at the exit-gas temperature minus that at the initial temperature, the low temperature state being taken as liquid water at fuel temperature.

2nd. Loss due to the water vapor formed in the burning of hydrogen. This loss results from the same causes as the first item and is represented by the difference in total heat of water vapor at exit temperature and liquid water at fuel temperature.

3rd. Loss due to the moisture in the air used for combustion. This moisture enters the furnace as water vapor and not as liquid water. The loss therefore is represented by the difference in total heat of the water vapor at exit-gas temperature and at the temperature of the air used in combustion. It should be noted that the low-temperature state is as vapor, and not as liquid water as in the case of items 1 and 2. The weight of moisture per pound of air supplied for combustion is determined from wet and dry bulb thermometer readings and psychrometric tables or charts. This loss is small and is frequently included with the unaccounted losses.

4th. Loss due to the heat carried away by the dry chimney gases. This loss results from the necessity of heating the dry products of combustion from the temperature of the air used in combustion to that of the exit gases. The weight of the dry products is the total weight including excess air.

5th. Loss due to unconsumed solid combustible matter. This loss is ordinarily based on the unconsumed combustible matter that is discharged with the furnace ash and refuse, but, for accuracy, should properly include such unconsumed combustible particles as may be carried over into the setting or up the stack. The latter portion of combustible can be determined only where the arrangement of the unit permits the collection of ash and cinders discharged in this manner, and is ordinarily not included in the heat balance. The solid combustible matter contained in the ash and refuse is usually assumed to be carbon, but where the percentage of combustible in the ash is high, the volatile constituent may be relatively large. The accuracy of the determination of this loss is primarily dependent upon the ash sampling.

6th. Loss due to unconsumed gaseous combustible matter discharged with the exit gases. This loss in the ordinary boiler trial is measured by the carbon monoxide constituent of the flue gases alone, though losses from unconsumed hydrocarbons may exist that are greater than that resulting from the presence of carbon monoxide. The ordinary gas analysis apparatus, however, does not carry the analysis further than the carbon monoxide constituent. The heat value used in the computation of this loss is that representing the difference in burning one pound of carbon to carbon monoxide and carbon dioxide.

7th. Radiation and unaccounted losses. While several methods of determining radiation losses have been offered, none is satisfactory, and such losses are ordinarily included with the unaccounted loss. As stated, this loss is represented by the difference between 100 per cent and the sum of the above items plus the boiler efficiency.

The methods of computing these losses, with an example in each instance, are given in the following:

(A). Loss due to moisture in the fuel. At temperatures below 1000 degrees and at vapor pressures below 3 pounds per square inch absolute, the total heat of saturated or superheated vapor depends only upon its temperature, and is given by the formula

$$H = 1{,}058.7 + 0.455T$$

where T is the actual temperature in degrees Fahrenheit. This formula is purely empirical, and the constant 0.455 is not to be confused with the specific heat of superheated steam.

The total heat of liquid water is

$$t - 32$$

where t is the temperature of the water.

The difference between these total heats is thus

$$[(1058.7 + 0.455T) - (t_f - 32)] = 1090.7 + 0.455T - t_f$$

The loss due to the moisture content of the fuel, therefore, is

$$\text{Loss, B.t.u. per pound} = W(1090.7 + 0.455T - t_f) \qquad (35)$$

where W = weight of moisture per pound of fuel

t_f = temperature of the fuel.

(B). Loss due to combustion of hydrogen. In combustion, one pound of hydrogen unites with eight pounds of oxygen to form water vapor. On the basis of (A), the loss from this source is then

$$\text{Loss, B.t.u. per pound} = 9H_2(1090.7 + 0.455T - t_f) \qquad (36)$$

where H_2 = weight of hydrogen per pound of fuel and T and t_f have the same values as in (A).

(C). Loss due to moisture in the air used for combustion. The difference in the total heat of this vapor at the temperatures of the exit gas and of the air for combustion is

$$[(1058.7 + 0.455T) - (1058.7 + 0.455t_a)] = 0.455(T - t_a)$$

where t_a = initial temperature of the air used for combustion.

The loss from this source is, then,

$$\text{Loss, B.t.u. per pound} = w \times m \times 0.455(T - t_a) \qquad (37)$$

where w = weight of air used for combustion per pound of fuel.

m = weight of moisture per pound of air,

(D). Loss in dry chimney gases. The weight of dry products of combustion per pound of fuel, W, has been shown by formula *(20)* to be

$$\frac{11CO_2 + 8O_2 + 7(CO + N_2)}{3(CO_2 + CO)}C$$

where C is the weight of carbon burned per pound of fuel and appearing in the products of combustion.

The loss from this source is, then,

$$\text{Loss, B.t.u. per pound} = W(T - t_a)0.24 \qquad (38)$$

where W = weight of dry products per pound of fuel,

T = temperature of exit gases,

t_a = temperature of air used for combustion,

0.24 = specific heat of products. (This value is not exact, as it depends upon temperature. It may be computed for actual temperatures if desired.)

(E). Loss due to unconsumed combustible in the ash. This loss may be expressed by

$$\text{Loss, B.t.u. per pound} = \frac{c \times C}{100}\ 14{,}600 \qquad (39)$$

where c = weight of ash per pound of fuel,

C = percentage of unconsumed combustible in the ash,

$c \times C$ = weight of unconsumed combustible in terms of total carbon per pound of fuel.

The value 14,600 is the heat value per pound of carbon, all of the combustible matter being assumed to be carbon. Such an assumption gives rise to an error, which, however, is so small as to be negligible.

(F). Loss due to unconsumed gaseous combustible matter. The weight of carbon monoxide in the exit gases is represented by the expression

$$\frac{CO}{CO_2 + CO}$$

where the symbols represent volumetric percentages. The loss from this source is

$$\text{Loss, B.t.u. per pound} = \frac{CO}{CO_2 + CO} \times C \times 10{,}160 \qquad (40)$$

where C = the weight of carbon burned and appearing in the products of combustion, per pound of fuel burned,

10,160 = the number of B.t.u. evolved in burning one pound of carbon *in carbon monoxide* to carbon dioxide.

(G). Radiation and unaccounted losses. If E is the heat absorbed by the steam-generating unit. expressed in B.t.u. per pound of fuel. or in terms of percentage. this loss is represented by

$$\text{Loss, B.t.u. per pound} = H.V. - (E + A + B + C + D + E + F) \qquad (41)$$

or

$$\text{Per cent loss} = 100 - (E + A + B + C + D + E + F) \qquad (41a)$$

the values of A, B, C, etc., being expressed in terms of B.t.u. or as percentages.

A heat balance may be given upon either a dry fuel or a fuel as fired basis, but care must be taken to have the numerical values used in the computation consistent with the basis used.

To illustrate the computation of the items of a heat balance as ordinarily reported, assume the data obtained in a boiler trial to be as given in Table 57. The heat balance as computed here is on a dry coal basis.

The heat absorbed per pound of dry coal is

$$11.12 \times 970.4 = 10{,}791 \text{ B.t.u.}$$

and the efficiency of the boiler

$$10{,}791 \div 14{,}000 = 77.08 \text{ per cent}$$

TABLE 57

BOILER TRIAL DATA

FOR COMPUTATION OF A HEAT BALANCE

Wet Bulb Thermometer, Degrees Fahrenheit		67
Temperature Boiler Room, Degrees Fahrenheit		73
Temperature Exit Gases, Degrees Fahrenheit		575
Ultimate Analysis Dry Coal	C, Per Cent	78.42
	H_2, Per Cent	5.56
	O_2, Per Cent	8.25
	N_2, Per Cent	1.09
	S, Per Cent	1.00
	Ash, Per Cent	5.68
Moisture in Coal, Per Cent		1.91
B.t.u. per Pound Dry Coal		14000
Ash and Refuse (Per Cent Dry Coal)		7.03
Unconsumed Carbon in Ash, Per Cent		31.50
Fuel Gas Analysis	CO_2, Per Cent	14.00
	O_2, Per Cent	5.50
	CO, Per Cent	0.42
	N_2, Per Cent	80.08
Evaporation from and at 212 Degrees per Pound Dry Coal, Pounds		11.12

(A). The moisture content of the coal, 1.91 per cent, in terms of dry coal, becomes

$$1.91 \div (100 - 1.91) = 1.95 \text{ per cent}$$

and the loss due to this moisture is, from (A),

$$0.0195 [1090.7 + (0.455 \times 575) - 73] = 25 \text{ B.t.u.}$$

or

$$25 \div 14,000 = 0.18 \text{ per cent}$$

(B). Loss due to burning of hydrogen

$$9 \times 0.0556 [1090.7 + (0.455 \times 575) - 73] = 640 \text{ B.t.u.}$$

or

$$640 \div 14,000 = 4.57 \text{ per cent}$$

(C). Loss due to moisture in air used for combustion.

From the wet and dry bulb thermometer readings and psychrometric tables, the weight of moisture in the air per pound of dry air supplied is 0.0127 pound.

The weight of dry air supplied per pound of dry coal is, from formula *(22)*

$$\frac{3.036 \times 80.08}{14.00 + 0.42} (0.7842 - 0.0221) = 12.849 \text{ pounds}$$

The weight of water vapor per pound of dry coal is

$$12.849 \times 0.0127 = 0.163 \text{ pound}$$

and the loss from this source is

$$0.163 \times 0.48 (575 - 73) = 39.3 \text{ B.t.u.}$$

or

$$39.3 \div 14.000 = 0.28 \text{ per cent}$$

(D). Loss in dry chimney gases. The weight of dry gases per pound of carbon burned, from formula *(19)*, is

$$\frac{11 \times 14.00 + 8 \times 5.50 + 7 (0.42 + 80.28)}{3 (14.00 + 0.42)} = 17.603 \text{ pounds}$$

and the weight per pound of fuel is

$$17.603 (0.7842 - 0.0221) = 13.415 \text{ pounds}$$

The loss due to the heat carried away in the dry chimney gases is then

$$13.415 \times 0.24 (575 - 73) = 1,616 \text{ B.t.u.}$$

or

$$1.616 \div 14.000 = 11.54 \text{ per cent}$$

(E). Loss due to unconsumed combustible matter in the ash and refuse. This loss is

$$\frac{0.0703 \times 31.5}{100} 14,600 = 323.3 \text{ B.t.u.}$$

or

$$323.3 \div 14,000 = 2.31 \text{ per cent}$$

(F). Loss due to unconsumed gaseous combustible matter. This loss is

$$\frac{0.42}{14.00 + 0.42} 0.7621 \times 10,160 = 225.5 \text{ B.t.u.}$$

or

$$225.5 \div 14,000 = 1.61 \text{ per cent}$$

(G). Radiation and unaccounted loss. This, expressed in terms of heat value of the fuel burned, is

$$14,000 - (10,791 + 25 + 640 + 39.3 + 1,616 + 323.3 + 225.5) = 14,000 - 13,660 = 340 \text{ B.t.u.}$$

or, in terms of percentage loss,

$$340 \div 14,000 = 2.43 \text{ per cent}$$

The heat balance as ordinarily tabulated is shown in Table 58.

TABLE 58

HEAT BALANCE

	B.t.u.	Per Cent
Heat absorbed by boiler	10,791	77.08
Loss due to moisture in coal	25	0.18
Loss due to moisture formed in burning H_2	642	4.58
Loss in dry chimney gases	1,616	11.54
Loss due to moisture in air	39	0.28
Loss due to incomplete combustion of C	227	1.62
Loss due to unconsumed C in ash	323	2.31
Radiation and unaccounted losses	337	2.41
Total	14,000	100.00

Application of the Heat Balance—A heat balance should be made in connection with any boiler test on which sufficient data for its computation are reported. This is particularly true where the boiler performance has been considered unsatisfactory, the distribution of the losses shown by the heat balance enabling any extraordinary loss to be detected. Even where accurate data for computing a heat balance are not available, such a computation based on certain assumptions is frequently of assistance in indicating unusual losses.

Where it is not practicable to weigh or measure the quantity of water evaporated or fuel burned, it is possible to determine the efficiency of the unit by a method of indirect heat balance. If a radiation and unaccounted loss is arbitrarily assumed, all the items of the ordinary heat balance with the exception of the boiler efficiency may be computed from fuel, ash and exit-gas analyses. The sum of these items and the assumed unaccounted loss deducted from 100 per cent give indirectly the efficiency of the unit. It is to be recognized that this method can give results that are only approximate, but with certain gaseous fuels, where it is frequently difficult or impossible to meter the quantity of gas burned, the method is of considerable use.

CAPACITY—The term capacity as applied to steam boilers in this country has come, through custom, to refer to boiler output expressed in terms of boiler horse-power. There is a decided tendency toward abandoning the use of the term horse-power as a measure of boiler capacity, and such tendency is entirely logical. Such use, however, has been so generally accepted and so thoroughly fixed, that it probably will be some time before it can be entirely abandoned and a more rational basis for the measure of boiler capacity be universally established. A discussion of a more rational basis is given later in the chapter.

Stationary boilers in this country are ordinarily rated today on a horse-power basis, though such a method frequently leads to a misunderstanding. A "horse-power," which is the unit of motive power in general use among steam engineers, is equivalent to 33,000 foot-pounds per minute. Work, as the term is used in mechanics, is the overcoming of resistance through space, while power is the rate of work or the amount done per unit of time. As the operation of a boiler in service implies no motion, it can produce no power in the sense of the term as understood in mechanics. Its operation is the generation of steam, which acts as a medium to convey the energy of the fuel which is in the form of heat to a prime mover in which that heat energy is converted into energy of motion or work, and power is developed.

If all engines developed the same amount of power from an equal amount of heat, a boiler might be designated as one having a definite horse-power, dependent upon the amount of engine horse-power its steam would develop. Such a statement of the rating of a boiler, though it would still be inaccurate if the term is considered in its mechanical sense, could, through custom, be interpreted to indicate that a boiler was of the exact capacity required to generate the steam necessary to develop a definite amount of horse-power in an engine. Such a basis of rating, however, is obviously impossible when the fact is considered that the amount of steam necessary to produce the same power in prime movers of different types and sizes varies over very wide limits.

To do away with the confusion resulting from an indefinite meaning of the term boiler horse-power, the Committee of Judges in charge of the boiler trials at the Centennial Exposition at Philadelphia in 1876 ascertained that a good engine of the type prevailing at the time required approximately 30 pounds of steam per hour per horse-power developed. In order to establish a relation between the engine power and the size of a boiler required to develop that power, they recommended that an evaporation of 30 pounds of water from an initial temperature of 100 degrees Fahrenheit to steam at 70 pounds gauge pressure be considered as one boiler horse-power. This recommendation, which was standardized by The American Society of Mechanical Engineers in 1889, has been, until recently, generally accepted as standard by American engineers, and when the term horse-power is used in connection with stationary boilers* throughout this country, without special definition, it is understood to have this meaning.

Inasmuch as an equivalent evaporation from and at 212 degrees Fahrenheit is the generally accepted basis of comparison, it is now customary to consider the

*Where the horse-power of marine boilers is stated, it generally refers to and is synonymous with the horse-power developed by the engines which they serve.

standard boiler horse-power as recommended by the Centennial Exposition Committee in terms of equivalent evaporation from and at 212 degrees. This will be 30 pounds multiplied by the factor of evaporation for 70 pounds gauge pressure and 100 degrees feed temperature, or 1.1494; $30 \times 1.1494 = 34.482$, or approximately 34.5 pounds. Hence, one boiler horse-power is equal to an evaporation of 34.5 pounds of water per hour from and at 212 degrees Fahrenheit. The term boiler horse-power, therefore, is clearly a measure of evaporation and not of power.

A method of basing the horse-power rating of a boiler adopted by boiler manufacturers is that of heating surfaces. Such a method is absolutely arbitrary and changes in no way the definition of a boiler horse-power just given. It is simply a statement by the manufacturer that his product, under ordinary operating conditions or conditions which may be specified, will evaporate 34.5 pounds of water from and at 212 degrees per definite amount of heating surface provided. The amount of heating surface that has been considered by manufacturers capable of evaporating 34.5 pounds from and at 212 degrees per hour has changed from time to time as the art has progressed. At the present time 10 square feet of heating surface is ordinarily considered the equivalent of one boiler horse-power among manufacturers of stationary water-tube boilers, and the horse-power for which a boiler is sold on such basis is ordinarily known as its normal rated capacity. In view of the arbitrary nature of such rating and of the widely varying rates of evaporation possible per square foot of heating surface with different boilers and different operating conditions, such a basis of rating has in reality no particular bearing on the question of horse-power and should be considered merely as a convenience.

The arguments to the effect that the terms capacity and horse-power as defined above are not definite and have no rational basis are unanswerable, except that they have been established by custom to an extent that will be difficult to overcome. The whole question of the unit of boiler capacity has been widely discussed, and though many methods have been proposed as offering a more definite and rational basis, none has ever met with universal approval.

It would seem most logical to rate boilers on size only, expressed in square feet of heating surface. The surface method of rating has been used in European countries for a number of years, and since the ultimate steam capacity, or evaporation per square foot of heating surface, is dependent solely upon the amount of fuel fired, such basis is suitable and does not involve capacity. The Committee on "Definitions and Values" of the Power Test Codes of The American Society of Mechanical Engineers has recommended this basis of rating boilers. This same Committee has recommended that boiler output, instead of being given in terms of boiler horse-power, be expressed:

(*a*) By the heat output in steam delivered by the unit per hour. as represented by the difference between the total heat in the steam delivered and in the feed water as fed to the unit.

(*b*) By the actual evaporation in pounds of steam per hour at the reported steam pressure and quality or temperature of steam, and the reported feed-water temperature, and,

(*c*) To do away with the objection resulting from (*a*) to the use of large numbers. that the total heat output be expressed in "units of output." such units being determined by the total heat ouput, (*a*), divided by 1000.

The efficiency of a boiler or steam-generating unit as a whole, and the maximum capacity, either in terms of maximum "rating" or maximum heat output for the particular combustion apparatus with which it is installed, can be determined accurately only by a boiler test. Standard methods of conducting such tests are recommended by the subcommittee on "Test Code for Stationary Steam Generating Units" of the Power Test Code Committee of The American Society of Mechanical Engineers, presented in 1923.* A résumé of this code is given in the following pages. The code as presented makes many references to other sections of the complete Power Test Code, which are too long to be included here, and it is suggested that before undertaking serious testing work, not only the Code for Stationary Steam Generating Units but also the General Instructions, the Code on Definitions and Values, the Code on Instruments and Apparatus, and the Code on Fuels, as given in the complete Power Test Codes of The American Society of Mechanical Engineers, be carefully studied. The résumé following, however, gives the principal recommendations and methods to be followed.

OBJECT AND PREPARATIONS

Definitely determine the object of the test and keep it continuously in mind. Take the dimensions, note the physical conditions, examine for leakage, install the testing appliances, etc., and make preparations for the test accordingly.

Determine the character of the fuel to be used.

In guarantee tests with fuel of a specified calorific value or other characteristics, there should be a clear understanding as to the permissible variation from the specified characteristics.

In guarantee tests of a waste-heat boiler, the performance of which is dependent upon gas weight and entering gas temperature, there should be a clear understanding as to the method to be used in the determination of the gas weight and the measurement of the gas temperature.

INSTRUMENTS AND APPARATUS

The instruments and apparatus required for boiler tests are:

(*a*) Scales for weighing coal, oil or other fuel, ashes, furnace refuse, etc.
(*b*) Graduated scales attached to the water glasses.
(*c*) Tanks or tanks and scales for volumetric or weight measurement of water.
(*d*) Meters or other apparatus for measuring gaseous fuels.
(*e*) Pressure gauges, thermometers, pyrometers and draft gauges.
(*f*) Calorimeter for determining the quality of the steam.
(*g*) Gas-sampling and analyzing apparatus.

As no two plants are alike, it is quite impracticable to do more than suggest how and where the various instruments should be located.

(*a*) Fuel-weighing apparatus should be near the point where the fuel is to be used, under direct observation of and convenient to the one in charge of the test. The weighing of refuse may be done at any convenient point.

(*b*) The water-glass scale should be so attached that the breaking of the glass will not disturb the scale.

*These recommendations are those of the second preliminary report of 1923 and have not, at the time of going to press, been finally adopted by the Society as a whole.

(*c*) The water-weighing or measuring apparatus, if other than a meter, should be arranged in an easily accessible place and made as convenient as practicable. It should also be under the direct observation of the one in charge of the test.

(*d*) Meters or other apparatus for measuring gaseous fuel may be placed in such locations as conditions dictate.

(*e*) Pressure gauges should be away from any disturbing influences, such as extreme heat, and in a position to be easily read. Thermometers for determining atmospheric temperatures should be placed so as to give an average indication away from cold or hot drafts, etc.

A thermometer well for measuring feed-water temperatures should be as close to the boiler inlet as possible, and the pipe between boiler and thermometer should be well protected with heat-insulating covering.

Saturated-steam temperatures may be measured at any point in the steam pipe where the pressure is the same as that where the temperature is desired, care being taken that the well is not cooled by condensate. The temperature and pressure of superheated steam should be measured as close to the outlet of the superheater as possible, eliminating all losses by pressure drop and radiation between the superheater and thermometer.

Pyrometers of any type must have the part on which the heat impinges so located that the temperature which it is desired to measure is actually obtained. Much experience and good judgment are essential in order to obtain even approximately reliable readings.

Draft at the gas outlet is usually measured, if nowhere else It is often desirable to know the draft at other points of the setting, in which cases the several locations should be chosen to suit the information wanted. Care should be exercised to see that the ends of the connecting pipes are not obstructed and that the open ends are not subject to other than static pressures. All joints of draft-gauge connections must be tight.

The sampling tube of the calorimeter should be as close to the point at which the quality of steam is desired as possible. All exposed piping in which condensation might influence the result must be well lagged. Ordinarily the quality of the steam leaving the boiler, or that entering the superheater, is desired.

Analysis of the exit gases is usually what is taken, but very frequently analyses are wanted from other points. Sample tubes must be so located that only the gas to be analyzed enters the tube and precautions taken against the possibility of inleakage of air anywhere in the line. The gas-analyzing apparatus should be in a readily accessible and protected place with good light and convenient facilities.

OPERATING CONDITIONS

Determine what the operating conditions and method of firing should be to conform with the object in view, and see that they prevail as nearly as possible throughout the trial.

The duration of tests to determine the efficiency of a steam-generating unit burning coal, either hand or stoker-fired, should preferably be 24 hours, but where operating conditions do not permit or other considerations make it advisable or necessary the length of test may be reduced to not less than 10 hours. When the rate of combustion is less than 25 pounds of coal per square foot of grate surface per hour, the test should be continued until a total of 250 pounds per square foot of grate area has been burned, except that in cases where a type of stoker is used that does not permit the quantity of fuel and (or) the condition of the fuel bed to be accurately estimated, the duration of test should not be reduced below that required to minimize the error.

In the case of a unit using pulverized fuel, the duration of test should be not less than 6 hours, and for liquid or gaseous fuel not less than 4 hours.

In the case of a waste-heat unit, where the operation of the industrial furnace is continuous and furnace conditions are constant, the duration of the test should be not less than 6 hours. Where the industrial-furnace operation is in cycles, the test should be of such duration as to cover at least one cycle of furnace operation.

In tests conducted under plant-operation conditions, where the service requires continuous operation night and day, with frequent shifts of firemen, the duration of test should be at least 24 hours. Likewise in such tests, either of a single boiler or of a plant of several boilers which operate regularly a certain number of hours and during the balance of the day are banked, the duration should be not less than 24 hours.

The duration of tests to determine the maximum evaporative capacity of a steam-generating installation, when the efficiency is not determined, should be not less than two hours, unless otherwise agreed upon.

STARTING AND STOPPING

Combustion, fuel, draft and temperature conditions, the water level, rate of feeding water, rate of steaming, and the steam pressure should be as nearly as possible the same at the end as at the beginning of the test.

To secure the desired equality of conditions with hand-fired boilers, the following method should be employed:

The furnace being well heated by a preliminary run at the same combustion rate that will prevail during the test and sufficiently long to thoroughly heat the setting, burn the fire low and thoroughly clean it, leaving enough live coal spread evenly over the grate (say, from 2 to 4 inches) to serve as a foundation for the new fire. Note quickly the thickness of the coal bed as nearly as it can be estimated or measured, also the water level, the steam pressure, and the time, and record the latter as the starting time. Fresh coal should then be fired from that weighed for the test, the ashpit thoroughly cleaned, and the regular work of the test proceeded with.

Before the end of the test the fire should again be burned low and cleaned in such a manner as to leave the same amount of live coal on the grate as at the start. When this condition is reached, observe quickly the water level, the steam pressure, and the time, and record the latter as the stopping time. If the water level is lower or higher than at the beginning, a correction should be made by computation, not by feeding additional water. Finally remove the ashes and refuse from the ashpit.

In a plant containing several boilers, where it is not practicable to clean them simultaneously, the fires should be cleaned one after the other as rapidly as may be, and each one after cleaning charged with enough coal to maintain a thin fire in good working condition.

After the last fire is cleaned and in working condition, burn all the fires low (say, 4 to 6 inches), note quickly the thickness of each, also the water levels, steam pressure, and time, which last is taken as the starting time. Likewise, when the time arrives for closing the test, the fires should be quickly cleaned one by one, and when this work is completed they should all be burned low the same as at the start, and the various final observations made as before stated.

In the case of a large boiler having several furnace doors requiring the fire to be cleaned in sections one after the other, the above directions pertaining to starting and stopping in a plant of several boilers may be followed.

To obtain the desired equality of conditions when a mechanical stoker is used, the above procedure should be modified as follows:

Stokers should be in operation at approximately the same rate that will prevail during the test for at least 12 hours before starting test. At the start and finish of the test, level the coal in stoker hopper. Make starting and stopping observations as in hand-fired tests.

(*a*) *Stokers of the Continuous-Dumping Type:* The desired operating conditions, *i. e.*, speed and stroke of coal-feeding mechanism, speed of grate, intensity of draft or blast, and rate of water feed should be maintained as nearly constant as possible for at least one hour, and preferably for two hours, before starting and stopping test.

(*b*) *Stokers of the Intermittent-Dumping Type:* Proceed as above except that stokers should be cleaned about one hour before starting and before stopping the test.

To obtain the desired equality of conditions when pulverized, liquid, or gaseous fuel is used, the following method should be employed:

The boiler or boilers to be tested should be operated before the start of the test under the fuel, furnace, and combustion conditions which are to be maintained throughout the test for a period of not less than three hours. Fuel temperature, fuel pressure and draft conditions should be kept as nearly constant as possible during this period and throughout the test. The references above for hand-fired tests as to starting and stopping observations hold also for pulverized, liquid, and gaseous-fuel tests.

In the case of a waste-heat boiler set in connection with an industrial furnace the operation of which is continuous, the rules governing the starting and stopping of tests with pulverized, liquid, or gaseous fuels will apply. Where the industrial furnace operates in cycles, the start and end of the test should be at the same point of two or more successive cycles. Starting and stopping observations should be in accordance with the previous rules.

RECORDS

Readings of the instruments at 15-minute intervals are usually sufficient. If there are sudden and wide fluctuations, the readings should be taken at intervals of such frequency as may be necessary to determine the true nature of the variations.

The approximate quantity of fuel needed each hour should be determined, and if the quantity required and inconvenience due to lack of room are not too great, the whole amount should be delivered on the firing floor at the beginning of each hour. If any is left at the end of the hour, the quantity should be estimated. In this way the amount of fuel used per hour can be determined and the approximate results be followed from hour to hour.

If the whole amount cannot be delivered at the beginning of the hour, convenient quantities may be weighed out at appropriate intervals so as to come out about even at the end of the hour. When hopper scales are used, receiving coal from bunkers and discharging directly into furnace hoppers, hourly quantities may be roughly determined by estimating furnace conditions. These hourly quantities should be properly noted on the log sheet. They are taken as a matter of convenience and guide, but only the totals are to be used in the final calculation.

The records should be such as to ascertain also the approximate consumption of feed water each hour, and thereby determine the degree of uniformity of evaporation. The maintenance of a uniformity of evaporation is greatly facilitated by the use of some form of graphic recording steam meter, which should be so placed as to keep continuously before the operator the rate of evaporation. Since the indications of this meter are not used in any of the test calculations, extreme accuracy is not essential.

Quality of Steam—If the boiler does not produce superheated steam, the quality of the steam should be determined by the use of a throttling or separating calorimeter. If the boiler has superheating surface, the temperature of the steam should be determined by the use of a thermometer inserted in a thermometer well.

Ashes and Refuse—The ashes and refuse withdrawn from the furnace and ashpit during the progress of and at the end of the test should be weighed, so far as possible, in a dry state. If wet, the amount of moisture should be ascertained and allowed for, a sample being taken and dried for this purpose. This sample may also serve for analysis.

Ash sampling is at best subject to larger errors, and every precaution should be taken to insure as representative a sample as possible If sufficiently hot to allow combustion to proceed. ash should be thoroughly quenched with water immediately after dumping into the ashpit. Enough time should elapse before weighing to permit the heat of the refuse to drive out most of this moisture. If possible. all the ash should then be passed through a crusher. thoroughly mixed. and reduced to a laboratory-size sample by successive quartering. If it is impracticable to crush all the ash, the clinkers and fines should be placed in separate piles, and each pile weighed and sampled separately, the two samples being combined in proportion to the relative weights of their respective piles.

Sampling Fuel—During the progress of the test fuel should be regularly sampled for purposes of analysis.*

Calorific Tests and Analyses of Fuels—The quality of the fuel should be determined by calorific tests and analyses of the fuel sample above referred to and of the furnace ash and refuse.

Analyses of Flue Gases—For approximate determinations of the composition of the flue gases, the Orsat apparatus or some modification thereof should be employed. Gas samples should preferably be taken continuously, but, if momentary samples are obtained, the analyses should be made as frequently as possible, noting the furnace and firing conditions at the time the sample is drawn. Where the firing is intermittent, gas samples should be taken at such intervals that the complete firing cycle will be covered by the average of individual readings. If the sample drawn is a continuous one, the intervals may be made longer. Where the unit includes an economizer, gas analyses should be made at both boiler and economizer outlet.

Smoke Observations—In tests requiring a determination of the amount of smoke produced, observations should be made regularly throughout the trial at intervals of five minutes, or if necessary one minute. For observations covering a period of one or more single firings with solid fuel, the intervals should be one-quarter minute or less.

Chart—In trials having for an object the determination and exposition of the complete boiler performance, the entire log of readings and data should be represented graphically.

It is recommended that a record be kept of the energy used by auxiliaries immediately connected with the steam-generating unit being tested and a specific note made thereof; no deductions, however, shall be made on the report forms or in computing results unless the object of the test so requires, in which case the report should specifically so state. This applies to steam or power used in driving stokers or other fuel-feeding apparatus, oil burners, fans, feed pumps, soot blowers, etc.

DATA AND RESULTS

The data and results should be reported in accordance with the tables given herewith, using the proper table for the class of fuel burned. If necessary, items of data not provided for may be added, or if certain items are not required such may be

*The matter of coal sampling is of so great importance in the determination of results that the standard method for sampling of coal as given by the American Society for Testing Materials, which is in agreement with the methods recommended by the Fuel Code of The American Society of Mechanical Engineers Power Test Code, is given in full at the end of this chapter

omitted as may conform to the object of the test. Unless otherwise indicated, the quantities should be the average of the observations.

DATA AND RESULTS OF TEST OF STATIONARY STEAM-GENERATING UNIT

GENERAL INFORMATION

1 Date of test . . . .
2 Location of plant . . . .
3 Owner . . . .
4 Maker of boiler . . . .
5 Test conducted by . . . .
6 Object of test . . . .

DESCRIPTIONS. DIMENSIONS, ETC.

7 Grate . . . .
8 Superheater . . . .
9 Economizer . . . .
10 Draft . . . .
11 Fuel . . . .
12 Boiler heating surface . . . . square feet
13 Superheating surface . . . . square feet
14 Grate area . . . . square feet
15 Volume of combustion space . . . . cubic feet
16 Furnace, center of grate to nearest heating surface . . . . feet
17 Furnace volume per square foot of heating surface . . . . cubic feet

FUEL AND GAS ANALYSES AND DATA

Fuel, Approximate Analysis

18 Volatile matter . . . . per cent
19 Fixed carbon . . . . per cent
20 Ash . . . . per cent
21 Moisture (as fired) . . . . per cent
22 B.t.u. per pound (as fired) . . . .
23 B.t.u. per pound, dry . . . .
24 Fusion temperature of ash . . . . degree Fahrenheit

Fuel, Ultimate Analysis

25 Carbon . . . . per cent
26 Hydrogen . . . . per cent
27 Oxygen . . . . per cent
28 Nitrogen . . . . per cent
29 Sulphur . . . . per cent
30 Ash . . . . per cent

Gas

31 Furnace-gas analysis:
Per cent of CO_2 . . O_2 . . N_2 . . CO . . SO_2 . .
32 Flue-gas analysis, boiler outlet:
Per cent of CO_2 . O_2 . . N_2 . . CO . . SO_2 . .
32a Flue-gas analysis, economizer outlet:
Per cent of CO_2 . . O_2 . . N_2 . . CO . . SO_2 . .
33 Dry gas per pound of fuel, furnace (as fired. dry) . . . . pounds
34 Dry gas per pound of fuel, boiler outlet (as fired, dry) . . . . pounds
34a Dry gas per pound of fuel, economizer outlet (as fired, dry) . . . . pounds
35 Dry gas per pound of fuel, theoretical . . . . pounds
36 Air supplied per pound of fuel (as fired, dry) . . . . pounds

TEST DATA AND RESULTS

Pressures and Drafts

No.	Item	Unit
37	Moisture in air	pound per pound of air
38	Steam pressure by gauge:	
	Boiler	pounds per square inch
	Superheater outlet	pounds per square inch
39	Air pressure in ashpit zone	inch of water
40	Draft in furnace	inch of water
41	Draft under boiler damper	inch of water
42	Draft at economizer outlet	inch of water

Temperatures

No.	Item	Unit
43	Steam temperature	degrees Fahrenheit
44	Moisture in steam	per cent
45	Superheat	degrees Fahrenheit
46	Temperature of feed water entering boiler	degrees Fahrenheit
47	Temperature of feed water entering economizer	degrees Fahrenheit
48	Furnace temperature	degrees Fahrenheit
49	Temperature of gases leaving boiler	degrees Fahrenheit
50	Temperature of gases leaving economizer	degrees Fahrenheit
51	Temperature of air for combustion	degrees Fahrenheit

Hourly Quantities

No.	Item	Unit
52	Duration of test	hours
53	Fuel as fired per hour	pounds
54	Dry fuel per hour	pounds
55	Fuel as fired per hour per square foot of grate	pounds
56	Fuel as fired per hour per retort	pounds
57	Dry fuel per square foot of grate per hour	pounds
58	Dry fuel per retort per hour	pounds
59	Combustion space per pound of coal per hour (as fired, dry)	cubic feet
60	Refuse per hour	pounds
61	Actual water per hour	pounds
62	Factor of evaporation	
63	Equivalent evaporation per hour	pounds

Refuse

No.	Item	Unit
64	Percentage of refuse in fuel (as fired, dry)	per cent
65	Percentage of combustible in refuse	per cent
66	Carbon burned per pound of fuel (as fired, dry)	pounds

Evaporation

No.	Item	Unit
67	Actual evaporation per pound of fuel as fired	pounds
68	Equivalent evaporation per pound of fuel as fired	pounds
69	Equivalent evaporation per pound of dry fuel	pounds
70	Equivalent evaporation per square foot of boiler heating surface per hour	pounds

Efficiency

No.	Item	Unit
71	Efficiency of boiler, superheater, furnace and grate	per cent
72	Efficiency including economizer	per cent

Heat Balance (based on dry fuel—fuel as fired)

No.	Item		
73	Heat absorbed by boiler and superheater	B.t.u.	per cent
74	Heat absorbed by economizer	B.t.u.	per cent
75	Loss due to moisture in fuel	B.t.u.	per cent
76	Loss due to moisture formed in burning H_2	B.t.u.	per cent
77	Loss due to moisture in air	B.t.u.	per cent
78	Loss due to chimney gases	B.t.u.	per cent
79	Loss due to incomplete combustion of carbon	B.t.u.	per cent
80	Loss due to unconsumed combustible in refuse	B.t.u.	per cent
81	Radiation and unaccounted for losses	B.t.u.	per cent

The forms for tests with liquid and with gaseous fuels are similar to the form above, with such items omitted or added as may be necessary because of methods of combustion used and different characteristics of the fuel. While tests with gaseous fuels may be reported on either a volumetric or weight basis, it is better, in order to make the test basis for all fuels in agreement, to convert volumetric fuel gas analyses to a weight basis and report results on such basis.

STANDARD METHOD OF SAMPLING COAL

(American Society for Testing Materials)

It is imperative that every sample be collected and prepared carefully and conscientiously and in strict accordance with the standard methods described herein, for if the sampling is improperly done, the sample will be in error, and it may be impossible or impracticable to take another sample; but if an analysis is in error, another analysis can easily be made of the original sample.

Gross samples of the quantities designated herein must be taken, whether the coal to be sampled consists of a few tons or several hundred tons, because of the following cardinal principle in sampling coal that must be recognized and understood: that is, the effect of the chance inclusion or exclusion of too many or too few pieces of slate or other impurities in what, or from what, would otherwise have been a representative sample will cause the analysis to be in error accordingly, regardless of the tonnage sampled. For example, the chance inclusion or exclusion of 10 pounds too much or too little of impurities in or from an otherwise representative sample of 100 pounds would cause the analysis to show an error in ash content and in heat units of approximately 10 per cent, whereas for a 1000-pound sample the effect would be approximately only 1 per cent, being the same whether the sample is collected from a 1-ton lot or from a lot consisting of several hundred tons.

When this method of sampling is to be employed, as a part of any contract or agreement, the following provisions shall be specifically agreed to by the parties to such contract or agreement:

(*a*) The place at which the coal is to be sampled (see Section 1).

(*b*) The approximate size of the sample required when the standard conditions do not apply (see Section 3).

(*c*) The number of samples to be taken or the amount of coal to be represented by each sample when the standard conditions do not apply (see Section 4).

For All Determinations Except Total Moisture

Time of Sampling.

1. The coal shall be sampled when it is being loaded into or unloaded from railroad cars, ships, barges, or wagons, or when discharged from supply bins, or from industrial railway cars, or grab buckets, or from any coal-conveying equipment, as the case may be. If the coal is crushed as received, samples usually can be taken advantageously after the coal has passed through the crusher. Samples collected from the surface of coal in piles or bins, or in cars, ships or barges are generally unreliable.

Size of Increments.

2. To collect samples, a shovel or specially designed tool or mechanical means shall be used for taking equal portions or increments. For slack or small sizes of

anthracite, increments as small as 5 to 10 pounds may be taken, but for run-of-mine or lump coal, increments should be at least 10 to 30 pounds.

Collection of Gross Samples.

3. The increments shall be regularly and systematically collected, so that the entire quantity of coal sampled will be represented proportionately in the gross sample, and with such frequency that a gross sample of the required amount shall be collected. The standard gross sample shall not be less than 1000 pounds, except that for slack coal and small sizes of anthracite in which the impurities do not exist in abnormal quantities or in pieces larger than ¾ inch, a gross sample of approximately 500 pounds shall be considered sufficient. If the coal contains an unusual amount of impurities, such as slate, and if the pieces of such impurities are very large, a gross sample of 1500 pounds or more shall be collected. The gross sample should contain the same proportion of lump coal, fine coal, and impurities as is contained in the coal sampled. When coal is extremely lumpy, it is best to break a proportional amount of the lumps before taking the various increments of a sample. Provision should be made for the preservation of the integrity of the sample.

Quality Represented.

4. A gross sample shall be taken for each 500 tons or less, or in case of larger tonnages, for such quantities as may be agreed upon.

Crushing.

5. After the gross sample has been collected, it shall be systematically crushed, mixed, and reduced in quantity to convenient size for transmittal to the laboratory. The sample may be crushed by hand or by any mechanical means, but under such conditions as shall prevent loss or the accidental admixture of foreign matter. Samples of the quantities indicated in the table shall be crushed so that no pieces of coal and impurities will be greater in any dimension, as judged by eye, than specified for the sample before division into two approximately equal parts.

Weight of Sample to be Divided, Pounds	Largest Size of Coal and Impurities Allowable in Sample before Division, Inches
1000 or over	1
500	¾
250	½
125	⅜
60	¼
30	$\frac{3}{16}$ or 4-mesh screen

The method of reducing by hand the quantity of coal in a gross sample shall be carried out as prescribed in Section 6, even should the initial size of coal and impurities be less than indicated in the table.

Hand Preparation.

6. The progressive reduction in the weight of the sample to the quantities indicated in the table shall be done by the following methods:

Mixing and Reduction by Discarding Alternate Shovelfuls.

(*a*) The alternate-shovel method of reducing the gross sample shall be repeated until the sample is reduced to approximately 250 pounds, and care shall be observed before each reduction in quantity that the sample has been crushed to the fineness prescribed in the table.

The crushed coal shall be shoveled into a conical pile by depositing each shovelful of coal on top of the preceding one, and then formed into a long pile in the following manner: The sampler shall take a shovelful of coal from the conical pile and spread it out in a straight line, having a width equal to the width of the shovel and a length of 5 to 10 feet. His next shovelful shall be spread directly over the top of the first shovelful, but in the opposite direction, and so on back and forth, the pile being occasionally flattened, until all the coal has been formed into one long pile. The sampler shall then discard half of this pile, proceeding as follows:

Beginning on one side of the pile, at either end, and shoveling from the bottom of the pile, the sampler shall take one shovelful and set it aside; advancing along the side of the pile a distance equal to the width of the shovel, he shall take a second shovelful and discard it; again advancing in the same direction one shovel width, he shall take a third shovelful and add it to the first. The fourth shall be taken in a like manner and discarded, the fifth retained, and so on, the sampler advancing always in the same direction around the pile so that its size will be gradually reduced in a uniform manner. When the pile is removed, about half of the original quantity of coal should be contained in the new pile formed by the alternate shovelfuls which have been retained.

Mixing and Reduction by Quartering.

(*b*) After the gross sample has been reduced by the above method to approximately 250 pounds, further reduction in quantity shall be by the quartering method. Before each quartering, the sample shall be crushed to the fineness prescribed in the table.

Quantities of 125 to 250 pounds shall be thoroughly mixed by coning and reconing; quantities less than 125 pounds shall be placed on a suitable cloth, measuring about 6 by 8 feet, mixed by raising first one end of the cloth and then the other, so as to roll the coal back and forth, and after being thoroughly mixed shall be formed into a conical pile by gathering together the four corners of the cloth. The quartering of the conical pile shall be done as follows:

The cone shall be flattened, its apex being pressed vertically down with a shovel or board, so that after the pile has been quartered, each quarter will contain the material originally in it. The flattened mass, which shall be of uniform thickness and diameter, shall then be marked into quarters by two lines that intersect at right angles directly under a point corresponding to the apex of the original cone. The diagonally opposite quarters shall then be shoveled away and discarded and the space that they occupied brushed clean. The coal remaining shall be successively crushed, mixed, coned and quartered until the sample is reduced to the desired quantity.

(*c*) The 30-pound quantity shall be crushed to $\frac{3}{16}$ inch or 4-mesh size, mixed, coned, flattened and quartered. The laboratory samples shall include all of one of the quarters, or all of two opposite quarters, as may be required. The laboratory sample shall be immediately placed in a suitable container and sealed in such a manner as to preclude tampering.

Mechanical Preparation.

7. Only such mechanical means as will give equally representative samples shall be used in substitution for the hand method of preparation herein standardized.

II. For the Determination of Total Moisture

Collection of Moisture Sample

8. The special moisture sample shall weigh approximately 100 pounds, and shall be accumulated by placing in a waterproof receptacle with a tight-fitting and waterproof lid small equal parts of freshly taken increments of the standard gross sample. The accumulated moisture sample shall be rapidly crushed and reduced mechanically or by hand to about a 5-pound quantity, which shall be immediately placed in a container and sealed air tight and forwarded to the laboratory without delay.

Use of Standard Gross Sample

9. Only when equally representative results will be obtained shall the standard gross sample be used instead of the special moisture sample for the determination of total moisture.

South Boston Station of the Boston Elevated Ry. Co., Boston, Mass. 15,600 Horse-power of Babcock & Wilcox Boilers and Superheaters Installed in this Station

THE SELECTION OF BOILERS WITH A CONSIDERATION OF THE FACTORS DETERMINING SUCH SELECTION

THE selection of steam boilers is a matter to which the most careful thought and attention may be well given. Within the last twenty years, radical changes have taken place in the methods and appliances for the generation and distribution of power. These changes have been made largely in the prime movers, both as to type and size, and are best illustrated by the changes in central station power-plant practice. It is hardly within the scope of this work to treat of power-plant design and the discussion will be limited to a consideration of the boiler end of the power plant.

As stated, the changes have been largely in prime movers, the steam-generating equipment having been considered more or less a standard piece of apparatus whose sole function is the transfer of the heat liberated from the fuel by combustion to the water stored or circulated in such apparatus. When the fact is considered that the cost of steam generation is roughly from 65 to 80 per cent of the total cost of power production, it may be readily understood that the most fruitful field for improvement exists in the boiler end of the power plant. The efficiency of the plant as a whole will vary with the load it carries, and it is in the boiler room where such variation is largest and most subject to control.

The improvements to be secured in the boiler-room results are not simply a matter of dictation of operating methods. The securing of perfect combustion, with the accompanying efficiency of heat transfer, while comparatively simple in theory, is difficult to obtain in practical operation. This fact is perhaps best exemplified by the difference between test results and those obtained in daily operation even under the most careful supervision. This difference makes it necessary to establish a standard by which operating results may be judged, a standard not necessarily that which might be possible under test conditions, but one which experiment shows can be secured under the very best operating conditions.

The study of the theory of combustion, draft, etc., as already given, will indicate that the question of efficiency is largely a matter of proper relation between fuel, furnace and generator. While the possibility of a substantial saving through added efficiency cannot be overlooked, the boiler design of the future must, even more than in the past, be considered particularly from the aspect of reliability and simplicity. A flexibility of operation is necessary as a guarantee of continuity of service.

In view of the above, before the question of the selection of boilers can be taken up intelligently, it is necessary to consider the subjects of boiler efficiency and boiler capacity, together with their relation to each other.

The criterion by which the efficiency of a boiler plant is to be judged is the cost of the production of a definite amount of steam. Considered in this sense, there must be included in the efficiency of a boiler plant the simplicity of operation, flexibility and reliability of the boiler used. The items of repair and upkeep cost are often high because of the nature of the service. The governing factor in these items is unquestionably the type of boiler selected.

West End Station of the Union Gas & Electric Company of Cincinnati, which has 20,27[illegible] Horse Power of Babcock & Wilcox Boilers
Equipped with Babcock & Wilcox Superheaters

The features entering into the plant efficiency are so numerous that it is impossible to make a statement as to a means of securing the highest efficiency which will apply to all cases. Such efficiency is to be secured by the proper relation of fuel, furnace and boiler heating surface, actual operating conditions, which allow the approaching of the potential efficiencies made possible by the refinement of design, and a systematic supervision of the operation assisted by a detailed record of performances and conditions. The question of supervision will be taken up later in the chapter on "Operation and Care of Boilers."

The possible number of combinations of different arrangements of boiler heating surface, different furnaces and different coals is so great and the efficiency varies so appreciably at different rates of evaporation that no table of efficiencies can be included here which would cover comprehensively all combinations and conditions. It is to be remembered that published boiler-performance figures are ordinarily performances under test conditions which are usually better than plant-operating conditions. Further, under test conditions it is frequently possible to operate for the duration of a test under conditions as to furnace temperature, etc., that would not represent the highest commercial efficiency in ordinary operation. For the highest thermal efficiency, the furnace temperature should be maintained at a maximum. Such a maximum temperature, however, combined with certain draft conditions that might be allowable for the duration of a test, might lead to excessive furnace brickwork maintenance, and under such conditions the maximum thermal efficiency would not represent the maximum commercial efficiency. The closeness with which operating conditions approach test conditions is dependent upon the character of the boiler-room supervision and the intelligence of the operating crew.

The bearing that the type of boiler has on the efficiency to be expected can only be realized from a study of the foregoing chapters.

CAPACITY—Capacity, as already defined, is the ability of a definite amount of boiler heating surface to generate steam. Boilers are ordinarily purchased under a manufacturer's specification, which rates a boiler at a nominal rated horse-power, usually based on 10 square feet of heating surface per horse-power. Such a builders' rating is absolutely arbitrary and implies nothing as to the limiting amount of water that his amount of heating surface will evaporate. It does not imply that the evaporation of 34.5 pounds of water from and at 212 degrees with 10 square feet of heating surface is the limit of the capacity of the boiler. Further, from a statement that a boiler is of a certain horse-power on the manufacturer's basis, it is not to be understood that the boiler is in any state of strain when developing more than its rated capacity.

Broadly stated, the evaporative capacity of a certain amount of heating surface in a well-designed boiler, that is, the boiler horse-power it is capable of producing, is limited only by the amount of fuel that can be burned under the boiler. While such a statement would imply that the question of capacity to be secured was simply one of making an arrangement by which sufficient fuel could be burned under a definite amount of heating surface to generate the required amount of steam, there are limiting features that must be weighed against the advantages of high capacity developed from small heating surfaces. Briefly stated, these factors are as follow:

1st. Efficiency. As the capacity increases, there will, in general, be a decrease in efficiency, this loss above a certain point making it inadvisable to try to secure more than a definite horse-power from a given boiler. This loss of efficiency with increased capacity is treated below in detail, in considering the relation of efficiency to capacity.

2nd. Grate Ratio Possible or Practicable. All fuels have a maximum rate of combustion, beyond which satisfactory results cannot be obtained, regardless of draft available or which may be secured by mechanical means. Such being the case, it is evident that with this maximum combustion rate secured, the only method of obtaining added capacity will be through the addition of grate surface. There is obviously a point beyond which the grate surface for a given boiler cannot be increased. This is due to the impracticability of handling grates above a certain maximum size, to the enormous loss in draft pressure through a boiler resulting from an attempt to force an abnormal quantity of gas through the heating surface and to innumerable details of design and maintenance that would make such an arrangement wholly unfeasible.

3rd. Feed Water. The difficulties that may arise through the use of poor feed water or that are liable to happen through the use of practically any feed water have already been pointed out. This question of feed is frequently the limiting factor in the capacity obtainable, for with an increase in such capacity comes an added concentration of such ingredients in the feed water as will cause priming, foaming or rapid scale formation. Certain waters which will give no trouble that cannot be readily overcome with the boiler run at ordinary ratings will cause difficulties at higher ratings entirely out of proportion to any advantage secured by an increase in the power that a definite amount of heating surface may be made to produce.

Where capacity in the sense of overload is desired, the type of boiler selected will play a large part in the successful operation through such periods. A boiler must be selected with which there is possible a furnace arrangement that will give flexibility without undue loss in efficiency over the range of capacity desired. The heating surface must be so arranged that it will be possible to install in a practical manner sufficient grate surface at or below the maximum combustion rate to develop the amount of power required. The design of boiler must be such that there will be no priming or foaming at high overloads and that any added scale formation due to such overloads may be easily removed. Certain boilers which deliver commercially dry steam when operated at about their normal rated capacity will prime badly when run at overloads, and this action may take place with a water that should be easily handled by a properly designed boiler at any reasonable load Such action is ordinarily produced by the lack of a well-defined, positive circulation.

Relation of Efficiency and Capacity—The statement has been made that in general the efficiency of a boiler will decrease as the capacity is increased. Considering the boiler alone, apart from the furnace, this statement may be readily explained.

Presupposing a constant furnace temperature, regardless of the capacity at which a given boiler is run; to assure equal efficiencies at low and high ratings, the exit temperature in the two instances would necessarily be the same. For this temperature at the high rating to be identical with that at the low rating, the rate of heat transfer from the gases to the heating surfaces would have to vary directly as the weight or volume of such gases. Experiment has shown, however, that this is not true, but that this rate of transfer varies as some power of the volume of gas less than one. As the heat transfer does not, therefore, increase proportionately with the volume of gases, the exit temperature for a given furnace temperature will be increased as the volume of gases increases. As this is the measure of the efficiency of the heating surface. the boiler efficiency will, therefore, decrease as the volume of gases increases or the capacity at which the boiler is operated increases.

Further, a certain portion of the heat absorbed by the heating surface is through direct radiation from the fire. Again, presupposing a constant furnace temperature; the heat absorbed through radiation is solely a function of the amount of surface exposed to such radiation. Hence, for the conditions assumed, the amount of heat absorbed by radiation at the higher ratings will be the same as at the lower ratings, but in proportion to the total absorption will be less. As the added volume of gas does not increase the rate of heat transfer sufficiently to offset the constant absorption by radiation, there are therefore two factors acting toward the decrease in the efficiency of a boiler with an increase in the capacity.

This increase in the efficiency of the boiler alone with the decrease in the rate at which it is operated will hold to a point where the radiation of heat from the boiler setting is proportionately large enough to be a governing factor in the total amount of heat absorbed.

The second reason given above for a decrease of boiler efficiency with increase of capacity, viz., the effect of radiant heat, is to a greater extent than the first reason dependent upon a constant furnace temperature. Any increase in this temperature will affect enormously the amount of heat absorbed by radiation, as this absorption will vary as the fourth power of the temperature of the radiating body. In this way it is seen that but a slight increase in furnace temperature will be necessary to bring the proportional part, due to absorption by radiation, of the total heat absorbed, up to its proper proportion at the higher ratings. This factor of furnace temperature more properly belongs to the consideration of furnace efficiency than of boiler efficiency. There is a point, however, in any furnace above which the combustion will be so poor as actually to reduce the furnace temperature and, therefore, the proportion of heat absorbed through radiation by a given amount of exposed heating surface.

Since it is thus true that the efficiency of the boiler considered alone will increase with a decreased capacity, it is evident that, if the furnace conditions are constant regardless of the load, the combined efficiency of boiler and furnace will also decrease with increasing loads. This fact was clearly proven in tests of the boilers at the Detroit Edison Company.* The furnace arrangement of these boilers and the great care with which the tests were run made it possible to secure uniformly good furnace conditions irrespective of load, and here the maximum efficiency was obtained at a point somewhat less than the rated capacity of the boilers.

In some cases, however, particularly where the combined boiler, furnace and stoker are designed for high capacities, the furnace efficiency will, up to a certain point, increase with an increase in power. This increase in furnace efficiency is ordinarily at a greater rate as the capacity increases than is the decrease in boiler efficiency, with the result that the combined efficiency of boiler and furnace will, to a certain point, increase with an increase in capacity. This makes the ordinary point of maximum combined efficiency somewhat above the rated capacity of the boiler, and in many cases the combined efficiency will be practically a constant over a considerable range of ratings. The features limiting the establishing of the point of maximum efficiency at a high rating are the same as those limiting the amount of grate surface

*See Transactions A.S.M.E., Volume XXXIII, 1912.

that can be installed under a boiler. The relative efficiency of different combinations of boilers and furnaces at different ratings depends so largely upon the furnace conditions that what might hold for one combination would not for another.

In view of the above, it is impossible to make a statement of the efficiency at different capacities of a boiler and furnace which will hold for any and all conditions.

Economical Loads—With the effect of capacity on economy in mind, the question arises as to what constitutes the economical load to be carried. In figuring on the economical load for an individual plant, the broader economy is to be considered, that in which, against the boiler efficiency, there are to be weighed the plant first cost, returns on such investment, fuel cost, labor, capacity, etc. This matter has been widely discussed, but unfortunately such discussion has been largely limited to central power station practice. The power generated in such stations, while representing an enormous total, is by no means the larger proportion of the total power generated throughout the country. The factors determining the economic load for the small plant, however, are the same as in the large, and in general the statements made relative to the question are equally applicable.

The economical rating at which a boiler plant should be run is dependent upon the load to be carried by that individual plant and the nature of such load. The economical rating is also influenced by the presence or absence of economizers. The economical load for each individual plant can be determined only from the careful study of each individual set of conditions or by actual trial.

The controlling factor in the cost of the plant, regardless of the nature of the load, is the capacity to carry the maximum peak load that may be thrown on the plant under any conditions.

While load conditions do, as stated, vary in every individual plant, in a broad sense all loads may be grouped into three classes: 1st, the approximately constant 24-hour load; 2nd, the steady 10- or 12-hour load usually with a noon-day period of no load; 3rd, the 24-hour variable load, found in central station practice. The economical load at which the boiler may be run will vary with these groups:

1st. For a constant load, 24 hours in the day, it will be found in most cases that, when all features are considered, the most economical load or that at which a given amount of steam can be produced the most cheaply will be considerably over the rated horse-power of the boiler. How much above the rated capacity this most economic load will be is dependent largely upon the cost of coal at the plant, but under ordinary conditions, the point of maximum economy will probably be found to be somewhere between 25 and 50 per cent above the rated capacity of the boilers. Where the boilers are equipped with economizers, the economical commercial capacity will be still higher than these ratings; how much higher is dependent upon the ratio of economizer surface installed to the boiler surface. The capital investment must be weighed against the coal saving through increased thermal efficiency, and the labor account, which increases with the number of units, must be given proper consideration. When the question is considered in connection with a plant already installed, the conditions are different from where a new plant is contemplated. In an old plant, where there are enough boilers to operate at low rates of capacity, the capital investment leads to a fixed charge, and it will be found that the most economical load at which boilers may be operated will be lower than where a new plant is under consideration.

OPERATION AND CARE OF BOILERS

THE general subject of boiler-room practice may be considered from two aspects. The first is that of the broad plant economy, with a suggestion as to the methods to be followed in securing the best economical results with the apparatus at hand and procurable. The second deals rather with specific recommendations which should be followed in plant practice, recommendations leading not only to economy but also to safety and continuity of service. Such recommendations are dictated from an understanding of the nature of steam-generating apparatus and its operation, as covered previously in this book.

It has already been pointed out that the attention given in recent years to steam-generating practice has come with a realization of the wide difference existing between the results being obtained in every-day operation and those theoretically possible. The amount of such attention and regulation given to the steam-generating end of a power plant, however, is comparatively small in relation to that given to the balance of the plant, but it may be safely stated that it is here that there is the greatest assurance of a return for the attention given. The statement holds to a greater extent in the small plant than in the larger stations which have come generally to a realization of the necessity for boiler-room economy.

In the endeavor to increase boiler-room efficiency, it is of the utmost importance that a standard basis be set by which average results are to be judged. With the theoretical efficiency obtainable varying so widely this standard cannot be placed at the highest efficiency that has been obtained regardless of operating conditions. It is better set at the best obtainable results for each individual plant under its conditions of installation and daily operation.

With an individual standard so set, present practice can only be improved by a systematic effort to approach this standard. The degree with which operating results will approximate such a standard will be found to be directly proportional to the amount of intelligent supervision given the operation. For such supervision to be given, it is necessary to have not only a full realization of what the plant can do under the best operating conditions but also a full and complete knowledge of what it is doing under all of the different conditions that may arise. What the plant is doing should be made a matter of continuous record so arranged that the results may be directly compared for any period or set of conditions, and where such results vary from the standard set, steps must be taken immediately to remedy the causes of such failure. Such a record is an important check on the losses in the plant.

As the size of the plant and the fuel consumption increase, such a check of losses and recording of results become a necessity. In the larger plants the saving of but a fraction of one per cent in the fuel bill represents an amount running into thousands of dollars annually, while the expense of the proper supervision to secure such saving is small. The methods of supervision followed in the large plants are necessarily elaborate and complete. In the smaller plants the same methods may be followed on a more moderate scale with a corresponding saving in fuel and an inappreciable increase in either plant organization or expense.

There has been within the last few years a great increase in the practicability and reliability of the various types of apparatus by which the records of plant operation may be secured. Much of this apparatus is ingenious and, considering the work

Portion of 4160 Horse-power Installation of Babcock & Wilcox Boilers at the Prudential Life Insurance Co. Building, Newark, N. J.

boilers are cut in to take care of increasing loads and peaks and placed again on bank when the peak periods have passed.

Other conditions of operation make it advisable to carry the load on a definite number of boiler units, operating these at slightly below their rated capacity during periods of light or low loads and securing the overload capacity during peaks by operating the same boilers at high ratings. In this method there are no boilers kept on banked fires, the spares being spares in every sense of the word.

A third method of handling widely varying loads which is coming somewhat into vogue is that of considering the plant as divided, one part to take care of what may be considered the constant plant load, the other to take care of the floating or variable load. With such a method, that portion of the plant carrying the steady load is so proportioned that the boilers may be operated at the point of maximum efficiency, this point being raised to a maximum through the use of economizers and the general installation of any apparatus leading to such results. The variable load will be carried on the remaining boilers of the plant under either of the methods just given, that is. at the high ratings of all boilers in service and banking others, or a variable capacity from all boilers in service.

This method is ordinarily followed to an extended degree where there are a number of plants interconnected in a single system. Under such conditions, one or more of the stations in the system are considered as base-load stations and are given such portion of the total load as will enable them to be operated with a high load factor. These base-load plants are equipped with highly efficient apparatus and are operated at a capacity at which the commercial operating efficiency closely approaches the maximum thermal efficiency. The older stations in the system or those not equipped for the highest practicable efficiencies carry the variations in the load.

The opportunity is again taken to indicate the very general character of any statements made relative to the economical load for any plant and to emphasize the fact that each individual case must be considered independently, with the conditions of operation applicable thereto.

With a thorough understanding of the meaning of boiler efficiency and capacity and their relation to each other, it is possible to consider more specifically the selection of boilers.

The foremost consideration is, without question, the adaptability of the design selected to the nature of the work to be done. An installation which is only temporary in its nature would obviously not warrant the first cost that a permanent plant would. If boilers are to carry an intermittent and suddenly fluctuating load, such as a hoisting load or a reversing mill load, a design would have to be selected that would not tend to prime with the fluctuations and sudden demand for steam. A boiler that would give the highest possible efficiency with fuel of one description would not of necessity give such efficiency with a different fuel. A boiler of a certain design which might be good for small-plant practice would not, because of the limitations in practicable size of units, be suitable for large installations. A discussion of the relative value of designs can be carried on almost indefinitely, but enough has been said to indicate that a given design will not serve satisfactorily under all conditions and that the adaptability to the service required will be dependent upon the fuel available, the class of labor procurable, the feed water that must be used, the nature of the plant's load, the size of the plant and the first cost warranted by the service the boiler is to fulfil.

TABLE 59

ACTUAL EVAPORATION FOR DIFFERENT PRESSURES AND TEMPERATURES OF FEED

WATER CORRESPONDING TO ONE HORSE-POWER (34½ POUNDS PER HOUR FROM AND AT 212 DEGREES FAHRENHEIT)

Temperature of Feed Degrees Fahrenheit	Pressure by Gauge—Pounds per Square Inch																				
	50	60	70	80	90	100	110	120	130	140	150	160	170	180	190	200	210	220	230	240	250
32	28.41	28.36	28.29	28.24	28.20	28.16	28.13	28.09	28.07	28.04	28.02	27.99	27.97	27.95	27.94	27.92	27.90	27.89	27.87	27.86	27.85
40	28.61	28.54	28.49	28.44	28.40	28.35	28.32	28.29	28.26	28.23	28.21	28.18	28.16	28.14	28.12	28.11	28.09	28.07	28.06	28.05	28.03
50	28.85	28.79	28.73	28.68	28.64	28.60	28.56	28.53	28.50	28.47	28.45	28.43	28.40	28.38	28.36	28.35	28.33	28.31	28.30	28.28	28.27
60	29.10	29.04	28.98	28.93	28.88	28.84	28.81	28.77	28.74	28.72	28.69	28.67	28.65	28.62	28.60	28.59	28.57	28.55	28.54	28.52	28.51
70	29.36	29.29	29.23	29.18	29.14	29.09	29.06	29.02	28.99	28.96	28.94	28.92	28.89	28.87	28.85	28.83	28.82	28.80	28.78	28.77	28.76
80	29.62	29.55	29.49	29.44	29.39	29.35	29.31	29.27	29.24	29.22	29.19	29.17	29.14	29.12	29.10	29.08	29.07	29.05	29.03	29.02	29.00
90	29.88	29.81	29.75	29.70	29.65	29.61	29.57	29.53	29.50	29.47	29.45	29.42	29.40	29.38	29.36	29.34	29.32	29.30	29.29	29.27	29.25
100	30.15	30.08	30.02	29.96	29.91	29.87	29.83	29.80	29.76	29.73	29.71	29.68	29.66	29.63	29.61	29.60	29.58	29.56	29.54	29.53	29.51
110	30.42	30.35	30.29	30.23	30.18	30.14	30.10	30.06	30.03	30.00	29.97	29.95	29.92	29.90	29.88	29.86	29.84	29.82	29.81	29.79	29.77
120	30.70	30.63	30.56	30.51	30.46	30.41	30.37	30.33	30.30	30.27	30.24	30.22	30.19	30.17	30.15	30.13	30.11	30.09	30.07	30.06	30.04
130	30.99	30.91	30.84	30.79	30.73	30.69	30.65	30.61	30.57	30.54	30.52	30.49	30.47	30.44	30.42	30.40	30.38	30.36	30.35	30.33	30.31
140	31.28	31.20	31.13	31.07	31.02	30.97	30.93	30.89	30.86	30.83	30.80	30.77	30.75	30.72	30.70	30.68	30.66	30.64	30.62	30.61	30.59
150	31.58	31.49	31.42	31.36	31.31	31.26	31.22	31.18	31.14	31.11	31.08	31.06	31.03	31.01	30.98	30.96	30.94	30.92	30.91	30.89	30.87
160	31.87	31.79	31.72	31.66	31.61	31.56	31.51	31.47	31.44	31.40	31.37	31.35	31.32	31.29	31.27	31.25	31.23	31.21	31.19	31.18	31.16
170	32.18	32.10	32.02	31.96	31.91	31.86	31.81	31.77	31.73	31.70	31.67	31.64	31.62	31.59	31.57	31.54	31.52	31.50	31.49	31.47	31.46
180	32.49	32.41	32.33	32.27	32.22	32.16	32.12	32.08	32.04	32.00	31.97	31.95	31.92	31.89	31.87	31.84	31.82	31.80	31.79	31.77	31.75
190	32.81	32.72	32.65	32.59	32.53	32.47	32.43	32.38	32.35	32.32	32.29	32.26	32.23	32.20	32.17	32.15	32.13	32.11	32.09	32.07	32.05
200	33.13	33.05	33.97	32.91	32.85	32.79	32.75	32.70	32.66	32.63	32.60	32.57	32.54	32.51	32.49	32.46	32.44	32.42	32.40	32.38	32.36
210	33.47	33.38	33.30	33.24	33.18	33.13	33.08	33.03	32.99	32.95	32.92	32.89	32.86	32.83	32.81	32.79	32.76	32.74	32.72	32.70	32.68

The proper consideration can be given to the adaptability of any boiler for the service in view only after a thorough understanding of the requirements of a good steam boiler, with the application of what has been said on the proper operation to the special requirements of each case. Of almost equal importance to the factors mentioned are the experience, the skill and responsibility of the manufacturer.

With the design of boiler selected that is best adapted to the service required, the next step is the determination of the boiler power requirements.

The amount of steam that must be generated is determined from the steam consumption of the prime movers. It has already been indicated that such consumption can vary over wide limits with the size and type of the apparatus used, but fortunately all types have been so tested that manufacturers are enabled to state within very close limits the actual consumption under any given set of conditions. It is obvious that conditions of operation will have a bearing on the steam consumption that is as important as the type and size of the apparatus itself. This being the case, any tabular information that can be given on such steam consumption, unless it be extended to an impracticable size, is only of use for the most approximate work, and more definite figures on this consumption should in all cases be obtained from the manufacturer of the apparatus to be used for the conditions under which it will operate.

To the steam consumption of the main prime movers there is to be added that of the auxiliaries. Again it is impossible to make a definite statement of what this allowance should be, the figure depending wholly upon the type and the number of such auxiliaries. Whatever figure is used should be taken high enough to be on the conservative side.

When any such figures are based on the actual weight of steam required, Table 59, which gives the actual evaporation for various pressures and temperatures of feed corresponding to one boiler horse-power (34.5 pounds of water per hour from and at 212 degrees), may be of service.

With the steam requirements known, the next step is the determination of the number and size of boiler units to be installed. This is directly affected by the capacity which a consideration of the economical load indicates is the best for the operating conditions which will exist. The other factors entering into such determination are the size of the plant and the character of the feed water.

The size of the plant has its bearing on the question from the fact that higher efficiencies are in general obtained from large units, that labor cost decreases with the number of units, the first cost of brickwork is lower for large than for small size units, a general decrease in the complication of piping, etc., and in general the cost per horse-power of any design of boiler decreases with the size of units. To illustrate this, it is only necessary to consider a plant of, say, 10,000 boiler horse-power, consisting of forty 250 horse-power units or seventeen 600 horse-power units.

The feed water available has its bearing on the subject from the other side, for it has already been shown that very large units are not advisable where the feed water is not of the best.

The character of an installation is also a factor. Where, say, 1000 horse-power is installed in a plant where it is known what the ultimate capacity is to be, the size of units should be selected with the idea of this ultimate capacity in mind rather than the amount of the first installation.

Boiler service, from its nature, is severe. All boilers have to be cleaned from time to time and certain repairs to settings, etc., are a necessity. This makes it necessary, in determining the number of boilers to be installed, to allow a certain number of units or spares to be operated when any of the regular boilers must be taken off the line. With the steam requirements determined for a plant of moderate size and a reasonably constant load, it is highly advisable to install at least two spare boilers where a continuity of service is essential. This permits taking off one boiler for cleaning or repairs and still allows a spare boiler in the event of some unforeseen occurrence, such as the blowing out of a tube or the like. Investment in such spare apparatus is nothing more nor less than insurance on the necessary continuity of service. In small plants of, say, 500 or 600 horse-power, two spares are not usually warranted in view of the cost of such insurance. A large plant is ordinarily laid out in a number of sections or panels and each section should have its spare boiler or boilers even though the sections are cross-connected. In central station work, where the peaks are carried on the boilers brought up from the bank, such spares are, of course, in addition to these banked boilers. From the aspect of cleaning boilers alone, the number of spare boilers is determined by the nature of any scale that may be formed. If scale is formed so rapidly that the boilers cannot be kept clean enough for good operating results, by cleaning in rotation, one at a time, the number of spares to take care of such proper cleaning will naturally increase.

In view of the above, it is evident that only a suggestion can be made as to the number and size of units, as no recommendation will hold for all cases. In general, it will be found best to install units of the largest possible size compatible with the size of the plant and operating conditions, with the total power requirements divided among such a number of units as will give proper flexibility of load, with such additional units for spares as conditions of cleaning and insurance against interruption of service warrant.

In closing the subject of the selection of boilers, it may not be out of place to refer to the effect of the builder's guarantee upon the determination of design to be used. Here, in one of its most important aspects, appears the responsibility of the manufacturer. Emphasis has been laid on the difference between test results and those secured in ordinary operating practice. That such a difference exists is well known and it is now pretty generally realized that it is the responsible manufacturer who, where guarantees are necessary, submits the conservative figures, figures which may readily be exceeded under test conditions and which may be closely approached under the ordinary plant conditions that will be met in daily operation.

OPERATION AND CARE OF BOILERS

THE general subject of boiler-room practice may be considered from two aspects. The first is that of the broad plant economy, with a suggestion as to the methods to be followed in securing the best economical results with the apparatus at hand and procurable. The second deals rather with specific recommendations which should be followed in plant practice, recommendations leading not only to economy but also to safety and continuity of service. Such recommendations are dictated from an understanding of the nature of steam-generating apparatus and its operation. as covered previously in this book.

It has already been pointed out that the attention given in recent years to steam-generating practice has come with a realization of the wide difference existing between the results being obtained in every-day operation and those theoretically possible. The amount of such attention and regulation given to the steam-generating end of a power plant, however, is comparatively small in relation to that given to the balance of the plant, but it may be safely stated that it is here that there is the greatest assurance of a return for the attention given. The statement holds to a greater extent in the small plant than in the larger stations which have come generally to a realization of the necessity for boiler-room economy.

In the endeavor to increase boiler-room efficiency, it is of the utmost importance that a standard basis be set by which average results are to be judged. With the theoretical efficiency obtainable varying so widely, this standard cannot be placed at the highest efficiency that has been obtained regardless of operating conditions. It is better set at the best obtainable results for each individual plant under its conditions of installation and daily operation.

With an individual standard so set, present practice can only be improved by a systematic effort to approach this standard. The degree with which operating results will approximate such a standard will be found to be directly proportional to the amount of intelligent supervision given the operation. For such supervision to be given, it is necessary to have not only a full realization of what the plant can do under the best operating conditions but also a full and complete knowledge of what it is doing under all of the different conditions that may arise. What the plant is doing should be made a matter of continuous record so arranged that the results may be directly compared for any period or set of conditions, and where such results vary from the standard set, steps must be taken immediately to remedy the causes of such failure. Such a record is an important check on the losses in the plant.

As the size of the plant and the fuel consumption increase, such a check of losses and recording of results become a necessity. In the larger plants, the saving of but a fraction of one per cent in the fuel bill represents an amount running into thousands of dollars annually. while the expense of the proper supervision to secure such saving is small. The methods of supervision followed in the large plants are necessarily elaborate and complete. In the smaller plants the same methods may be followed on a more moderate scale with a corresponding saving in fuel and an inappreciable increase in either plant organization or expense.

There has been within the last few years a great increase in the practicability and reliability of the various types of apparatus by which the records of plant operation may be secured. Much of this apparatus is ingenious and, considering the work

Portion of 4100 Horse-power Installation of Babcock & Wilcox Boilers at the Prudential Life Insurance Co. Building, Newark, N. J.

to be done, is remarkably accurate. From the delicate nature of some of the apparatus, the liability to error necessitates frequent calibration, but even where the accuracy is known to be only within limits of, say, five per cent either way, the records obtained are of the greatest service in considering relative results. Some of the records desirable and the apparatus for securing them are given below.

Inasmuch as the ultimate measure of the efficiency of the boiler plant is the cost of steam generation, the important records are those of steam generated and fuel consumed. Records of temperature, analyses, draft and the like serve as a check on this consumption, indicating the distribution of the losses and affording a means of remedying conditions where improvement is possible.

Coal Records—There are many devices on the market for conveniently weighing the coal used. These are ordinarily accurate within close limits, and where the size or nature of the plant warrants the investment in such a device, its use is to be recommended. The coal consumption should be recorded by some other method than from the weights of coal purchased. The total weight gives no way of dividing the consumption into periods and it will unquestionably be found to be profitable to put into operation some scheme by which the coal is weighed as it is used. In this way, the coal consumption, during any specific period of the plant's operation, can be readily seen. The simplest of such methods which may be used in small plants is the actual weighing on scales of the fuel as it is brought into the fire room and the recording of such weights.

Aside from the actual weight of the fuel used, it is often advisable to keep other coal records, coal and ash analyses and the like, for the evaporation to be expected will be dependent upon the grade of fuel used and its calorific value, fusibility of its ash, and like factors.

The highest calorific value for unit cost is not necessarily the indication of the best commercial results. The cost of fuel is governed by this calorific value only when such value is modified by local conditions of capacity, labor and commercial efficiency. One of the important factors entering into fuel cost is the consideration of the cost of ash handling and the maintenance of ash-handling apparatus if such be installed. The value of a fuel, regardless of its calorific value, is to be based only on the results obtained in every-day plant operation.

Coal and ash analyses used in connection with the amount of fuel consumed are a direct indication of the relation between the results being secured and the standard of results which has been set for the plant. The methods of such analyses have already been described. The apparatus is simple and the degree of scientific knowledge necessary in using it is only such as may be readily mastered by plant operatives.

The ash content of a fuel, as indicated from a coal analysis checked against ash weights as actually found in plant operation, acts as a check on grate efficiency. The effect of any saving in the ashes, that is, the permissible ash to be allowed in the fuel purchased, is determined by the point at which the cost of handling, combined with the falling off in the evaporation, exceeds the saving of fuel cost through the use of poorer coal.

Water Records—Water records with the coal consumption form the basis for judging the economic production of steam. The methods of securing such records are of later introduction than for coal, but great advances have been made in the apparatus to be used. Here, possibly, to a greater extent than in any recording device, are the records of value in determining relative evaporation, that is, an error is rather allowable provided such an error be reasonably constant.

The apparatus for recording such evaporation is of two general classes: Those measuring water before it is fed to the boiler and those measuring the steam as it leaves. Of the first, the venturi meter is perhaps the best known, though recently there has come into considerable vogue an apparatus utilizing a weir notch for the measuring of such water. Both methods are reasonably accurate and apparatus of this description has an advantage over one measuring steam in that it may be calibrated much more readily. Of the steam-measuring devices, the one in most common use is the steam-flow meter. Provided the instruments are selected for a proper flow, etc., they are of inestimable value in indicating the steam consumption. Where such instruments are placed on the various engine-room lines, they will immediately indicate an excessive consumption for any one of the units. With a steam-flow meter placed on each boiler, it is possible to fix relatively the amount produced by each boiler, and, considered in connection with some of the "check" records described below, clearly to indicate whether its portion of the total steam produced is up to the standard set for the over-all boiler-room efficiency.

Flue-gas Analysis—The value of a flue-gas analysis as a measure of furnace efficiency has already been indicated. There are on the market a number of instruments by which a continuous record of the carbon dioxide in the flue gases may be secured and in general the results so recorded are accurate. The limitations of an analysis showing only CO_2 and the necessity of completing such an analysis with an Orsat, or like apparatus, and in this way checking the automatic device, have already been pointed out, but where such records are properly checked from time to time and are used in conjunction with a record of flue temperatures, the losses due to excess air or incomplete combustion and the like may be directly compared for any period. Such records act as a means for controlling excess air and also as a check on individual firemen.

Where the size of a plant will not warrant the purchase of an expensive continuous CO_2 recorder, it is advisable to make analyses of samples for various conditions of firing and to install an apparatus whereby a sample of flue gas covering a period of, say, eight hours, may be obtained and such a sample afterwards analyzed.

Temperature Records— Flue-gas temperatures, feed-water temperatures and steam temperatures are all taken with recording thermometers, any number of which will, when properly calibrated, give accurate results.

A record of flue temperatures is serviceable in checking stack losses and, in general, the cleanliness of the boiler. A record of steam temperatures, where superheaters are used, will indicate excessive fluctuations and lead to an investigation of their cause. Feed temperatures are valuable in showing that the full benefit of the exhaust steam is being derived.

Draft Regulation—As the capacity of a boiler varies with the combustion rate and this rate with the draft, an automatic apparatus satisfactorily varying this draft with the capacity demands on the boiler will obviously be advantageous.

As has been pointed out, any fuel has some rate of combustion at which the best results will be obtained. In a properly designed plant where the load is reasonably steady, the draft necessary to secure such a rate may be regulated automatically.

Automatic apparatus for the regulation of draft has recently reached a stage of perfection which, in the larger plants at any rate, makes its installation advisable. The installation of a draft gauge or gauges is strongly to be recommended and a record of such drafts should be kept as being a check on the combustion rates.

An important feature to be considered in the installing of all recording apparatus is its location. Thermometers, draft gauges and flue-gas sampling pipes should be so located as to give as nearly as possible an average of the conditions, the gases flowing freely over the ends of the thermometers, couples and sampling pipes. With the location permanent, there is no certainty that the samples may be considered an average, but in any event comparative results will be secured which will be useful in plant operation. The best permanent location of apparatus will vary considerably with the design of the boiler.

It may not be out of place to refer briefly to some of the shortcomings found in boiler-room practice, with suggestions as to means of overcoming them.

1st. It is sometimes found that the operating force is not fully acquainted with the boilers and apparatus. Probably the most general of such shortcomings is the fixed idea in the heads of the operatives that boilers run above their rated capacity are operating under a state of strain and that by operating at less than their rated capacity the most economical service is assured; whereas, by determining what a boiler will do, it may be found that the most economical rating under the conditions of the plant will be considerably in excess of the builder's rating. Such ideas can be dislodged only by demonstrating to the operatives what maximum load the boilers can carry, showing how the economy will vary with the load and the determining of the economical load for the individual plant in question.

2nd. Stokers. With stoker-fired boilers, it is essential that the operators know the limitations of their stokers as determined by their individual installation. A thorough understanding of the requirements of efficient handling must be insisted upon. The operatives must realize that smokeless stacks are not necessarily the indication of good combustion, for, as has been pointed out, absolute smokelessness is oftentimes secured at an enormous loss in efficiency through excess air.

Another feature in stoker-fired plants is in the cleaning of fires. It must be impressed upon the operatives that before the fires are cleaned they should be put into condition for such cleaning. If this cleaning is done at a definite time, regardless of whether the fires are in the best condition for cleaning, good fuel will be lost.

3rd. It is necessary that in each individual plant there be a basis on which to judge the cleanliness of a boiler. From the operative's standpoint, it is probably more necessary that there be a thorough understanding of the relation between scale and tube difficulties than between scale and efficiency. It is, of course, impossible to keep boilers absolutely free from scale at all times, but experience in each individual plant determines the limit to which scale can be allowed to form before tube difficulties will begin or a perceptible falling off in efficiency will take place. With such a limit of scale formation fixed, the operatives should be impressed with the danger of allowing it to be exceeded.

4th. The operatives should be instructed as to the losses resulting from excess air due to leaks in the setting and as to losses in efficiency and capacity due to the by-passing of gases through the setting, that is, not following the path of the baffles as originally installed. In replacing tubes and in cleaning the heating surfaces, care must be taken not to dislodge baffle brick or tile.

5th. That an increase in the temperature of the feed reduces the amount of work demanded from the boiler has been shown. The necessity of keeping the feed temperature as high as the quantity of exhaust steam will allow should be thoroughly understood. As an example of this, there was a case brought to our attention where a large amount of exhaust steam was wasted simply because the feed pump showed a tendency to leak if the temperature of feed water was increased above 140 degrees. The amount wasted was sufficient to increase the temperature to 180 degrees but was not utilized simply because of the slight expense necessary to overhaul the feed pump.

The highest return will be obtained when the speed of the feed pumps is maintained reasonably constant, for should the pumps run very slowly at times, there may be a loss of the steam from other auxiliaries by blowing off from the heaters.

6th. With a view to checking steam losses through the useless blowing of safety valves, the operative should be made to realize the great amount of steam that it is possible to get through a pipe of a given size. Oftentimes the fireman feels a sense of security from objections to a drop in steam simply because of the blowing of safety valves, not considering the losses due to such a cause and makes no effort to check this flow either by manipulation of dampers or regulation of fires.

The shortcomings outlined above, which may be found in many plants, are almost entirely due to lack of knowledge on the part of the operating crew as to the conditions existing in their own plants and the better performances being secured in others. Such shortcomings can be overcome only by the education of the operatives, the showing of the defects of present methods, and instruction in better methods. Where such instruction is necessary the value of records is obvious. There is fortunately a tendency toward the employment of a better class of labor in the boiler room, a tendency which is becoming more and more marked as the realization of the possible saving in this end of the plant increases.

The second aspect of boiler-room management, dealing with specific recommendations as to the care and operation of the boilers, is dictated largely by the nature of the apparatus. Some of the features to be watched in considering this aspect follow.

Before placing a new boiler or one which has been idle over an extended period in service a careful and thorough inspection should be made of the pressure parts and the setting. In the case of new boilers, baffles should be checked for correct location from the drawings of the particular installation, the setting should be checked to make sure that the pressure parts are free to expand, and all brick and mortar cleaned off the setting and pressure parts.

In the case of new boilers, since firing the units with green walls will invariably crack the brickwork, this should be properly dried out. To start this drying process, as soon as the brickwork is completed the damper and ashpit doors should be blocked open to maintain a circulation of air through the setting. Whenever possible this should be done for several days before firing. When ready for firing, wood should be used for a light fire, gradually building it up until the setting is thoroughly warmed. If the precautions discussed below have been followed, coal may then be fired and the boiler put on light load for several days and then into regular operation.

The following precautions apply both to new boilers and those that have been out of service for an appreciable length of time.

Internally the boiler should be inspected to assure the absence of dirt, waste, oil and tools. If there is oil or paint inside the boiler, soda ash should be placed in the drum, the boiler filled with water to its normal level and a slow fire started. After twelve hours the fire should be allowed to die out, the boiler cooled slowly, then opened and washed out thoroughly. This will remove oil and grease from the interior of the setting and prevent foaming when the boiler is placed in service.

The combustion chamber and gas passages should be clean and in good repair.

Dampers and stoker and fan drives should be operated and shown to be ready for service. Outlet dampers should be left open.

After the boiler is closed, the blow-off valves, water-column and water-glass drains, try cocks and feed valves should be examined for proper operating condition and closed. The vent valves and the valves to the steam-pressure gauge should be opened, and the boiler filled. The stem of the main stop valve should be eased without lifting it from its seat to make sure that it is not stuck.

If coal is the fuel a thin layer should be placed on the grate and a light fire started. In the case of oil, gas or pulverized fuel the back damper should be partly opened, and with a light suction at the burners a torch should be placed near the burners to obtain quick ignition and a light fire started by admitting fuel slowly. With these fuels care must be taken to avoid possible flare-backs, and a torch should be used in lighting all burners until the furnace has been brought to its working temperature, and even then when all burners have been out.

In oil-burning plants the oil system and burners should be inspected before starting, thoroughly cleaned and shown to be in proper working condition.

In new settings, tie rods and buckstays should be set up snug and then slacked slightly until the setting has been thoroughly warmed after the first firing.

The fires should be operated so that a steam pressure is not generated until approximately one half hour before the boiler is to be cut into the line. When the water has become heated, the water level should be checked by the try cocks and the blow-off valves examined for leaks. When steam has escaped from the vent valves for a few minutes, these may be closed.

Steam pressure should be raised slowly and when it approaches the operating pressure the boiler should be examined for leaks, the pressure-gauge connections blown out sufficiently to assure cleanliness, the safety valve tested by hand, and the boiler blown down.

Where superheaters are installed these should be thoroughly drained before the boiler is fired, and protected during the period of steam raising by letting a small quantity of steam pass through them. As soon as there is any pressure on the boiler, the drains on both inlet and outlet superheater headers should be opened. That on the inlet header should then be closed and that on the outlet header left open until the boiler is put on the line. Just before cutting in, the drain on the inlet header should be opened momentarily to assure no water in that header which might be carried over.

Before cutting the boiler into the line, the connections between it and the steam main should be thoroughly drained. Where there is no non-return check valve or stop-and-check valve installed, the stop valve should be opened very slowly as the

PORTION OF 2600 HORSE-POWER INSTALLATION OF BABCOCK & WILCOX BOILERS EQUIPPED WITH BABCOCK & WILCOX CHAIN GRATE STOKERS AT THE PETER SCHOENHOFEN BREWING CO., CHICAGO, ILL.

pressure in the boiler reaches the pressure in the main. If the main stop valve is equipped with a by-pass, this should be opened very slowly, and in the event of any water hammer closed at once. Where a non-return check or stop-and-check valve is installed, the stop valve should be slightly opened when the pressure within the boiler is within from 10 to 15 pounds of that in the main and the non-return check allowed to operate automatically.

In regular operation, the safety valve and the steam gauge should be checked daily. If practicable the steam pressure should be raised until the safety valve blows. If not, the valve should be gently raised from its seat by hand. If the valve sticks, the pressure should be lowered and the defect causing the sticking remedied. When the valves blow, the steam gauge should be examined. If it shows a different pressure from that at which the valves are set the gauge should be tested. If such test shows the gauge to be correct, the safety valve should be adjusted by some competent person.

When safety-valve springs become weak, care should be taken in screwing down that there is no restriction in the relieving capacity as called for by The American Society of Mechanical Engineers' Rules for the Construction of Power Boilers.

The connections to the steam-pressure gauge should be blown down when raising steam, when testing safety valves and gauge, and when foaming occurs.

Water glasses should be kept clean and unscratched. The water column should be blown down thoroughly at least once on every shift, and the height of water as indicated by the glass checked by use of the try cocks on the column. If steam is allowed to leak from the water column or its connections, the gauge glass will not record a true water level.

In the larger stations the amount and frequency of blowing down are ordinarily determined from a chemical analysis of the water within the boiler. Where such analyses are not made, the boiler should be blown down at least once in every twenty-four hours, the amount of blow-down depending upon the class of feed water, the number of hours a day the boiler is in service, and the capacity at which the boiler is operated.

In blowing down, the blow-off cock should be opened first and the blow-off valve second; in closing, the valve should be closed first and the cock second. The valves should be opened carefully and slowly. Any valve leaks that develop should be repaired as soon as discovered.

In a distinct case of low water, resulting either from carelessness or unforeseen operating conditions, the essential object is to extinguish the fire in the quickest possible manner. The fuel supply should be shut off, the air supply to the fuel closed, and where practicable the fires should be covered with wet ashes. keeping the grates in motion. Under certain conditions it is feasible to put out the fire with a heavy stream of high-pressure water from a fire hose, and where this method can be followed it should be used. Such a method is dangerous, to an extent, but is justifiable, particularly in the case of large units, from the standpoint of saving the boiler. The feed valve should be closed and under no circumstances should water be fed to the boiler. The safety valves should not be opened nor the opening of the main steam valve be changed. In general, nothing should be done that would tend to alter the stresses acting on the boiler.

The boiler in which low water has occurred should be cut out of the line and a thorough examination be made to determine what damage, if any, has been done before placing it back in service.

If foaming occurs, the fuel supply should be stopped and the steam outlet valve closed long enough to establish the true water level. If the water level is sufficiently high, the boiler should be blown down and fresh water fed. The alternate blowing and feeding should be repeated several times, and if the foaming does not stop, the fires should be drawn, and the alternating blowing down and feeding continued until the boiler is cool enough to be opened and the cause of the priming determined. This cooling of the boiler should be done slowly, as described hereafter in the discussion of the methods of taking a boiler off the line. If priming is a result of the presence of oil in the boiler, the methods of removal are those described in the case of a new boiler. If due to other water conditions, the methods of removing the cause are those given in the chapter on feed-water treatment.

The efficiency and capacity of a boiler depend to an extent very much greater than is ordinarily appreciated upon its cleanliness internally and externally, and systematic cleaning should be included as a regular feature in the operating of any steam plant.

The outer surfaces of the tubes should be blown free from soot with automatic soot blowers or a steam lance at regular intervals. The frequency of such blowing depends upon the class of fuel burned, but should be at least once a day. Before blowing, the supply lines to the blowing elements or the steam lance should be thoroughly drained. Steam should not be used for dusting boilers not under pressure.

When permanent soot blowers are installed, care must be taken to see that the nozzles of the soot-blowing elements are maintained in their proper position relative to the boiler tubes. If these be improperly placed or become displaced in such manner that there is a direct impingement of the steam on the tubes, erosion of the metal will result. In blowing with a steam lance, no one portion of the heating surface should be blown for too long a period or similar erosion will occur.

With coals having a tendency toward the formation of slag on the boiler surfaces adjacent to the furnace, slag-cleaning doors should be installed and used at such intervals as the draft loss and exit temperatures indicate are necessary for proper commercial efficiency.

Boilers should be taken off the line at regular intervals for internal inspection, cleaning and repairs. In cutting out a boiler, the fuel supply should be stopped and when the fuel is coal the air supply shut off and the dampers closed. When the boiler no longer requires any feed and the non-return check valve, if installed, has closed, the main steam valve should be shut. It is best to clean the heating surfaces externally before the fire has died out. The boiler and setting should be allowed to cool slowly and when possible should be allowed to stand twenty-four hours after the fires have been drawn before emptying and opening. Any attempt to hasten the cooling process by allowing cooler air to pass through the setting will result in brickwork difficulties. While the boiler is off for cleaning, a careful inspection should be made externally and internally, and all leaks of steam, water or air through the setting stopped. Internally the tubes should be cleaned from scale and sludge which will accumulate, due to the concentration of solids present in practically any boiler-feed water. This internal cleaning can best be accomplished by the use of an air or water-driven turbine. The illustration on page 347 shows a turbine cleaner which has been found to give satisfactory results. The wheel cutter head should be used except where the scale has been allowed to form to an excessive thickness.

When scale has been allowed to accumulate to an excessive thickness the work of removing it is difficult. Where the scale is of sulphate formation its removal may be facilitated by filling the boiler with water in which there has been placed fifty pounds of soda ash to each 1000 gallons of water, starting a slow fire and allowing the water to boil for twenty-four hours without allowing any pressure on the boiler. It should then be cooled slowly, drained and the turbine cleaner used immediately, as the action of the air tends to harden the scale. While the use of chemicals in feed water is permissible with a view to preventing the formation of scale, such an agent should not be introduced into the boiler while it is in operation with a view to softening or loosening any scale that may already be present in the boiler.

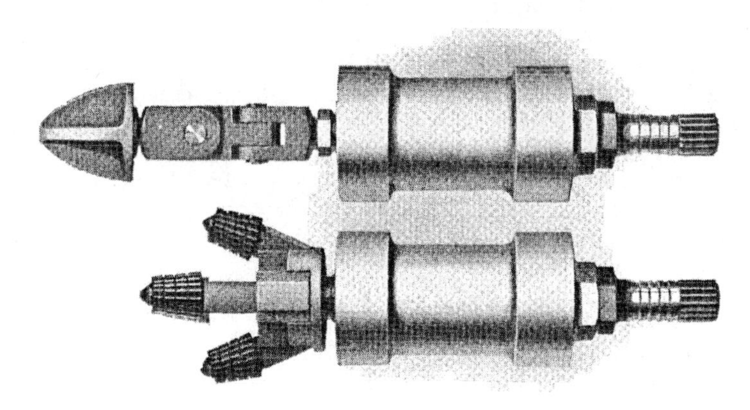

Turbine Tube Cleaner with Two Types of Head

Aside from the aspect of efficiency and capacity, a clean interior of boiler heating surfaces insures protection from burning. In the absence of a blow-pipe action of the flames, it is impossible to burn a metal surface when water is in intimate contact with that surface. Any formation of scale on the interior surfaces of a boiler will keep the water from those surfaces and increase their tendency to burn. Particles of loose scale which may have become detached will lodge at certain points in the tubes and act at such points in the same manner as a continuous coating of scale except that the tendency to burn is localized. If oil is allowed to enter the boiler with the feed water, its action will be the same as that of scale in keeping the water from the metal of the tubes, in this way increasing their liability to burn.

It has been proven beyond doubt that a very large percentage of tube losses is due to the presence of scale which, in many instances, has been so thin as to be considered of no moment, and the importance of maintaining the interior of boiler heating surfaces in a clean condition cannot be emphasized too strongly.

If pitting or corrosion is noted, the parts affected should be carefully cleaned and painted with white zinc. The cause of such action should be determined immediately and steps taken to see that a proper remedy is applied.

When making an internal inspection of a boiler or when cleaning the interior of the heating surfaces, great care must be taken to guard against the possibility of steam entering the boiler in question from any other boilers on the line through open blow-off valves or through the careless opening of the boiler stop valve. Bad cases of scalding have resulted from neglect of this precaution.

When boilers are taken out of service they should be allowed to cool slowly and when possible allowed to stand twelve hours after the fires are drawn before opening. The cooling process should not be hastened by causing cold air to rush through the setting, for this will cause difficulties with the setting brickwork. While the boiler is off for cleaning, a careful examination should be made of its condition, both internally and externally, and all leaks of steam, water and air through the setting should be promptly stopped.

If a boiler is to remain idle for some time it is liable to deteriorate much faster than when in service. If the period for which it is to be laid off is not long it may be filled with water while out of service. The boiler should be thoroughly cleaned, internally and externally, all soot and ashes being removed from the setting and any accumulation of scale removed from the interior surfaces. It should then be filled with water to which about four buckets of soda ash have been added, a very light fire started to drive the air from the water, the fire then allowed to die out and the boiler pumped full. When boilers are idle and filled, the water within should be kept cold, as lukewarm water tends toward a maximum corrosion. This necessitates tight feed and steam connections.

If the boiler is to be out of service for an extended period, it should be emptied, cleaned and thoroughly dried. A tray of quicklime should be placed in each drum, the boiler closed up, the grates covered and a quantity of quicklime placed on these. Special care must be taken to prevent air, steam or water leaks into the setting or onto the pressure parts, to obviate danger of corrosion.

When a boiler is to be out of service for a period of three months or more it should be given a protective external coat of some material such as red lead, black japan or tar paint to prevent external corrosion.

BRICKWORK BOILER SETTINGS

THE consideration of brickwork for stoker-fired furnaces may be divided into three parts, namely, furnace design, quality of brick used, and workmanship in the laying up of brick. The question as here considered is limited to the furnace proper, and deals, therefore, only with fire brick.

DESIGN—The design of the furnace is obviously of the greatest importance. Such design, however, varies so widely with different types of boilers and stokers, different fuels, and different operating conditions that no general statement that will apply to all cases may be made as to what constitutes a proper furnace design. The number of possible combinations of boilers, stokers, fuels and operating conditions is so very large that no attempt will be made here to suggest a furnace design. Each individual installation of boiler and stoker should be considered by itself and the furnace design based upon experience as to what has given the most satisfactory service for a similar set of conditions.

QUALITY OF FIRE BRICK—The modern tendency toward high overloads has greatly increased the severity of the service under which furnace brickwork is called upon to stand, and to a very great extent the life of the furnace is dependent upon the quality of fire brick entering into its construction.

Excluding the workmanship in building a furnace, and considering only the quality of the fire brick itself, failures in service can be traced to a small number of comparatively simple reasons. These reasons are given approximately in the order of their relative importance, the most important first:

Plastic Deformation—When a brick is subjected to successively higher temperatures, it ordinarily changes shape by expansion. As the temperature increases, there may be a period in which shrinkage is taking place, and at some temperature plastic deformation occurs, even under no greater load than its own weight. Expansion and shrinkage are to be distinguished from plastic deformation, because the two former leave the brick geometrically similar to its original shape, while the latter does not, and is the result of an externally applied stress, such as gravity. Fire-clay brick of the best quality generally obtainable begins to show plastic deformation under a load of 20 pounds per square inch at from 2200 to 2400 degrees Fahrenheit. Reduction in the load to 10 pounds per square inch will increase this limit about 200 degrees Fahrenheit. The importance of this quality in fire-clay brick in boiler furnaces appears when it is recognized that furnace temperatures as high as 2400 degrees Fahrenheit are frequent in hand-firing with coal, and may reach 2700 degrees Fahrenheit with stokers and 3000 degrees Fahrenheit with oil. It is evident, since the load due to the design of the furnace may be as high as 15 pounds per square inch, that should the temperature of the brickwork reach the furnace temperature for any considerable depth in the brickwork, fire-clay brick could not be used at all for boiler furnaces. These figures also show the necessity for protecting furnace arches from full furnace temperatures on both sides of the arch and for avoidance of pressures in the furnace above atmosphere.

A method of testing brick for this characteristic is given in the Technologic Paper No. 7 of the Bureau of Standards dealing with "The Testing of Clay Refractories, with Special Reference to Their Load-carrying Capacity at Furnace

Temperatures." Referring to the test for this specific characteristic, this publication recommends the following: "When subjected to the load test in a manner substantially as described in this bulletin, at 1350 degrees Centigrade (2462 degrees Fahrenheit), and under a load of 50 pounds per square inch, a standard fire brick tested on end should show no serious deformation and should not be compressed more than 1 inch, referred to the standard length of 9 inches."

Activity—Under this general word may be grouped the causes of failure resulting from expansion and shrinkage. If expansion or shrinkage occurs slowly (that is to say, if for a given percentage of volume change a considerable temperature range is required and the rate of change is more or less continuous), then the failure of the brick structure comes about from dislocation of the furnace structure as a whole; that is, the results are evident in bulging and cracking of walls. Expansion of this sort may occur every time the brick passes through a large range of temperature, the brick returning about to its original volume when the temperature returns to its starting point. Unfortunately, the brick structure under such action does not return to its original shape. Shrinkage of the slow and continuous type referred to is more apt to be permanent, and on this account is not apt to recur each time the brick passes through a cycle of temperature change. For this reason, if a brick structure does not have trouble from shrinkage the first time it is brought up to its working temperature and back again, it is not likely that trouble from this cause will ever be experienced.

Where expansion or shrinkage occurs within a very narrow temperature range, if the change of volume is also considerable, then disintegration of the brick by " spalling " is apt to occur. The reason for spalling under these conditions is double, since a sudden volume change within a close temperature range usually corresponds to a marked change in the chemical structure of the brick, and since, also, the effect of the temperature gradient through the brick becomes greater. Various tests for determining this quality of brick are used. They are all relative tests, as distinct from measurements. One test consists in heating one face of the brick so rapidly that the brick for a small depth below the face is at a temperature of 2400 to 3000 degrees Fahrenheit when the remainder of the brick is comparatively cold. Any test of this sort yields information from which the spalling quality of a brick in service can be predicted with a good deal of certainty if a sufficient number of various other sorts of brick have undergone the test and if sufficient is known from actual experience as to how they perform in service.

Tests for expansion and shrinkage lend themselves to measurements with a fair degree of convenience, and a limit for first-class fire-clay brick of 1 per cent linear at 3000 degrees Fahrenheit for the best quality of fire-clay brick for boilers should not be exceeded. A corresponding limit for shrinkage is also satisfactory.

Fusion Point—Under ordinary furnace operating conditions in boilers, the interior surface of the brick will approach quite closely the temperature of the furnace (flame temperature), so that fire brick to be satisfactory in this respect should have a fusion point above the flame temperature of the furnace by a distinct margin, probably not less than 100 degrees Fahrenheit and preferably 200 degrees Fahrenheit. At first sight, the fusion point of a fire brick would appear to be its most important characteristic, but in practice, failure in a brick structure will ordinarily take place from plastic deformation before failure by fusion appears. At small areas in a furnace, failure from fusion may take place without the furnace as a whole showing evidence of deformation, but such a difficulty is usually controllable by design. First-class fire-clay brick should show a fusion point above 3100 degrees Fahrenheit.

Slagging—Failure from the effect of slagging is ordinarily met with in the region of the fuel bed, but can occur under furnace conditions where particles of molten slag are brought by the gases into contact with the brickwork. The resistance of a given brick to slagging is the result of its chemical composition, and its physical structure as well. Fire-clay bricks, as a class, chemically, are reasonably resistant to slags such as are ordinarily formed in boiler furnaces from the usual varieties of coal. They are by no means as resistant as kaolin brick, though difficulties in the manufacture of kaolin brick have prevented as yet their coming into general use. The variation in the amount of trouble experienced with various fire-clay brick from slagging is to be explained chiefly by their macro-structure, which, in turn, is influenced by the sieve characteristics of the flint clay, the nature of the bond and similar details of manufacture. There is no exact measurement test for this characteristic of a brick, the test ordinarily used being either that given in the American Society for Testing Materials, Specification No. C20, or one of a very similar character.

A fire-clay brick for boiler furnaces may safely be said to be of the best quality generally obtainable for the purpose if it meets the limiting figures given above for plastic deformation, fusion, expansion and shrinkage, and if its action under the spalling test classifies it with other fire-clay bricks that experience has shown work well in practice in that particular.

Nothing has been so far said as to the information that chemical analysis, density and porosity measurements, hardness, flint-clay nodule size, color, etc., may reveal. The reason for the omission of these characteristics of the fire-clay brick for boiler service from the causes of failure of such brick is that, though they are all of effect in failures, they are either difficult to determine or misleading in the information that they give. Proper chemical composition will not insure good brick, and the opposite is true within limits, that is, a brick that is not of specially good chemistry, may, by proper sizing, bonding, burning and other manufacturing methods, actually give better results in practice than a less well-made brick of better chemistry. In general, fire-clay brick found to be especially good for high-temperature conditions will be found in the middle range of hardness, rather above the average in porosity and below it in density, and the brick will be, when fractured, rough in texture, with the flint-clay nodule areas prominent. The color of the fracture will be from yellow toward white, with absence of red, the latter being indicative of iron. It is to be borne in mind that a given amount of iron will show color principally in proportion to its dispersion, so that any judgment as to the amount of iron present from the reddish color of the brick is extremely uncertain.

Fire-brick Cements—The brickwork of a furnace is its most expensive component, and although in many cases the best of brick is used the results secured are poor unless the workmanship is of the highest grade and the best possible materials are used in laying the brick.

The functions of a fire-brick cement are:

1. To act as a bond with fire brick without shrinking, cracking or expanding.
2. To have a melting point higher than the maximum working temperature of the furnace.

3. To burn to a density sufficient to prevent the penetrating of flames, slag, vapors, gases, etc.
4. To be of such character that it will vitrify by the application of heat.
5. To be of such character that it will not disintegrate upon being heated and cooled repeatedly.

Joints are usually subjected to heat on one side only, and under this condition a refractory material showing a high shrinkage has a tendency to disintegrate before it reaches the first stages of vitrification. Cements should be so compounded that they will permit compression without disintegration while in the process of vitrification. This quality is very necessary, as usually there are stresses imposed upon the joints during the first firing of the furnace.

Where the cement shows a high shrinkage or a high expansion, cracking of the brickwork will be the result. If the cement used disintegrates or fuses and allows the penetration of heat, slag, gases, etc., between the bricks, each brick then becomes attacked from several sides, thereby shortening greatly the time of destruction of the brickwork. In cases where fusion of the cement occurs, the fused cement may join in the attack on the brick.

Other sources of trouble from the use of a poor cement are the exposure of broken or cracked portions of brick to the cutting action of flames and slag, the opportunity for clinkers to cling to the joints, and the infiltration of cold air from outside sources.

There are two very distinct types of jointing materials, namely, natural fire clay and prepared cement.

Fire clay is cheaper and, although not so suitable, can sometimes be made to compare favorably with cement if the furnace temperature is low. If ordinary fire clay is used, it should be chemically as like the brick as possible and very finely ground.

Fire clay ordinarily acts simply as a filler to take up irregularities of the faces of the brick in the joint, and until it vitrifies it has no binding strength. Its shrinkage is very much greater than that common to prepared cements, and is so great that very often disintegration or cracking of the brickwork takes place first. For this reason fire clay should always be used in as small quantities as possible and at low temperatures and only where furnace conditions are not severe.

The superiority of prepared cements over fire clay can be covered only in a general way unless specific materials are considered, there being much variation in the chemical composition and physical characteristics of both groups. Taking the average drying shrinkage of a number of fire clays, it is found to be 6 per cent; and the average shrinkage from dry to 2400 degrees Fahrenheit is found to be 5 per cent, making a total of 11 per cent. A high-grade refractory cement will show not more than 2 per cent shrinkage upon drying and a total of not more than 4 per cent at 2400 degrees Fahrenheit.

By examining the necessary properties of an ideal joint, the proper selection of a cement can easily be made. If a prepared cement is used, its point of fusion should be as high as the fusion point of the brick and its initial point of vitrification (which, roughly speaking, is the point at which its cement action comes into play) should be low in order that its holding power can be developed throughout the entire joint and not simply at its exposed surface.

The distinction between fire clays and cements has only recently been recognized. Results with cements, however, over a considerable length of service and under widely varying furnace conditions, are now available, and such results apparently show the superior merit of the proper cements over fire clays. On the other hand, there are a number of cements offered that are wholly unsatisfactory and in comparing the relative values it is necessary to understand the real properties of a proper cement, otherwise the advantages of the latter may be obscured.

LAYING UP—With the grade of brick and fire clay or "cement" selected best suited to the service of the boiler to be set, the other factor affecting the life of the boiler setting is the laying. While it is probable that more setting difficulties arise from improper workmanship in the laying up of the brick than from poor material, on the other hand no written instructions, specifications or description of the methods to be followed in this work will compensate for the lack of an elementary knowledge of good and poor brick and proper and improper joints.

Unfortunately, fire brick and even red brick in any local district are not of the same sizes. This makes it difficult if not impossible to tie fire brick into red-brick backing unless special provision is made to insure that the brick are of the correct size. If the red brick that are used in backing a fire-brick wall are of the correct size, all joints will be broken uniformly throughout the wall, all joints can be made of a minimum thickness, and both inside and outside faces of the wall can be left a true plane. All fire, insulating, and red brick, when laid up in walls, should not exceed 23½ inches for every nine courses in height. This will insure tight bed joints and proper bonding in the transverse and longitudinal sections of walls.

It is not expected, in backing a fire-brick wall with red brick, that the joints can be held as thin or as uniform in the red-brick backing as in the fire-brick lining. It is, however, recognized that, if proper care is used in the selection of the brick and the workmanship in laying up, the joints of the red-brick backing can be held down to a thickness that will not affect the stability of the structure or the life of the wall. No joints in a red-brick wall should exceed $\frac{5}{16}$ inch.

Fire brick, when laid for any purpose, should always be dipped in thin fire clay or "cement," and under no circumstance should be "buttered." As each course of the wall is laid, it should be grouted with very thin fire clay, and after the grouting has been brushed into the joints and all joints filled, the surplus should be entirely removed from the top of that course.

In the red-brick backing, the same method of laying up should be followed, the only difference being that the mortar for red brick should be spread to a thickness that will hold the upper face of the red-brick course in the same plane as the corresponding course of fire brick, and that a lime and cement grout should be used to slush the red brick instead of fire clay.

The pointing up of joints in fire-brick work is solely to give the finished appearance of a new job. The boiler wall should be so laid up that the pointing up is reduced to only the smoothing of the visible joints, and should not be a filling of any joints that should have been filled from the inside during the laying of any course.

All 4½-inch fire-brick walls should be three stretchers and one header. All fire-brick linings 9 inches or over should be laid up of three courses of headers and one stretcher backed with a full header course of fire brick. Furnace center-walls should be built entirely of fire brick. If the center of such walls is built of red brick, they will often melt down and cause the failure of the wall as a whole.

Fire-brick arches should be constructed of selected brick which are smooth, straight and uniform. The forms on which such arches are built, called arch centers, should be constructed of batten strips not over 2 inches wide. The brick should be laid on these in courses, not in rings. each joint being broken with a bond equal to the length of half a brick. A large 9-inch brick should be used at each end of the arch on every other course to break the bond. Soap brick should never be used in arch work for breaking the bond. Each course should be tried in place dry and should be checked with a straight edge to insure all the brick of one course being the same thickness. Each brick should be dipped on one side and one edge only and rubbed into place. Wedge-brick courses should be used only where necessary to keep the bottom faces of the straight-brick course in even contact with the centers. When such contact cannot be exactly secured by the use of wedge brick, the straight brick should lean away from the center of the arch rather than toward it. When the arch is approximately two-thirds completed. a trial ring should be laid to determine whether the key course will fit. When cutting is necessary to secure such a fit, it should be done on two or more of the adjacent courses on the side of the brick away from the key. It is necessary that the keying course be a true fit from top to bottom, and after it has been dipped and driven it should not extend below the surface of the arch. but preferably should have its lower edge ¼ inch above this surface. After fitting, the keys should be dipped, replaced loosely, and the whole course driven uniformly into place by means of a heavy hammer and a piece of wood extending the full length of the keying course. Such a driving in of this course should raise the arch as a whole from the center. The center should be so constructed that it may be dropped free from the arch when the key course is in place and removed from the furnace without being burned out.

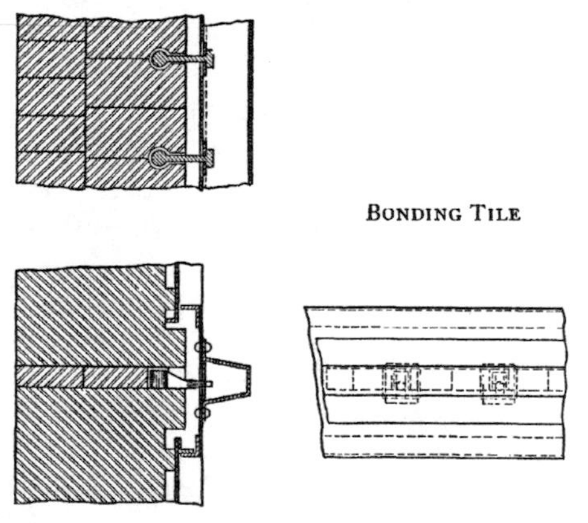
Bonding Tile

Bonding Tile—The walls of boiler furnaces subjected to high temperatures, and particularly with the larger furnace volumes and increased boiler capacities that are coming into general use, tend to bulge inward and, in some instances, fail by falling inward. A method that has been successfully employed for holding furnace walls in alignment is shown in the accompanying illustration. This construction, which is patented, consists of introducing horizontal rows of bonding tile in the walls, the tile being attached by members to horizontal buckstays outside the walls in such manner that the wall can expand freely both vertically and horizontally and is still given stability and prevented from bulging either inwardly or outwardly.

Relation of Draft to Setting Brickwork—The bearing that the draft available has upon the boiler setting and particularly the furnace setting is a factor that, in general, has only recently been given its proper consideration. Such a relation is to be distinguished from that of draft and combustion rates.

The draft available should be such as to provide a suction throughout all parts of the boiler setting at all times and under all conditions of operation. Where such a suction does not exist and a back pressure is found at any point in the setting, there is a tendency to force the gases of combustion outward through the boiler setting and to overheat the brickwork and access and inspection doors. This overheating, which will increase as the gases are hotter or as the boiler furnace is approached, will naturally cause a rapid deterioration of the setting and warping of the doors and frames. Where the products of combustion are not carried away from the boiler furnace and through the setting by an ample draft suction, the cost of upkeep of the setting will be excessive and will be greatest where such a suction does not exist in the boiler furnace. Here the highest temperatures are found and, if the hot gases are not removed promptly, the "soaking up" of the heat by the furnace walls and arches cannot but be harmful from the standpoint of length of life.

With a natural-draft stoker, the fact that there is sufficient draft in the furnace to burn the necessary amount of coal to develop the rating at which the boiler is being operated is ordinarily a safe indication that there is a draft suction throughout all parts of the setting and that the gases are being properly carried away from the furnace. This statement. of course, refers to those instances in which there is no undue loss in draft in passing from the furnace proper to the point at which the gases encounter the boiler heating surface.

With forced-draft stokers, on the other hand, the blast is ordinarily relied upon to give the required combustion rates. With this class of apparatus, therefore, the function of the stack is simply to remove the products of combustion from the furnace and it is in such cases that the question of suction throughout *all* parts of the setting is to be watched.

It may be readily conceived and, in fact, the condition is frequently found in practice, that a draft suction in the boiler furnace does not necessarily indicate that such a suction exists throughout all parts of the setting. For instance, in a boiler with vertical or semi-vertical passes for the gases, a suction may be found in the furnace while at the top of such a pass a slight back pressure may exist. Such a condition is due to the effect of the column of heated gases passing upward, which acts in the same way as the gases in a chimney, in so far as providing a draft at the bottom is concerned.

In determining stack sizes for forced-draft stokers, as for all boiler work, the diameter is a function of the amount of gases to be handled and should be made such as to give no undue frictional resistance to the gases because of insufficient area. The height is purely a function of the draft that must be supplied. With natural-draft stokers, as with hand firing, it must be sufficient to provide in the boiler furnace ample draft to give the combustion rate necessary to develop the maximum capacity at which the boiler is to be operated, proper attention being given to losses in draft due to length of flues. turns, resistance offered in the passage through the boiler, etc.

With forced-draft stokers the stack height must be such that a draft suction is assured throughout all portions of the setting under all conditions of operation, regardless of the intensity of the blast supplied to give the necessary combustion rates, and the same attention must be given to factors causing draft losses.

In the early days of forced-draft stoker work manufacturers of this class of apparatus had a tendency to overcarry the mark so far as the reduction in stack sizes was concerned, taking a stand that their product required practically no stack The importance of furnace and setting upkeep cost, however, is now appreciated by such manufacturers and they are insisting that sufficient stack be provided to maintain a draft suction throughout the boiler under all conditions.

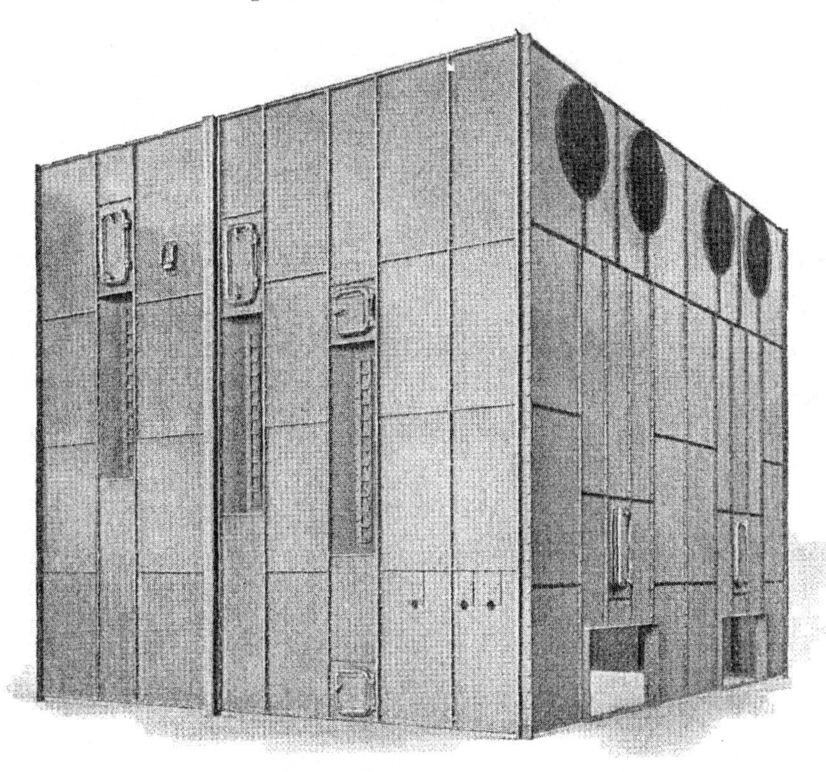

STEEL CASING FOR BABCOCK & WILCOX BOILERS

STEEL CASINGS—In the chapter dealing with the losses operating against high efficiencies as indicated by the heat balance it has been shown that a considerable portion of such losses is due to radiation and to air infiltration into the boiler setting. These losses have been variously estimated at 2 to 10 per cent, depending upon the condition of the setting and the amount of radiation surface, the latter in turn being dependent upon the size of the boiler used. In the modern efforts after the highest obtainable plant efficiencies, much has been done to reduce such losses by the use of an insulated steel casing covering the brickwork. In an average size boiler unit, the use of such a casing when properly installed, will reduce radiation losses from 1 to 2 per cent over what can be accomplished with the best brick setting without such a casing and, in addition, prevent the loss due to the infiltration of air, which may amount to an additional 5 per cent, as compared with brick settings that are not maintained in good order Steel plate, or steel plate backed by asbestos mill-board, while acting as a preventive against the infiltration of air through the boiler setting, is not as effective from the standpoint of decreasing radiation losses as a casing properly insulated from the brick portion of the setting by magnesia block and asbestos mill-board. A casing which has been found to give excellent results in eliminating air leakage and in the reduction of radiation losses is illustrated on page 356.

Many attempts have been made to use some material other than brick for boiler settings, but up to the present nothing has been found that may be considered successful or which will give as satisfactory service under severe conditions as properly laid brickwork.

3000 Horse-power Installation of Babcock & Wilcox Boilers in the Main Power Plant, Chicago & Northwestern Ry. Depot, Chicago, Ill.

One of Six Babcock & Wilcox Cross-drum Boilers, Aggregating 8060 Horse-power, Equipped with Babcock & Wilcox Superheaters, in the West End Station of the Union Gas and Electric Co., Cincinnati

FLOW OF STEAM THROUGH PIPES AND ORIFICES

VARIOUS formulæ for the flow of steam through pipes have been advanced. all having their basis upon Bernoulli's theorem of the flow of water through circular pipes with the proper modifications made for the variation in constants between steam and water. The loss of energy due to friction in a pipe is given by Unwin (based upon Weisbach) as

$$E_f = \frac{f\,2\,v^2\,WL}{gd} \qquad (42)$$

where E is the energy loss in foot-pounds due to the friction of W units of weight of steam passing with a velocity of v feet per second through a pipe d feet in diameter and L feet long; g represents the acceleration due to gravity (32.2) and f the coefficient of friction.

Numerous values have been given for this coefficient of friction, f, which, from experiment, apparently varies with both the diameter of pipe and the velocity of the passing steam. There are no authentic data on the rate of this variation with velocity and, as in all experiments, the effect of change of velocity has seemed less than the unavoidable errors of observation, the coefficient is assumed to vary only with the size of the pipe.

Unwin established a relation for this coefficient for steam at a velocity of 100 feet per second,

$$f = K\left(1 + \frac{3}{10d}\right) \qquad (43)$$

where K is a constant experimentally determined, and d the internal diameter of the pipe in feet.

If h represents the loss of head in feet, then

$$E_f = Wh = \frac{f\,2\,v^2\,WL}{gd} \qquad (44)$$

$$\text{and } h = \frac{f\,2\,v^2\,L}{gd} \qquad (45)$$

If D represents the density of the steam or weight per cubic foot, and p the loss of pressure due to friction in pounds per square inch, then

$$p = \frac{hD}{144} \qquad (46)$$

and from equations (38), (40) and (41),

$$p = \frac{Dv^2L}{72gd} \times K\left(1 + \frac{3}{10d}\right) \qquad (47)$$

To convert the velocity term and to reduce to units ordinarily used, let d_1 = the diameter of pipe in inches = $12d$, and w = the flow in pounds per minute; then

$$w = 60v\frac{\pi}{4}\left(\frac{d_1}{12}\right)^2 D$$

$$\text{and } v = \frac{9.6w}{\pi d_1^2 D}$$

Substituting this value and that of d in formula (47),

$$p = 0.04839\,K\left(1 + \frac{3.6}{d_1}\right)\frac{w^2L}{Dd_1^5} \qquad (48)$$

Some of the experimental determinations for the value of K are:

$K = 0.005$ for water (Unwin).
$K = 0.005$ for air (Arson).
$K = 0.0028$ for air (St. Gothard tunnel experiments).
$K = 0.0026$ for steam (Carpenter at Oriskany).
$K = 0.0027$ for steam (G. H. Babcock)

The value 0.0027 is apparently the most nearly correct, and substituting in formula *(48)* gives,

$$p = 0.000131 \left(1 + \frac{3.6}{d_1}\right)\frac{w^2L}{Dd_1^5} \qquad (49)$$

$$\text{or } w = 87\left(\frac{pDd_1^5}{\left(1 + \frac{3.6}{d_1}\right)L}\right)^{\frac{1}{2}} \qquad (50)$$

Where w = the weight of steam passing in pounds per minute,
p = the difference in pressure between the two ends of the pipe in pounds per square inch,
D = density of steam or weight per cubic foot,*
d_1 = internal diameter of pipe in inches,
L = length of pipe in feet.

This formula is the most generally accepted for the flow of steam in pipes. Table 60 is calculated from this formula and gives the amount of steam passing per minute that will flow through straight smooth pipes having a length of 240 diameters from various initial pressures with one pound difference between the initial and final pressures.

To apply this table for other lengths of pipe and pressure losses than those assumed, let L = the length and d the diameter of the pipe, both in inches; l, the loss in pounds; Q, the weight under the conditions assumed in the table, and Q_1, the weight for the changed conditions.

For any length of pipe, if the weight of steam passing is the same as given in the table, the loss will be,

$$l = \frac{L}{240d} \qquad (51)$$

If the pipe length is the same as assumed in the table but the loss is different, the quantity of steam passing per minute will be,

$$Q_1 = Ql^{\frac{1}{2}} \qquad (52)$$

For any assumed pipe length and loss of pressure, the weight will be,

$$Q_1 = Q\left(\frac{240d\,l}{L}\right)^{\frac{1}{2}} \qquad (53)$$

Example: Find the weight of steam at 100 pounds initial gauge pressure, which will pass through a 6-inch pipe 720 feet long with a pressure drop of 4 pounds.

***D, the density, is taken as the mean of the density at the initial and final pressures.**

Under the conditions assumed in the table, 287 pounds would flow per minute; hence, $Q = 287$, and

$$Q_1 = 287 \left(\frac{240 \times 6 \times 4}{720 \times 12}\right)^{1/2} = 234.3 \text{ pounds.}$$

Elbows, globe valves and a square-ended entrance to pipes all offer resistance to the passage of steam. It is customary to measure the resistance offered by such construction in terms of the diameter of the pipe. Many formulæ have been advanced for computing the length of pipe in diameters equivalent to such fittings or valves which offer resistance. These formulæ, however, vary widely and for ordinary purposes it will be sufficiently accurate to allow for resistance at the entrance of a pipe a length equal to 60 times the diameter; for a right-angle elbow, a length equal to 40 diameters; for a globe valve, a length equal to 60 diameters.

TABLE 60

FLOW OF STEAM THROUGH PIPES

LENGTH OF PIPE = 240 DIAMETERS

Initial Gauge Pressure. Pounds per Square Inch	DIAMETER OF PIPE IN INCHES														Density*
	¾	1.0	1½	2.0	2½	3.0	4.0	5.0	6.0	8.0	10.0	12.0	15.0	18.0	
	WEIGHT OF STEAM PER MINUTE IN POUNDS WITH ONE POUND LOSS IN PRESSURE														
1	1.13	2.04	5.49	9.85	14.8	24.4	45.0	74.2	105	202	329	482	771	1127	0.0397
10	1.39	2.54	6.82	12.3	18.5	30.3	55.9	92.3	138	248	409	599	958	1404	0.0595
20	1.63	2.98	8.02	14.4	21.9	35.6	65.7	108	163	295	481	705	1128	1645	0.0823
30	1.84	3.37	9.07	16.2	24.5	40.2	74.1	122	183	333	542	795	1272	1858	0.1047
40	2.03	3.71	9.95	17.9	26.9	44.2	81.6	135	202	366	596	874	1388	2089	0.1267
50	2.19	4.02	10.8	19.4	29.2	47.9	88.3	146	219	397	646	948	1515	2214	0.1486
60	2.35	4.30	11.5	20.7	31.2	51.3	93.7	156	234	424	691	1014	1623	2372	0.1703
70	2.50	4.56	12.3	22.0	33.1	54.4	100	166	248	451	734	1075	1722	2517	0.1919
80	2.63	4.81	12.9	23.2	34.9	57.4	106	175	262	475	774	1135	1814	2653	0.2134
90	2.76	5.05	13.6	24.3	36.6	60.2	111	183	275	499	812	1192	1893	2782	0.2347
100	2.88	5.27	14.2	25.4	38.3	62.8	116	191	287	520	848	1243	1990	2908	0.2560
120	3.10	5.69	15.3	27.4	41.3	67.9	125	207	310	562	915	1342	2148	3138	0.2985
150	3.38	6.26	16.8	30.2	45.5	74.7	138	228	341	619	1007	1477	2361	3450	0.3616
200	3.88	6.96	19.1	34.3	51.5	84.8	156	258	387	703	1144	1677	2685	3919	0.4664
250	4.33	7.86	21.1	37.9	57.1	93.7	173	285	428	776	1264	1855	2962	4337	0.5695
300	4.68	8.56	23.0	41.2	62.1	102	188	310	466	845	1376	2019	3226	4717	0.6748
350	5.03	9.22	24.7	44.5	66.8	110	202	334	501	908	1478	2168	3467	5060	0.7788

*From Marks and Davis Tables

The flow of steam of a higher toward a lower pressure increases as the difference in pressure increases to a point where the external pressure becomes 58 per cent of the absolute initial pressure. Below this point the flow is neither increased nor decreased by a reduction of the external pressure, even to the extent of a perfect vacuum. The lowest pressure for which this statement holds when steam is discharged into the atmosphere is 25.37 pounds. For any pressure below this figure, the atmospheric pressure, 14.7 pounds, is greater than 58 per cent of the initial pressure. Table 61, by D. K. Clark, gives the velocity of outflow at constant density, the actual velocity of outflow expanded (the atmospheric pressure being taken as 14.7 pounds absolute, and the ratio of expansion in the nozzle being 1.624), and the corresponding discharge per square inch of orifice per minute.

Napier deduced an approximate formula for the outflow of steam into the atmosphere which checks closely with the figures just given. This formula is:

$$W = \frac{pa}{70} \tag{54}$$

Where W = the pounds of steam flowing per second,
p = the absolute pressure in pounds per square inch,
and a = the area of the orifice in square inches.

TABLE 61

FLOW OF STEAM INTO THE ATMOSPHERE

Absolute Initial Pressure per Square Inch Pounds	Velocity of Outflow at Constant Density Feet per Second	Actual Velocity of Outflow Expanded Feet per Second	Discharge per Square Inch of Orifice per Minute Pounds	Horse-power per Square Inch of Orifice if Horse-power =30 Pounds per Hour
25.37	863	1401	22.81	45.6
30.	867	1408	26.84	53.7
40.	874	1419	35.18	70.4
50.	880	1429	44.06	88.1
60.	885	1437	52.59	105.2
70.	889	1444	61.07	122.1
75.	891	1447	65.30	130.6
90.	895	1454	77 94	155.9
100.	898	1459	86.34	172.7
115.	902	1466	98.76	197.5
135.	906	1472	115.61	231.2
155.	910	1478	132.21	264.4
165.	912	1481	140.46	280.9
215.	919	1493	181.58	363.2

In some experiments made by Professor C. H. Peabody, on the flow of steam through pipes from ¼ inch to 1½ inches long and ¼ inch in diameter, with rounded entrances, the greatest difference from Napier's formula was 3.2 per cent excess of the experimental over the calculated results.

For steam flowing through an orifice from a higher to a lower pressure where the lower pressure is greater than 58 per cent of the higher, the flow per minute may be calculated from the formula:

$$W = 1.9AK\sqrt{(P-d)d} \tag{55}$$

Where W = the weight of steam discharged in pounds per minute,
A = area of orifice in square inches,
P = the absolute initial pressure in pounds per square inch,
d = the difference in pressure between the two sides in pounds per square inch,
K = a constant = 0.93 for a short pipe, and 0.63 for a hole in a thin plate or a safety valve.

HEAT TRANSFER

THE rate at which heat is transmitted from a hot gas to a cooler metal surface over which the gas is flowing has been the subject of a great deal of investigation, from both the experimental and theoretical sides. A more or less complete explanation of this process is necessary for a detailed analysis of the performance of steam boilers. Such information at the present is almost entirely lacking and for this reason a boiler, as a physical piece of apparatus, is not as well understood as it might be. This, however, has had little effect in its practical development and it is hardly possible that a more complete understanding of the phenomena discussed will have any radical effect on the present design.

The amount of heat that is transferred across any surface is usually expressed as a product. of which one factor is the slope or linear rate of change in temperature and the other is the amount of heat transferred per unit difference in temperature in unit length. In Fournier's analytical theory of the conduction of heat, this second factor is taken as a constant and is called the " conductivity " of the substance. Following this practice, the amount of heat absorbed by any surface from a hot gas is usually expressed as a product of the difference in temperature between the gas and the absorbing surface into a factor which is commonly designated the " transfer rate." There has been considerable looseness in the writings of even the best authors as to the way in which the gas temperature difference is to be measured. If the gas varies in temperature across the section of the channel through which it is assumed to flow, and most of them seem to consider that this would be the case, there are two mean gas temperatures, one the mean of the actual temperatures at any time across the section, and the other the mean temperature of the entire volume of the gas passing such a section in any given time. Since the velocity of flow will of a certainty vary across the section, this second mean temperature, which is one tacitly assumed in most instances, may vary materially from the first. The two mean temperatures are only approximately equal when the actual temperature measured across the section is very nearly a constant. In what follows it will be assumed that the mean temperature measured in the second way is referred to. In English units the temperature difference is expressed in Fahrenheit degrees and the transfer rate in B.t.u. per hour per square foot of surface. Pecla, who seems to have been one of the first to consider this subject analytically. assumed that the transfer rate was constant and independent both of the temperature differences and the velocity of the gas over the surface. Rankine, on the other hand, assumed that the transfer rate, while independent of the velocity of the gas, was proportional to the temperature difference, and expressed the total amount of heat absorbed as proportional to the square of the difference in temperature. Neither of these assumptions has any warrant in either theory or experiment and they are only valuable in so far as their use determines formulæ that fit experimental results. Of the two, Rankine's assumption seems to lead to formulæ that more nearly represent actual conditions. It has been quite fully developed by William Kent in his " Steam Boiler Economy." Professor Osborne Reynolds in a short paper reprinted in Volume I of his " Scientific Papers " suggests that the transfer rate is proportional to the product of the density and velocity of the gas, and it is to be assumed that he had in mind the mean velocity, density and temperature over the section of the channel through which the gas was assumed to flow. Contrary

Portion of 6[illegible] Horse-Power Installation of Babcock & Wilcox Boilers Equipped with Babcock & Wilcox Chain Grate Stokers at the Campbell Street Plant of the Louisville Railway Co., Louisville, Ky. This Company Operates a Total of 10,000 Horse Power of Babcock & Wilcox Boilers

to prevalent opinion, Professor Reynolds gave neither a valid experimental nor a theoretical explanation of his formula and the attempts that have been made since its first publication to establish it on any theoretical basis can hardly be considered of scientific value. Nevertheless, Reynolds' suggestion was really the starting point of the scientific investigation of this subject and while his formula cannot in any sense be held as completely expressing the facts, it is undoubtedly correct to a first approximation for small temperature differences if the additive constant, which in his paper he assumed as negligible, is given a value.*

Experimental determinations have been made during the last few years of the heat transfer rate in cylindrical tubes at comparatively low temperatures and small temperature differences. The results at different velocities have been plotted and an empirical formula determined expressing the transfer rate with the velocity as a factor. The exponent of the power of the velocity appearing in the formula, according to Reynolds, would be unity. The most probable value, however, deduced from most of the experiments makes it less than unity. After considering experiments of his own, as well as experiments of others, Dr. Wilhelm Nusselt† concludes that the evidence supports the following formula:

$$a = b\frac{\lambda_w}{d^{1-u}}\left(\frac{wc_p\delta}{\lambda}\right)^u$$

Where a is the transfer rate in calories per hour per square meter of surface per degree Centigrade difference in temperature,

u is a physical constant, equal to 0.786 from Dr. Nusselt's experiments,

b is a constant which, for the units given below, is 15.90.

w is the mean velocity of the gas in meters per second,

c_p is the specific heat of the gas at its mean temperature and pressure in calories per kilogram,

δ is the density in kilograms per cubic meter,

λ is the conductivity at the mean temperature and pressure in calories per hour per square meter per degree Centigrade temperature drop per meter,

λ_w is the conductivity of the steam at the temperature of the tube wall.

d is the diameter of the tube in meters.

If the unit of time for the velocity is made the hour, and in the place of the product of the velocity and density is written its equivalent, the weight of gas flowing per hour divided by the area of the tube, this equation becomes:

$$a = .0255\,\frac{\lambda_w}{d^{.214}}\left(\frac{Wc_p}{A\lambda}\right)^{.786}$$

where the quantities are in the units mentioned, or, since the constants are absolute constants, in English units,

a is the transfer rate in B.t.u. per hour per square foot of surface per degree difference in temperature,

W is the weight in pounds of the gas flowing through the tube per hour,

A is the area of the tube in square feet,

d is the diameter of the tube in feet,

*H. P. Jordan, "Proceedings of The Institute of Mechanical Engineers," 1909.

†"Zeitschrift des Vereines Deutscher Ingenieure," 1909, page 1750.

c_p is the specific heat of the gas at constant pressure,

λ is the conductivity of the gas at the mean temperature and pressure in B.t.u. per hour per square foot of surface per degree Fahrenheit drop in temperature per foot,

λ_w is the conductivity of the steam at the temperature of the wall of the tube.

The conductivities of air, carbonic acid gas and superheated steam, as affected by the temperature, in English units, are:

Conductivity of air . . . . $0.0122 (1 + 0.00132 T)$
Conductivity of carbonic acid gas $0.0076 (1 + 0.00228 T)$
Conductivity of superheated steam $0.0119 (1 + 0.00261 T)$

where T is the temperature in degrees Fahrenheit.

Nusselt's formulæ can be taken as typical of a number of other formulæ proposed by German, French and English writers.* Physical properties, in addition to the density, are introduced in the form of coefficients from a consideration of the physical dimensions of the various units and of the theoretical formulæ that are supposed to govern the flow of the gas and the transfer of heat. All assume that the correct method of representing the heat transfer rate is by the use of one term, which seems to be unwarranted and probably has been adopted on account of the convenience in working up the results by plotting them logarithmically. This was the method Professor Reynolds used in determining his equation for the loss in head in fluids flowing through cylindrical pipes, and it is now known that the derived equation cannot be considered as anything more than an empirical formula. It, therefore, is well for anyone considering this subject to understand at the outset that the formulæ discussed are only of an empirical nature and applicable to limited ranges of temperature under conditions approximately the same as those surrounding the experiments from which the constants of the formula were determined.

It is not probable that the subject of heat transfer in boilers will ever be on any other than an experimental basis until the mathematical expression connecting the quantity of fluid which will flow through a channel of any section under a given head has been found, and some explanation of its derivation obtained. Taking the simplest possible section, namely, a circle, it is found that at low velocities the loss of head is directly proportional to the velocity and the fluid flows in straight stream lines or the motion is direct. This motion is in exact accordance with the theoretical equations of the motion of a viscous fluid and constitutes almost a direct proof that the fundamental assumptions on which these equations are based are correct. When, however, the velocity exceeds a value which is determinable for any size of tube, the direct or stream-line motion breaks down and is replaced by an eddy or mixing flow. In this flow the head loss by friction is approximately, although not exactly, proportional to the square of the velocity. No explanation of this has ever been found in spite of the fact that the subject has been treated by the best mathematicians and physicists for years back. It is to be assumed that the heat transferred during the mixing flow would be at a much higher rate than with the direct or stream-line flow, and Professors Croker and Clement† have demonstrated that this is true, the increase in the transfer

* Heinrich Gröber—Zeit. d. Ver. Ing., March, 1912, December, 1912. Leprince-Ringuet—Revue de Mécanique, July, 1911 John Perry—"The Steam Engine." T E. Stanton—Philosophical Transactions, 1897. Dr J T. Nicholson—Proceedings Institute of Engineers & Shipbuilders in Scotland, 1910. W. E. Dally—Proceedings Institute of Mechanical Engineers, 1909.

† "Proceedings Royal Society," Vol. LXXI

being so marked as to enable them to determine the point of critical velocity from observing the rise in temperature of water flowing through a tube surrounded by a steam jacket.

The formulæ given apply only to a mixing flow and inasmuch as, from what has just been stated, this form of motion does not exist from zero velocity upward, it follows that any expression for the heat transfer rate that would make its value zero when the velocity is zero, can hardly be correct. Below the critical velocity, the transfer rate seems to be little affected by change in velocity and Nusselt,* in another paper which mathematically treats the direct or stream-line flow, concludes that, while it is approximately constant as far as the velocity is concerned in a straight cylindrical tube, it would vary from point to point of the tube, growing less as the surface passed over increased.

It should further be noted that no account in any of this experimental work has been taken of radiation of heat from the gas. Since the common gases absorb very little radiant heat at ordinary temperatures, it has been assumed that they radiate very little at any temperature. This may or may not be true, but certainly a visible flame must radiate as well as absorb heat. However this radiation may occur, since it would be a volume phenomenon rather than a surface phenomenon it would be considered somewhat differently from ordinary radiation. It might apply as increasing the conductivity of the gas which, however independent of radiation, is known to increase with the temperature. It is, therefore, to be expected that at high temperatures the rate of transfer will be greater than at low temperatures. The experimental determinations of transfer rates at high temperatures are lacking.

Although comparatively nothing is known concerning the heat radiation from gases at high temperatures, there is no question but what a large proportion of the heat absorbed by a boiler is received directly as radiation from the furnace. Experiments show that the lower row of tubes of a Babcock & Wilcox boiler absorb heat at an average rate per square foot of surface between the first baffle and the front headers equivalent to the evaporation of from 50 to 75 pounds of water from and at 212 degrees Fahrenheit per hour. Inasmuch as in these experiments no separation could be made between the heat absorbed by the bottom of the tube and that absorbed by the top, the average includes both maximum and minimum rates for those particular tubes and it is fair to assume that the portion of the tubes actually exposed to the furnace radiation absorb heat at a higher rate. Part of this heat was, of course, absorbed by actual contact between the hot gases and the boiler heating surface. A large portion of it, however, must have been due to radiation. Whether this radiant heat came from the fire surface and the brickwork and passed through the gases in the furnace with little or no absorption, or whether, on the other hand, the radiation was absorbed by the furnace gases and the heat received by the boiler was a secondary radiation from the gases themselves and at a rate corresponding to the actual gas temperature, is a question. If the radiation is direct, then the term "furnace temperature," as usually used, has no scientific meaning, for obviously the temperature of the gas in the furnace would be entirely different from the radiation temperature, even were it possible to attach any significance to the term "radiation temperature," and it is not possible to do this unless the radiation is what is known as "full radiation" from a so-called "black body." If furnace radiation takes place in this

*"Zeitschrift des Vereines Deutscher Ingenieure," 1910, page 1154

manner, the indications of a pyrometer placed in a furnace are hard to interpret and such temperature measurements can be of little value. If the furnace gases absorb the radiation from the fire and from the brickwork of the side walls and in their turn radiate heat to the boiler surface, it is scientifically correct to assume that the actual or sensible temperature of the gas would be measured by a pyrometer and the amount of radiation could be calculated from this temperature by Stefan's law, which is to the effect that the rate of radiation is proportional to the fourth power of the absolute temperature, using the constant with the resulting formula that has been determined from direct experiment and other phenomena. With this understanding of the matter, the radiation absorbed by a boiler can be taken as equal to that absorbed by a flat surface covering the portion of the boiler tubes exposed to the furnace and at the temperature of the tube surface, when completely exposed on one side to the radiation from an atmosphere at the temperature in the furnace. With this assumption, if S' is the area of the surface, T the absolute temperature of the furnace gases, t the absolute temperature of the tube surface of the boiler, the heat absorbed per hour measured in B.t.u. is equal to

$$1600\left(\left(\frac{T}{1000}\right)^4-\left(\frac{t}{1000}\right)^4\right)S'$$

In using this formula, or in any work connected with heat transfer, the external temperature of the boiler heating surface can be taken as that of saturated steam at the pressure under which the boiler is working, with an almost negligible error, since experiments have shown that with a surface clean internally, the external surface is only a few degrees hotter than the water in contact with the inner surface, even at the highest rates of evaporation. Further than this, it is not conceivable that in a modern boiler there can be much difference in the temperature of the boiler in the different parts, or much difference between the temperature of the water and the temperature of the steam in the drums which is in contact with it.

If the total evaporation of a boiler measured in B.t.u. per hour is represented by E, the furnace temperature by T_1, the temperature of the gas leaving the boiler by T_2, the weight of gas leaving the furnace and passing through the setting per hour by W, the specific heat of the gas by C, it follows from the fact that the total amount of heat absorbed is equal to the heat received from radiation plus the heat removed from the gases by cooling from the temperature T_1 to the temperature T_2, that

$$E=1600\left(\left(\frac{T}{1000}\right)^4-\left(\frac{t}{1000}\right)^4\right)S'+WC(T_1-T_2)$$

This formula can be used for calculating the furnace temperature when E, t and T_2 are known, but it must be remembered that an assumption which is probably, in part at least, incorrect is implied in using it or in using any similar formula. Expressed in this way, however, it seems more rational than one proposed a few years ago by Dr. Nicholson,* where, in place of the surface exposed to radiation, he uses the grate surface and assumes the furnace gas temperature as equal to the fire temperature.

If the heat transfer rate is taken as independent of the gas temperature and the heat absorbed by an element of the surface in a given time is equated to the heat given out from the gas passing over this surface in the same time, a single integration gives

$$(T-t)=(T_1-t)\,e^{-\frac{Rs}{WC}}$$

* "Proceedings Institute of Engineers and Shipbuilders," 1910.

where s is the area of surface passed over by the gases from the furnace to any point where the gas temperature T is measured, and the rate of heat transfer is R. As written, this formula could be used for calculating the temperature of the gas at any point in the boiler setting. Gas temperatures, however, calculated in this way are not to be depended upon, as it is known that the transfer rate is not independent of the temperature. Again, if the transfer rate is assumed as varying directly with the weight of the gases passing, which is Reynolds' suggestion, it is seen that the weight of the gases entirely disappears from the formula and as a consequence if the formula was correct, as long as the temperature of the gas entering the surface from the furnace was the same, the temperatures throughout the setting would be the same. This is known also to be incorrect. If, however, in place of T is written T_2 and in place of s is written S, the entire surface of the boiler, and the formula is re-arranged, it becomes:

$$R = \frac{WC}{S} \log_e \left(\frac{T_1 - t}{T_2 - t}\right)$$

This formula can be considered as giving a way of calculating an average transfer rate. It has been used in this way for calculating the average transfer rate from boiler tests in which the capacity has varied from an evaporation of a little over 3 pounds per square foot of surface up to 15 pounds. When plotted against the gas weights, it was found that the points were almost exactly on a line. This line, however, did not pass through the zero point but started at a point corresponding approximately to a transfer rate of 2. Checked out against many other tests, the straight line law seems to hold generally, and this is true even though material changes are made in the method of calculating the furnace temperature. The inclination of the line, however, varied inversely as the average area for the passage of the gas through the boiler. If A is the average area between all the passes of the boiler, the heat transfer rate in Babcock & Wilcox type boilers with ordinary clean surfaces can be determined to a rather close approximation from the formula:

$$R = 2.00 + 0.0014 \frac{W}{A}$$

The manner in which A appears in this formula is the same as it would appear in any formula in which the heat transfer rate was taken as depending upon the product of the velocity and the density of the gas jointly, since this product, as pointed out above, is equivalent to $W \div A$. Nusselt's experiments, as well as those of others, indicate that the ratio appears in the proper way.

While the underlying principles from which the formula for this average transfer rate was determined are questionable and at best only approximately correct, it nevertheless follows that, assuming the transfer rate as determined experimentally, the formula can be used in an inverse way for calculating the amount of surface required in a boiler for cooling the gases through the range of temperature covered by the experiments and it has been found that the results bear out this assumption. The practical application of the theory of heat transfer, as developed at present, seems consequently to rest on these last two formulæ, which from their nature are more or less empirical.

Through the range in the production of steam met with in boilers now in service, which in the marine type extends to the average evaporation of 12 to 15 pounds of water from and at 212 degrees Fahrenheit per square foot of surface, the constant 2 in the approximate formula for the average heat transfer rate constitutes quite a large proportion of the total. The comparative increase in the transfer rate due to a change in weight of the gases is not as great consequently as it would be if this constant were zero. For this reason, with the same temperature of the gases entering the boiler surface, there will be a gradual increase in the temperature of the gases leaving the surface as the velocity or weight of flow increases and the proportion of the heat contained in the gases entering the boiler which is absorbed by it is gradually reduced. It is, of course, possible that the weight of the gases could be increased to such an amount, or the area for their passage through the boiler could be so reduced by additional baffles, that the constant term in the heat transfer formula would be relatively unimportant. Under such conditions, as pointed out previously, the final gas temperature would be unaffected by a further increase in the velocity of the flow and the fraction of the heat carried by the gases removed by the boiler would be constant. Actual tests of waste heat boilers, in which the weight of gas per square foot of sectional area for its passage is many times more than in ordinary installations, show, however, that this condition has not been attained and it will probably never be attained in any practical installation. It is for this reason that the conclusions of Dr. Nicholson in the paper referred to and of Messrs. Kreisinger and Ray in the pamphlet, "The Transmission of Heat into Steam Boilers," published by the Department of the Interior in 1912, are not applicable without modification to boiler design.

In superheaters the heat transfer is effected in two different stages; the first transfer is from the hot gas to the metal of the superheater tube and the second transfer is from the metal of the tube to the steam on the inside. There is, theoretically, an intermediate stage in the transfer of the heat from the outside to the inside surface of the tube. The conductivity of steel is sufficient, however, to keep the temperatures of the two sides of the tube very nearly equal to each other so that the effect of the transfer in the tube itself can be neglected. The transfer from the hot gas to the metal of the tube takes place in the same way as with the boiler tubes proper, regard being paid to the temperature of the tube which increases as the steam is heated. The transfer from the inside surface of the tube to the steam is the inverse of the process of the transfer of the heat on the outside and seems to follow the same laws. The transfer rate, therefore, will increase with the velocity of the steam through the tube. For this reason, internal cores are quite often used in superheaters and actually result in an increase in the amount of superheat obtained from a given surface. The average transfer rate in superheaters based on a difference in mean temperature between the gas on the outside of the tubes and the steam on the inside of the tubes is, if R is the transfer rate from the gas to the tube and r the rate from the tube to the steam:

$$\frac{Rr}{R+r}$$

and is always less than either R or r. This rate is usually greater than the average transfer rate for the boiler as computed in the way outlined in the preceding paragraphs Since, however, steam cannot, under any imagined set of conditions, take up more heat from a tube than would water at the same average temperature, this fact supports the contention made that the actual transfer rate in a boiler must increase quite rapidly with the temperature. The actual transfer rate in superheaters is affected by so many conditions that it has not been possible so far to evolve any formula of practical value.

MISCELLANEOUS TUBE AND PIPE DATA

TABLE 62

TUBE, WIRE AND SHEET-METAL GAUGES

(DIAMETERS OR THICKNESSES IN DECIMAL PARTS OF AN INCH)

Gauge No.	Birmingham Wire Gauge (B. W. G.): Stubs Iron Wire Gauge *For steel tubes, wire and sheets*	U. S. Standard Gauge for Sheet Metal. *For iron and steel sheets.*	American Wire Gauge: Brown & Sharpe Gauge. *For copper wire.*	Steel Wire Gauge: Washburn & Moen and Roebling Gauge. *For steel wire.*	Stubs Steel Wire Gauge	British Imperial Standard Wire Gauge (S. W. G).	Trenton Iron Company Gauge	Standard Birmingham Sheet and Hoop Gauge (B. G.)
0	0.340	0.312	0.325	0.3065	. . .	0.324	0.305	0.3964
1	0.300	0.281	0.289	0.2830	0.227	0.300	0.285	0.3532
2	0.284	0.266	0.258	0.2625	0.219	0.276	0.265	0.3147
3	0.259	0.250	0.229	0.2437	0.212	0.252	0.245	0.2804
4	0.238	0.234	0.204	0.2253	0.207	0.232	0.225	0.2500
5	0.220	0.219	0.182	0.2070	0.204	0.212	0.205	0.2225
6	0.203	0.203	0.162	0.1920	0.201	0.192	0.190	0.1981
7	0.180	0.188	0.144	0.1770	0.199	0.176	0.175	0.1764
8	0.165	0.172	0.128	0.1620	0.197	0.160	0.160	0.1570
9	0.148	0.156	0.114	0.1483	0.194	0.144	0.145	0.1398
10	0.134	0.141	0.102	0.1350	0.191	0.128	0.130	0.1250
11	0.120	0.125	0.091	0.1205	0.188	0.116	0.1175	0.1113
12	0.109	0.109	0.081	0.1055	0.185	0.104	0.105	0.0991
13	0.095	0.094	0.072	0.0915	0.182	0.092	0.0925	0.0882
14	0.083	0.078	0.064	0.0800	0.180	0.080	0.080	0.0785
15	0.072	0.070	0.057	0.0720	0.178	0.072	0.070	0.0699
16	0.065	0.062	0.051	0.0625	0.175	0.064	0.061	0.0625
17	0.058	0.056	0.045	0.0540	0.172	0.056	0.0525	0.0556
18	0.049	0.050	0.040	0.0475	0.168	0.048	0.045	0.0495
19	0.042	0.0438	0.036	0.0411	0.164	0.040	0.040	0.0440
20	0.035	0.0375	0.032	0.0348	0.161	0.036	0.035	0.0392

TABLE 63

STANDARD OPEN-HEARTH OR LAP-WELDED STEEL TUBES

External Diameter Inches	Birmingham Wire Gauge Number	Thickness Inch	Internal Diameter Inches	Circumference Inches		Transverse Area Square Inches		Square Feet of External Surface per Foot of Length	Length in Feet per Square Foot of External Surface	Nominal Weight Pounds per Foot
				External	Internal	External	Internal			
1½	10	0.134	1.232	4.712	3.870	1.7671	1.1921	0.392	2.546	1.955
1½	9	0.148	1.204	4.712	3.782	1.7671	1.1385	0.392	2.546	2.137
1½	8	0.165	1.170	4.712	3.676	1.7671	1.0751	0.392	2.546	2.353
2	10	0.134	1.732	6.283	5.441	3.1416	2.3560	0.523	1.909	2.670
2	9	0.148	1.704	6.283	5.353	3.1416	2.2778	0.523	1.909	2.927
2	8	0.165	1.670	6.283	5.246	3.1416	2.1904	0.523	1.909	3.234
3¼	11	0.120	3.010	10.210	9.456	8.2958	7.1157	0.850	1.175	4.011
3¼	10	0.134	2.982	10.210	9.368	8.2958	6.9840	0.850	1.175	4.459
3¼	9	0.148	2.954	10.210	9.280	8.2958	6.8535	0.850	1.175	4.903
4	10	0.134	3.732	12.566	11.724	12.566	10.939	1.047	0.954	5.532
4	9	0.148	3.704	12.566	11.636	12.566	10.775	1.047	0.954	6.000
4	8	0.165	3.670	12.566	11.530	12.566	10.578	1.047	0.954	6.758

Dimensions are nominal and except where noted are in inches.

TABLE 64

DIMENSIONS OF STANDARD WROUGHT-IRON AND STEEL PIPE

Size, Inches	Diameters Inches		Nominal Thickness Inch	Circumference Inches		Transverse Areas Square Inches			Linear Feet of Pipe per Square Foot of		Length of Pipe Containing One Cubic Foot, Feet	Nominal Weight per Foot Pounds		Number of Threads per Inch of Screw
	External	Approximate Internal		External	Internal	External	Internal	Metal	External Surface	Internal Surface		Plain Ends	Threaded and Coupled	
1/8	0.405	0.269	0.068	1.272	0.845	0 129	0.057	0.072	9.431	14.199	2,533.775	0.244	0.245	27
1/4	0.540	0.364	0.088	1.696	1.144	0.229	0.104	0.125	7.073	10.493	1,383.789	0.424	0.425	18
3/8	0.675	0.493	0.091	2.121	1.549	0.358	0.191	0.167	5.658	7.747	754.360	0.567	0.568	18
1/2	0.840	0.622	0.109	2.639	1.954	0.554	0.304	0.250	4.547	6.141	473.906	0.850	0.852	14
3/4	1.050	0.824	0.113	3.299	2.589	0.866	0.533	0.333	3.637	4.635	270.034	1.130	1.134	14
1	1.315	1.049	0.133	4.131	3.296	1.358	0.864	0.494	2.904	3.641	166.618	1.678	1.684	11 1/2
1 1/4	1.660	1.380	0.140	5.215	4.335	2.164	1.495	0.669	2.301	2.767	96.275	2.272	2.281	11 1/2
1 1/2	1.900	1.610	0.145	5.969	5.058	2.835	2.036	0.799	2.010	2.372	70.733	2.717	2.731	11 1/2
2	2.375	2.067	0.154	7.461	6.494	4.430	3.355	1.075	1.608	1.847	42.913	3.652	3.678	11 1/2
2 1/2	2.875	2.469	0.203	9.032	7.757	6.492	4.788	1.704	1.328	1.547	30.077	5.793	5.819	8
3	3.500	3.068	0.216	10.996	9.638	9.621	7.393	2.228	1.091	1.245	19.479	7.575	7.616	8
3 1/2	4.000	3.548	0.226	12.566	11.146	12.566	9.886	2.680	0.954	1.076	14.565	9.109	9.202	8
4	4.500	4.026	0.237	14.137	12.648	15.904	12.730	3.174	0.848	0.948	11.312	10.790	10.889	8
4 1/2	5.000	4.506	0.247	15.708	14.156	19.635	15.947	3.688	0.763	0.847	9.030	12.538	12.642	8
5	5.563	5.047	0.258	17.477	15.856	24.306	20.006	4.300	0.686	0.756	7 198	14.617	14.810	8
6	6.625	6.065	0.280	20.813	19.054	34.472	28.891	5.581	0.576	0.629	4.984	18.974	19.185	8
7	7.625	7.023	0.301	23.955	22.063	45.664	38.738	6.926	0.500	0.543	3.717	23.544	23.769	8
8	8.625	8.071	0.277	27.096	25.356	58.426	51.161	7.265	0.442	0.473	2.815	24.696	25.000	8
8	8.625	7.981	0.322	27.096	25.073	58.426	50.027	8.399	0.442	0.478	2.878	28.554	28.809	8
9	9.625	8.941	0.342	30.238	28.089	72.760	62.786	9.974	0.396	0.427	2.294	33.907	34.188	8
10	10.750	10.192	0.279	33.772	32.019	90.763	81.585	9.178	0.355	0.374	1.765	31.201	32.000	8
10	10.750	10.136	0.307	33.772	31.843	90.763	80.691	10.072	0.355	0.376	1.785	34.240	35.000	8
10	10.750	10.020	0.365	33.772	31.479	90.763	78.855	11.908	0.355	0.381	1.826	40.483	41.132	8
11	11.750	11.000	0.375	36.914	34.558	108.434	95.033	13.401	0.325	0.347	1.515	45.557	46.247	8
12	12.750	12.090	0.330	40.055	37.982	127.676	114.800	12.876	0.299	0.315	1.254	43.773	45.000	8
12	12.750	12.000	0.375	40.055	37.699	127.676	113.097	14.579	0.299	0.318	1.273	49.562	50.706	8

TABLE 65

DIMENSIONS OF EXTRA STRONG WROUGHT-IRON AND STEEL PIPE

Nominal Internal Diameter, Inches	Diameters Inches		Nominal Thickness Inches	Circumference Inches		Transverse Areas Square Inches			Linear Feet of Pipe per Square Foot of		Length of Pipe Containing One Cubic Foot. Feet	Nominal Weight per Foot, Plain Ends, Pounds
	External	Approximate Internal		External	Internal	External	Internal	Metal	External Surface	Internal Surface		
1/8	0.405	0.215	0.095	1.272	0.675	0.129	0.036	0.093	9.431	17.766	3,966.392	0.314
1/4	0.540	0.302	0.119	1.696	0.949	0.229	0.072	0.157	7.073	12.648	2,010.290	0.535
3/8	0.675	0.423	0.126	2.121	1.329	0.358	0.141	0.217	5.658	9.030	1,024.689	0.738
1/2	0.840	0.546	0.147	2.639	1.715	0.554	0.234	0.320	4.547	6.995	615.017	1.087
3/4	1.050	0.742	0.154	3.299	2.331	0.866	0.433	0.433	3.637	5.147	333.016	1.473
1	1.315	0.957	0.179	4.131	3.007	1.358	0.719	0.639	2.904	3.991	200.193	2.171
1 1/4	1.660	1.278	0.191	5.215	4.015	2.164	1.283	0.881	2.301	2.988	112.256	2.996
1 1/2	1.900	1.500	0.200	5.969	4.712	2.835	1.767	1.068	2.010	2.546	81.487	3.631
2	2.375	1.939	0.218	7.461	6.092	4.430	2.953	1.477	1.608	1.969	48.766	5.022
2 1/2	2.875	2.323	0.276	9.032	7.298	6.492	4.238	2.254	1.328	1.644	33.976	7.661
3	3.500	2.900	0.300	10.996	9.111	9.621	6.605	3.016	1.091	1.317	21.801	10.252
3 1/2	4.000	3.364	0.318	12.566	10.568	12.566	8.888	3.678	0.954	1.135	16.202	12.505
4	4.500	3.826	0.337	14.137	12.020	15.904	11.497	4.407	0.848	0.998	12.525	14.983
4 1/2	5.000	4.290	0.355	15.708	13.477	19.635	14.455	5.180	0.763	0.890	9.962	17.611
5	5.563	4.813	0.375	17.477	15.120	24.306	18.194	6.112	0.686	0.793	7.915	20.778
6	6.625	5.761	0.432	20.813	18.099	34.472	26.067	8.405	0.576	0.663	5.524	28.573
7	7.625	6.625	0.500	23.955	20.813	45.664	34.472	11.192	0.500	0.576	4.177	38.048
8	8.625	7.625	0.500	27.096	23.955	58.426	45.663	12.763	0.442	0.500	3.154	43.388
9	9.625	8.625	0.500	30.238	27.096	72.760	58.426	14.334	0.396	0.442	2.464	48.728
10	10.750	9.750	0.500	33.772	30.631	90.763	74.662	16.101	0.355	0.391	1.929	54.735
11	11.750	10.750	0.500	36.914	33.772	108.434	90.763	17.671	0.325	0.355	1.587	60.075
12	12.750	11.750	0.500	40.055	36.914	127.676	108.434	19.242	0.299	0.325	1.328	65.415

TABLE 66

MAXIMUM ALLOWABLE WORKING PRESSURES FOR TUBES FOR WATER-TUBE BOILERS FOR DIFFERENT DIAMETERS AND GAUGES OF TUBES

(FROM THE BOILER CODE OF THE AMERICAN SOCIETY OF MECHANICAL ENGINEERS)

Outside Diameter of Tube Inches	Birmingham Wire Gauge Number, and Thickness of Tube Wall												
	17	16	15	14	13	12	11	10	9	8	7	6	5
D	$t=$ 0.058	$t=$ 0.065	$t=$ 0.072	$t=$ 0.083	$t=$ 0.095	$t=$ 0.109	$t=$ 0.120	$t=$ 0.134	$t=$ 0.148	$t=$ 0.165	$t=$ 0.180	$t=$ 0.203	$t=$ 0.220
½	434	686	938	1334	. .	. .	. .	.	. .	. .	. .	.	.
¾	206	374	542	806	1094	. .	. .	.	. .	.	. .	. .	. .
1	. .	218	344	542	758	1010	. .	. .	. .	.	. .	. .	. .
1⅛		166	278	454	646	870	1046	. .	. .	.	. .	. .	.
1¼		124	225	383	557	758	916	1118	. .	.	. .	.	.
1½			146	278	422	590	722	890	1058	.	.		.
1¾				203	326	470	583	727	871	1046	.	.	.
2				146	254	380	479	605	731	884	1019	.	.
2¼	. .	. .	. .	.	198	310	398	510	622	758	878	1062	.
2½	. .	. .	.	. .	153	254	333	434	535	657	765	931	1053
2¾	.	. .	.	. .	117	208	280	372	464	575	673	824	935
3	. .	. .	. .	.	.	170	236	320	404	506	596	734	836
3¼	. .	. .	.	. .	.	. .	199	276	354	448	531	658	752
3½	. .	.	.	. .	.	. .	167	238	310	398	475	594	681
3¾	.	. .	. .	. .	.	. .	139	206	273	355	427	537	619
4	. .	.	. .	. .	.	. .	. .	178	240	317	385	488	565
4½	. .	. .	. .	. .	.	. .	. .	. .	186	254	314	406	474
5	. .	. .	. .	. .	. .	.	. .	.	142	204	258	340	402

$$P = \left(\frac{t - 0.039}{D}\right) 18000 - 250$$

Where P = Maximum allowable working pressure, pounds per square inch
t = Thickness of tube wall, inches
D = Outside diameter of tube, inches

NOTE—Maximum allowable working pressures for superheater tubes shall be the same as for boiler tubes.

TABLE 67

RADIATION FROM COVERED AND UNCOVERED STEAM PIPES

CALCULATED FOR 160 POUNDS PRESSURE AND 60 DEGREES TEMPERATURE

Pipe Inches	Thickness of Covering	½ inch	¾ inch	1 inch	1¼ inch	1½ inch	Bare
2	B.t.u. per lineal foot per hour . .	149	118	99	86	79	597
	B.t.u. per square foot per hour . .	240	190	161	138	127	959
	B.t.u. per square foot per hour per one degree difference in temperature	0.770	0.613	0.519	0.445	0.410	3.198
4	B.t.u. per lineal foot per hour . . .	247	193	160	139	123	1085
	B.t.u. per square foot per hour . .	210	164	136	118	104	921
	B.t.u. per square foot per hour per one degree difference in temperature	0.677	0.592	0.439	0.381	0.335	2.970
6	B.t.u. per lineal foot per hour .	352	269	221	190	167	1555
	B.t.u. per square foot per hour . . .	203	155	127	110	96	897
	B.t.u. per square foot per hour per one degree difference in temperature	0.655	0.500	0.410	0.355	0.310	2.89
8	B.t.u. per lineal foot per hour . .	443	337	276	235	207	1994
	B.t.u. per square foot per hour . . .	196	149	122	104	92	883
	B.t.u. per square foot per hour per one degree difference in temperature	0.632	0.481	0.394	0.335	0.297	2.85
10	B.t.u. per lineal foot per hour . .	549	416	337	287	250	2468
	B.t.u. per square foot per hour . . .	195	148	120	102	89	877
	B.t.u per square foot per hour per one degree difference in temperature	0.629	0.477	0.387	0.329	0.287	2.83

Covering—Magnesia, canvas covered.

For calculating radiation for pressure and temperature other than 160 pounds and 60 degrees, use B.t.u. figures for one degree difference.

INDEX

*Illustration

*Illustration

*Illustration

*Illustration

*Illustration

*Illustration

Printed in the United States
49853LVS00003B/130

9 780766 194595